# Practical Handbook of

# GENETIC ALGORITHMS

## New Frontiers
## Volume II

Edited by Lance Chambers



CRC

CRC Press
Boca Raton New York

**Library of Congress Cataloging-in-Publication Data**

Chambers, Lance.
Practical handbook of genetic algorithms : new frontiers / Lance Chambers.
p. cm.
Includes bibliographical references (p. - ) and index.
ISBN 0-8493-2529-3 (v. 2 : alk. paper)
Genetic algorithms. I. Title.
QA402.5.C44 1995
005.1--dc20 95-17139
CIP

This book contains information obtained from authentic and highly regarded sources. Reprinted material is quoted with permission, and sources are indicated. A wide variety of references are listed. Reasonable efforts have been made to publish reliable data and information, but the author and the publisher cannot assume responsibility for the validity of all materials or for the consequences of their use.



Direct all inquiries to CRC Press, LLC, 2000 Corporate Blvd., N.W., Boca Raton, Florida 33431.


International Standard Book Number 0-8493-2529-3
Library of Congress Card Number 95-17139
Printed in the United States of America 2 3 4 5 6 7 8 9 0
Printed on acid-free paper

# PREFACE

This is the second volume of the *Practical Handbook of Genetic Algorithms*. In the first volume we covered a number of GA applications. This required code segments for a range of differing applications and problems encountered in the coding of GAs. These code examples were offered in a number of differing languages so that people with differing language proficiencies could benefit.

This second volume covers an array of new areas for the applications of GAs, arenas not seen before or if seen only glimpsed occasionally. The reasons for selecting these particular undertakings we so that problems particular to these areas could be canvassed. These new applications each have problems that are unique and require unique approaches. This book covers these approaches and offer a springboard for the theorist or practitioner to continue their endeavors.

The field of genetic algorithms is growing and developing so fast that it is impossible to remain up-to-date in all areas in which it is applied. This volume also demonstrates some of the leading applications in the field as of the date of publication. With this information on hand it becomes possible for researchers and practitioners to gain a glimpse of the wide array of problem types within which GAs can successfully operate and to consolidate any thoughts they may have had as to the value of continueing down a GA-centered path.

This collection, like the last, demonstrates the significant ability of GAs to solve new and different problems in ever more elegant ways. It is hoped that this work will, as has the first volume, prove to be of great value to all those who read and refer to it.

I feel emboldened by the people I have met and communicated with during the development of this work. We all share a common language, understanding, and faith in the capacity of GAs, either alone or in a hybridized environment, to generate significant real value in our world.

We are still, and will continue to be, confronted with significant quantitative and qualitative problems that need to be solved to further improve our quality of life. It may sound rather far-fetched and grandiose to attribute further advances of great significance to the study and use of genetic algorithms but I don't believe that these statements are at all at odds with what will become reality. We are still at the frontiers of understanding the impacts of Evolutionary Computation in all its forms and manifestations.

The new sciences of which Evolutionary Computation and hence GAs is one, from the select fields of Nanotechnology, AI, Fuzzy Systems, Complexity, Catastrophe, and Chaos Theories, are all in the early gestation stages. If there were ever a new light for science today it is, in my opinion, led by those on this short list.

Read the books, talk to the people, and imagine what is possible.

We can achieve a brilliant future if we can develop and apply the right tools to the right problems. I hope this book will assist in setting people onto a path that will allow each to contribute what they are able towards achieving that hoped for future.

Any errors, omissions, and mistakes are purely my fault and no other party can or should be held accountable.

Have fun reading it, I know I did in editing it.

Lance Chambers
140 Treasure Road
Queens Park
Perth
West Australia 6107

e-mail: stratthink@dot.wa.gov.au

May 25, 1995

# TABLE OF CONTENTS

## Chapter 0

**V. Rao Vemuri and Walter Cedeño**
Department of Applied Science
University of California, Davis and
Lawrence Livermore National Laboratory
Livermore, CA 94550

(vemuri@icdc.llnl.gov and wcedeno@llnl.gov )

# Multi-Niche Crowding for Multi-Modal Search



## Abstract

Multi-Niche Crowding (MNC) is a new genetic algorithm designed for locating multiple peaks in a multimodal function. The validity of this method is demonstrated by applying it first to a variety of test functions. Then the method is applied to solve the problem of assembling the so-called restriction fragments obtained from partial digestion of DNA molecules.

0-8493-2529-3/95/$0.00 + $.50

## 0.1 Introduction

Searching for extrema in a multi-modal space is different from locating the extremum of a unimodal function. When a search technique proven to be useful for unimodal functions is applied to multi-modal functions, the method tends to converge to an optimum in the local neighborhood of the first guess. One can use methods like simulated annealing to escape from a local optimum and locate the global optimum. However, there are many applications where the location of "k best extrema" of a multi-modal function are of interest. Searching for these locations goes by the name "multi-modal optimization." In this chapter we describe a genetic algorithm (GA) suitable for multi-modal function optimization.

A GA is a mathematical search technique based on the principles of natural selection and genetic recombination [Holland, 1975]. A possible solution to a problem is referred to as an *individual.* An individual is represented by a computational data structure called a *chromosome*, which is encoded using a fixed length alphabet. *Species* are individuals with a common characteristic and *niches* are subdomains of the search space. By encouraging niching and speciation, GAs can facilitate simultaneous convergence to more than one optimum in a multi-modal search space.

Section 2 of this chapter describes multi-modal function optimization using the multi-niche crowding (MNC) model. In Section 3 the MNC method is applied to a number of test functions. In Section 4, the MNC method is used to solve the so-called DNA restriction fragment assembly problem, an important and challenging problem from biotechnology. In the last two sections of the chapter we included some discussion on the suitability of this method to the solve the restriction fragment assembly problem.

## 0.2 Genetic Algorithms for Multi-Modal Search

There are many versions of genetic algorithms, one differing from another in some detail. In a nutshell, all genetic algorithms have two basic steps: during the *selection step*, a decision is made as to who in the population is allowed to produce offspring, and during the *replacement step* another decision is made as to which of the members from one generation are forced to perish (or vacate a slot) in order to make room for an offspring to compete (or, occupy a slot).

The Simple GA (or SGA), which will be used as a point of departure for presenting the method discussed here, starts by randomly generating a population of $N$ individuals, that is, individual solutions. These individuals are evaluated for their fitness. Individuals with higher fitness scores are selected, with replacement, to create a mating pool of size $N$. This method of selection is called *fitness proportionate reproduction* (FPR). The genetic operators of crossover and mutation are applied at this stage in a probabilistic manner which results in some individuals from the mating pool to reproduce. The assumption here is that each pair of parents produces only one pair of offspring through the crossover operation. Now the population pool contains some individuals who never got a chance to reproduce and the offspring of those who got a chance to reproduce.

The procedure continues until a suitable termination condition is satisfied. All other versions of GAs differ from this basic method in some detail or another.

The steady-state GA (or SSGA) differs from SGA mainly in the replacement step, and to a lesser extent on the way the genetic operators are applied [Whitley, 1988; Syswerda, 1989]. The SSGA selects two individuals using FPR and allows them to mate to produce two offspring. This selection step is identical to the corresponding step of SGA. However, in SSGA, the offspring are inserted into the population, thus replacing two individuals, soon after they are generated whereas the SGA generates *N* offspring prior to replacing the entire population. In other words SGA uses *simultaneous* replacement strategy whereas the SSGA uses the *successive* replacement strategy. Thus SGA and SSGA are analogous, respectively, to the Jacobi and Gauss-Seidal methods of solving systems of algebraic equations.

Both SGA and SSGA suffer from the possibility of premature convergence to a local minimum, primarily due to the *selection pressure* exerted by the FPR rule. Simply assigning an exponential number of mating trials to those members of the population that exhibit above average survival traits, as the FPR rule does, is not a good strategy for a thorough exploration of complex search spaces with multiple peaks. Due to this problem as well as the *deceptiveness* exhibited by muti-modal search spaces [Goldberg et al., 1992], SGA and SSGA are not suitable for multi-modal search and optimization.

### 0.2.1 Multi-Modal Search: Crowding-Based Methods:

The GA model described here, called *multi-niche crowding* (or, MNC), has the ability to converge to multiple solutions at the same time by encouraging competition between individuals within the same locally optimal group (Cedeño and Vemuri, 1992, 1994). In MNC, both the selection and replacement steps of the SGA are modified with the introduction of some form of *crowding* in order to render it suitable for searching spaces characterized by multiple peaks or niches. So it is essential that the concept of crowding is briefly reviewed here.

Crowding (De Jong, 1975) is a generalization of *preselection* (Cavicchio, 1970). In *crowding*, selection and reproduction are the same as in the SGA; but replacement is different. For concreteness, it is assumed that two parents are selected to produce two offspring. In order to make room for these offspring, it is necessary to identify two members of the population for replacement. The policy of replacing a member of the present generation by an offspring is carried out as follows, in two steps. First, a group of C individuals is selected at random from the population. C, called the *crowding factor*, indicates the size of the group. A value of C = 2 or 3 appears to work well for De Jong. Second, the bit strings in the offspring chromosomes are compared with those of the C individuals in the group using Hamming distance as a measure of similarity. The group member that is most similar to the offspring is now replaced by the offspring.This procedure is repeated for the other offspring as well. This second offspring can conceivably replace its own sibling that just entered the population pool, although such a scenario is rather unlikely. In any event, crowding is essentially a successive replacement strategy. This strategy maintains the diversity in the population and postpones premature convergence. Crowding cannot maintain stable subpopulations due to the selection pressure imparted by FPR. Summarizing, in crowding offspring replace similar individuals from the population. Crowding slows down premature convergence of the traditional GA and in most cases can find the global optimum in a multi-modal search space. On the other hand it does not converge to multiple solutions and after many generations one of the peaks takes over.

In *deterministic crowding* (Mahfound, 1992) selection pressure is eliminated and preselection is introduced to obtain a GA suitable for multi-modal function optimization. This version appears to have minimal overhead, thus contributing to its efficiency. In this method selection pressure is eliminated by allowing individuals to mate at random with any other individual in the population. Pressure is applied, however, during replacement step using *preselection.* Toward this goal, each of the two offspring is first paired with one of the parents; this pairing is not done randomly, rather the pairings are done in such a manner the offspring is paired with the most similar parent. Then each offspring is compared with its paired parent and the individual with the higher *fitness* is allowed to stay in the population and the other is eliminated. It is not clear if multiple solutions can be maintained for many generations using this method, although it appears that multiple solutions are sustained for more generations than when crowding is used alone.

### 0.2.2 Multi-Modal Search: Sharing-Based Methods

Goldberg and Richardson [1987] used the *sharing* concept of Holland [1975] as a way of reducing the selection pressure caused by FPR. In sharing, the fitness values of an individual are adjusted downward in accordance with the number of individuals in its neighborhood or niche. The more individuals there are in a niche, the more pressure they create on each other. The downward adjustment of fitness value is done with the help of a suitably defined *sharing function,* a function that takes into account the similarity (i. e., physical proximity) between two individuals. This approach allows multiple solutions to converge in parallel but it is very computing intensive. Deb and Goldberg [1989] applied mating restriction to sharing methods. Mating restriction only allows individuals within the same niche to mate. The method is too restrictive and the user must have an idea of the search space in order to define a suitable sharing function. Yin and Germay [1993] introduced cluster analysis to sharing to reduce its complexity and to group similar members naturally. Here the user must provided special attention to the parameters for the clustering algorithm to select the correct ones for the problem at hand.

Beasley *et al.* [1993] applied traditional GAs to multi-modal function optimization by using a fitness derating function to prevent convergence to a known local optima. In their approach the GA is applied iteratively to the problem and every solution found in previous iterations is used to derate the fitness of individuals near them. The time complexity is similar to that of sharing functions. Like in sharing the user must have an idea of the search space beforehand. Derating a large region may eliminate a possible solution.

### 0.2.3 Multi-Niche Crowding (MNC)

In multi-niche crowding (MNC), both the selection and replacement steps are modified with some type of crowding. The idea is to eliminate the selection pressure caused by FPR while allowing the population to maintain some diversity. This objective is achieved, in part, by encouraging mating and replacement within members of the same niche while allowing for some competition for population slots among the niches. The result is an algorithm that (a) maintains stable subpopulations within different niches, (b) maintains diversity throughout the search, and (c) converges to different local minima.

In MNC, the FPR selection is replaced by what we call *crowding selection.* In crowding selection, each individual in the population has the same chance for mating in every generation. Application of this selection rule takes place in two steps. First, an individual A is selected for mating. This selection can be either sequential or random. Second, its mate M is selected, not from the entire population, but from a group of individuals of size $c_s$, picked at random from the population. The mate M thus chosen must be the one who is the most "similar" to A. The similarity metric used here is not a *genotypic metric* such as the Hamming distance, but a suitably defined *phenotypic distance metric*. Crowding Selection promotes mating between individuals from the same niche while allowing matings between individuals from different niches.

During the replacement step, MNC uses a replacement policy called *worst among the most similar.* The goal of this step is to pick an individual from the population for replacement by an offspring. Implementation of this policy follows these steps. First, $c_f$ groups are created by randomly picking *s* individuals per group from the population. These groups are called *crowding factor groups.* Second, one individual from each group that is most similar to the offspring is identified. This gives $c_f$ individuals that are candidates for replacement by virtue of their similarity to the offspring that will replace them. From this group of most similar individuals, we pick the one with the lowest *fitness* to die and that slot is filled with the offspring.

A similar technique called *enhanced crowding* (Goldberg 1989) has been used before in classifier systems, but there the most similar individual out of a group of worst candidates is replaced. Figure 0.1 is an example of the worst among most similar replacement strategy.

After the offspring becomes part of the population it competes for survival with other individuals when the next offspring is inserted in the population. In *worst among most similar replacement* offspring are likely to replace low fitted individuals from the same niche. It can also happen that it replaces a higher fitted individual from the same niche or an individual from another niche. This allows a more diverse population to exist throughout the search. At the same time estimulates competition between members of the same niche and between members belonging to different niches as well.

The following pseudo-code summarizes the salient features of the method:

```
1 Generate initial population of N individuals
2 For gen = 1 to MAX_GEN
3         For i = 1 to N
4                 Use crowding selection to find mate for individual i
5                 Mate and mutate
6                 Insert offspring in the population using
                      worst among most similar replacement.
```

If we use FPR in Step 4 shown above and replace the lowest fitted individuals in the population (Step 6) with the newly generated offspring, this model corresponds to a steady-state GA. In contrast with the most common generational GA, offspring are available for mating as soon as they are generated, and good individuals can survive for many generations. For the purposes of this paper, a generation is every *N*/2 mating operations, where *N* is the population size.

MNC GA converges consistently to the global optimum. The complexity added by the selection and replacement operators to the GA is dependent on the values of *Cs* and *Cf s*. Normally *Cf* is a value in the interval [1, 4], and *Cs* and *s* are values ranging from 1% to 15% of the population size.

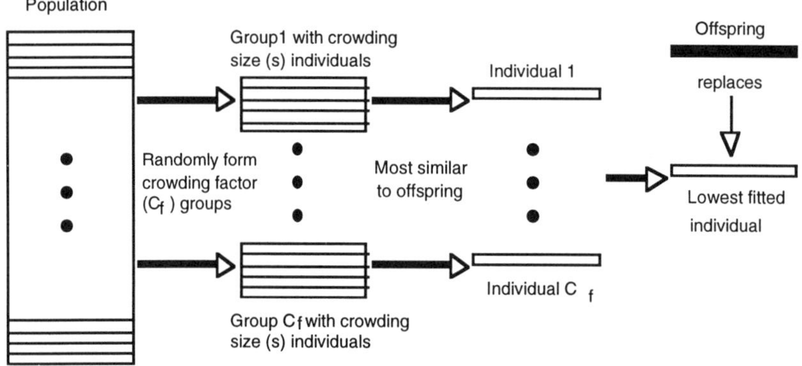


Figure 0.1: Schematic showing crowding factor groups created during the replacement step.

## 0.3 Application of MNC to Multi-Modal Test Functions

To evaluate the performance of the MNC model we used five different test functions. The first three of these functions have been used earlier by other investigators. Functions *F1(x)* and *F2(x)* defined by

$$F_1(x) = \sin^6(5.1\pi x + 0.5),$$

$$F_2(x) = \exp^{-4(\ln 2)(x-0.0667)^2/0.64} \sin^6(5.1\pi x + 0.5),$$

are shown on Figure 0.2. They correspond to the five-optima sine functions used by Goldberg in his work on sharing. Both functions were maximized using binary chromosome encoding for numbers in the interval [0, 1]. In both cases the GA with sharing was able to maintain stable subpopulations and diversity in the population. Convergence was good, but not all the peaks in $F_1$ had the individuals distributed close to the top. It is not clear that in $F_2$ sharing will be able to maintain a proportional number of individuals for larger number of generations. Function *F3(x,y)*, called "Shekel's foxholes" with twenty five optima, was used by De Jong in his work with crowding.

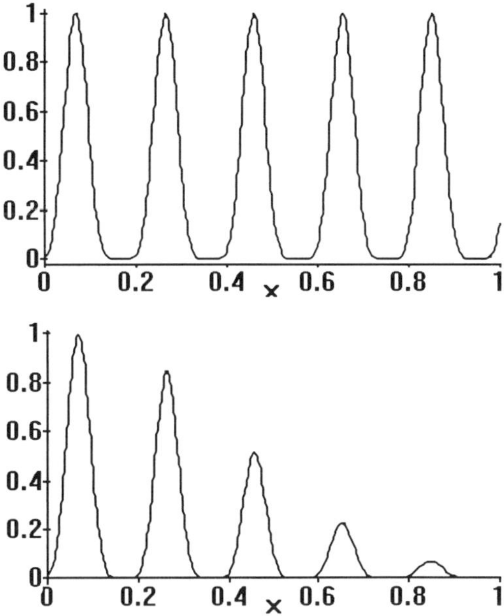


Figure 0.2: Test functions *F1(x)* (top) and *F2(x)* (bottom).

We also considered other functions not exhibiting the symmetry present in the above functions. Function *F4(x,y)*, shown on the left in Figure 0.4, contains two global optima with the same height and width but located far apart. Function *F5(x,y)*, the sample shown on the right in Figure 0.4, contains five optima with height, width, and location chosen at random in every run. Both of these functions are defined by

$$\sum_{i=1}^{p} \frac{A_i}{1 + W_i\left((x - X_i)^2 + (y - Y_i)^2\right)},$$

where $p$ indicates the number of peaks in the function, (*Xi*, *Yi*) the coordinates of peak *i*, *Ai* the height of peak *i*, and*Wi* determines how narrow or wide is the base of peak *i*.

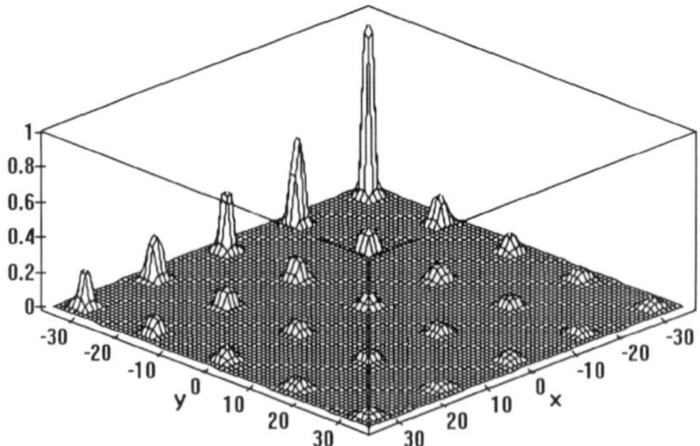


Figure 0.3: Test function *F3*(x), Shekel's foxholes.

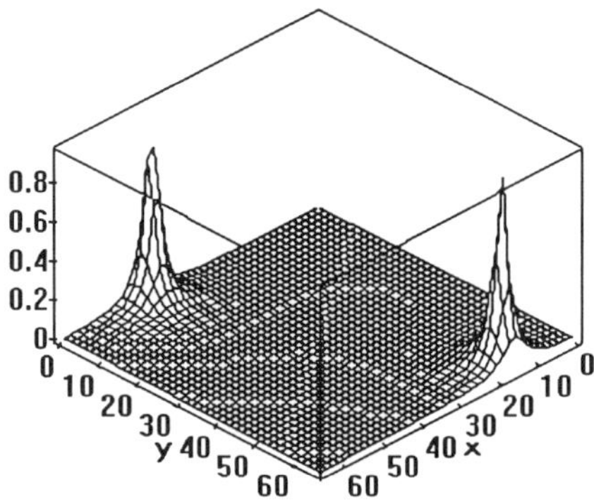


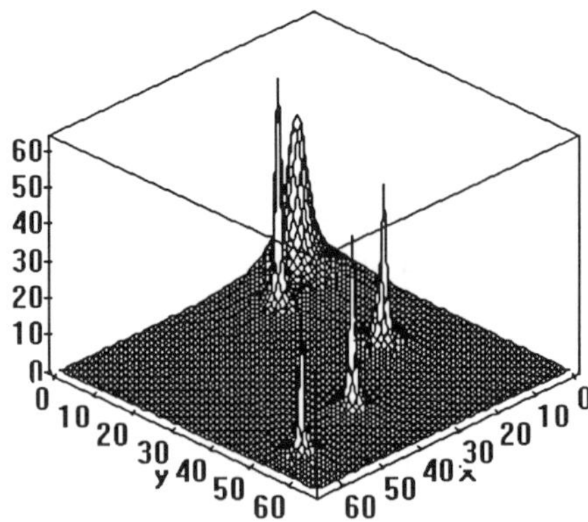


Figure 0.4: Test functions *F4* (top) and *F5* (bottom).

The simulations were done in a 486/33MHz PC. The variables $x$ and $y$ were encoded using two 30 bit chromosomes. To generate the initial population the search space was divided into $n$ (population size) equally sized regions. One individual was chosen at random in each region. The crossover probability ($pc$) was set at 0.95 and mutation probability ($pm$) at 0.001. Similarity between two individuals was determined by the Euclidean distance between the two points. The MNC GA was executed for 100 generations in each run. Other parameters are summarized in Table 0.1.

| | *F1 & F2* | *F3* | *F4* | *F5* |
|---|---|---|---|---|
| Population size ($n$): | 100 | 500 | 100 | 200 |
| Number of chromosomes: | 1 | 2 | 2 | 2 |
| Crowding selection size ($Cs$ ): | 15 | 75 | 5 | 15 |
| Crowding factor ($Cf$ ): | 2 | 2 | 3 | 3 |
| Crowding size ($s$): | 15 | 75 | 5 | 15 |

Table 0.1: Function specific parameters used in the MNC GA.

For all test cases the MNC GA was able to maintain stable subpopulations in all the higher peaks without exhibiting premature convergence. Only very small peaks in *F3* and *F5* did not have any significant number of individuals in the last generation even though some were present during the initial generations. The number of individuals at the lower peaks decreased as the fitness of the individuals in other peaks increased. This can be observed in Figure 0.5 for the two peaks located near (0,0) in function *F5*. The plot on the left has the average fitness of each peak for every generation. The plot on the right has the number of individuals in each peak for every generation. During the initial ten generations the wider peak had more individuals in the population. As the fitness of the individuals in the other peak improved the number of individuals increased. After about 20 generations the number of individuals in the skinnier peak was greater than those in the wider peak. There are of course other factors that contribute to this pattern and that needs to be investigated further.

We also observed that the number of individuals in a peak is related to more than just its height. In function *F3* where the optima are located on a 5x5 grid, peaks along the same $x$ and $y$ axis as the global optima had more individuals than other peaks with higher values. Some of the extra individuals can be attributed to mutation since a bit change in one of the chromosomes will cause an individual to move along the $x$ or $y$ axis. We ran the same test with mutation set at 0.0 and no major changes were observed. More tests are needed to determine other factors affecting the number of individuals in a peak.

The functions *F1*, *F2*, and *F4* were not hard for the GA. After 100 generations all five global optima in function *F1* were found including the local optimum at $x$ = 1.0. All five optima were also found for function *F2*. Both the peaks at *F4* were also successfully found. The distance between the peaks did not cause any problem for the GA.

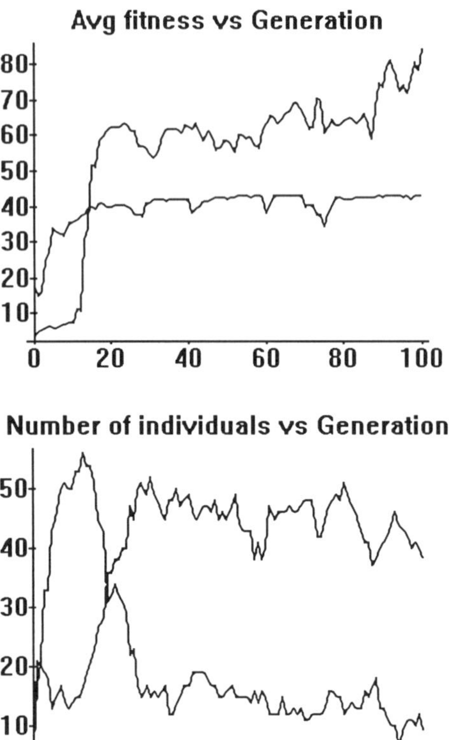


Figure 0.5: Average fitness and individuals in population for two peaks in function *F5*.

Overall, the properties exhibited by the MNC GA are very encouraging. A more rigorous analysis is necessary before statements can be made about the validity of this method in a more general context.

## 0.4 Application to DNA Restriction Fragment Map Assembly

The genetic material contained in the chromosomes of a cell is collectively called the genome. Chromosomes are essentially DNA molecules which are made out of four distinct types of molecules called nucleotides. Although these four nucleotides, denoted here by the letters A, C, G, T, can theoretically form sixteen pairs, only AT, TA, CG and GC pairings are allowed. It is estimated that the human genome is comprised of about three billion of these "base-pairs." The ambitious goal of the Human Genome Project is to decipher or map the exact sequence of these three billion nucleotide molecules.

### 0.4.1 Restriction Fragment Data

The experimental process of gathering DNA data is too intricate to describe here. However, a thumbnail sketch is sufficient to describe the broad outlines of what really takes place in the laboratory. First, the DNA molecule is cut to a manageable size using a so-called restriction enzyme. If total digestion is desired, then the enzymes are allowed to act on the restriction sites (of multiple copies) for a sufficiently long time and the result will be several identical pieces of DNA with non-overlapping base-pair sequences. However, under partial digestion, not all restriction sites experience the same rate of reaction. Furthermore, under partial digestion, the same restriction site on different copies of a DNA may behave differently. At some point, when the reaction is stopped, one finds many DNA fragments with overlapping base-pair sequences. Both complete digestion and partial digestion have roles to play in DNA sequencing studies.

Typically, the hierarchy of the fragmentation process goes somewhat like this. The DNA under study is first divided, using complete digestion into non-overlapping islands, called *contigs,* whose lengths may range between 150K to 200K base-pairs. This step is purely for the convenience of working with shorter pieces of a chromosome of manageable length. These contigs are then inserted into a type of viral DNA called a cloning vector. The type of cloning vector used in our study are called *cosmids*. These cosmid cloning vectors, containing a piece of the human DNA, are then used to infect bacterial cells at the rate of one per bacterial cell. The bacteria reproduce rapidly, and in doing so produce many copies of the piece of human DNA that was inserted into the vector. The identical copies of the contigs thus obtained are called cosmid clones. These clones are now subjected to partial digestion with restriction enzymes. This process yields several copies of cosmid clones, with overlapping base-pair segments. That is, identical base-pair segments from the original parent contig may appear in two cosmid clones, say Cosmid clone A and Cosmid clone B. Typically, each cosmid clone is approximately 40K base-pairs long. Deciding the relative position of a cosmid clone on the parent contig by inspecting these 40,000 base-pairs is a tortuous task. To render the problem more manageable, each cosmid clone is further divided into the so-called *restriction fragments* using complete digestion with a single restriction enzyme, such as EcoRI. These restriction fragments can be physically separated by size when they are placed in a porous gel. When an electrical current is applied to the gel, fragments of different sizes line up as bands and this pattern of bands is called the *fingerprint* of the cosmid clone. These fragments typically range in length from 0.5K to 15K base-pairs. A pictorial view of this process is shown in Figure 0.6. During this process information about the relative location of the islands, cosmid clones, and fragments in the original DNA is lost.

The DNA Restriction Fragment Map Assembly problem then is comprised of reassembling the cosmid clones in order to reestablish the correct sequence of base-pairs in the original contig. This reassembly is not a trivial exercise. First, there are only four base-pairs, repeated thousands of times along the length of a DNA with no predictable pattern. So when a DNA is cut down to the size of fragments, it is conceivable that one finds many fragments that are similar. Second, due to experimental difficulties, the restriction enzymes do not always succeed in cutting the fragments. Due to these reasons, assembling the contigs from fragment data continues to be a challenging problem.

To establish the relative position of each cosmid clone on a contig, we determine the possible locations of the fragments in each cosmid clone in such a manner that the fragment overlap among the clones is maximized while a suitably defined total error between the overlapped fragments is minimized. There are other constraints such as the total length of the assembly should be equal to the contig's original size. The problem is one of assembling cosmid clone sequences as shown in Figure 0.6. The problem is complicated further by the uncertainty in the data, the possibility of data loss (fragments of the same size are hard to distinguish during fingerprinting), and the known fact that data related to *corner fragments* (i.e., fragments near the fragment boundaries) are almost always unreliable. Problems of this type are known to be hard [Opatrny, 1979]. For example, the case with 10 cosmid clones, there are 10!/2 possible clone sequences.

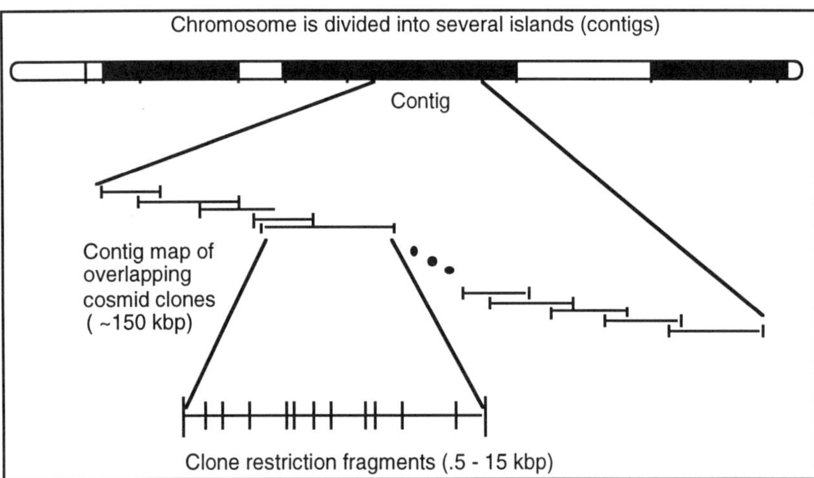


Figure 0.6: Physical map for island using fragments from a set of overlapping clones.

### 0.4.2 Problem Representation

Before going into the details about the use of genetic operators, it is important to show how the data for the problem is presented to the GA. Figure 0.7 shows fragment sizes obtained from fingerprinting for a set of overlapping cosmid clones.

| ALLELE NUMBER | CLONE ID | FRAGMENT SIZES (in thousands of base-pairs) | | | | | | | | |
|---|---|---|---|---|---|---|---|---|---|---|
| C0 | 5154 | 16.55 | 4.4 | 1.68 | 1.07 | 4.81 | 8.5 | | | |
| C1 | 7442 | 0.79 | 0.79 | 2.6 | 4.35 | 8.24 | 2.7 | 6.9 | 5.16 | |
| C2 | 21230 | 0.96 | 1.68 | 1.08 | 4.77 | 8.47 | 1.44 | 2.37 | 6.29 | 0.62 |

| | | |
|---|---|---|
| C3 | 8131 | 0.92 3.73 19.8 4.43 1.69 1.25 4.68 5.63 |
| C4 | 18993 | 0.96 6.31 5.48 8.61 7.29 0.81 0.81 2.6 4.36 1.92 |
| C5 | 5435 | 2.89 8.24 2.7 6.9 5.14 5.14 2.89 1.54 |
| C6 | 7255 | 1.04 8.21 2.69 6.89 5.12 5.12 2.88 1.94 2.42 1.37 3.33 |
| C7 | 12282 | 4.52 5.13 5.13 2.87 1.94 2.42 1.39 3.35 5.41 |
| C8 | 27714 | 6.69 5.07 5.41 2.88 1.92 2.32 1.4 3.35 5.46 17.65 1 10.49 0.58 1.74 |
| C9 | 10406 | 2.03 1.43 2.34 6.28 5.46 8.58 7.27 |

Figure 0.7: Cosmid clones with fragment data.

For example, cosmid clone with the ID number 5154 which is also labelled as Allele Number C0, is known to be comprised of six fragments, containing 16550, 4400, 1680, 1070, 4810 and 8500 base-pairs, in that order. Also, cosmid clone with the ID number 8131 which is also labelled as Allele Number C3, is known to be comprised of eight fragments, containing 920, 3730, 19800, 4430, 1690, 1250, 4680 and 5630 base-pairs, in that order. By comparing these fragment sequences one can surmise that the third and fourth fragments of Allele C0 are probably the same as the fifth and sixth fragments of Allele C3 mainly because the fragment lengths are so nearly equal to each other. If this is true then Allele C3 can be "aligned" below Allele C0 in such a manner that the fifth and sixth fragments of Allele C3 fall right below the third and fourth fragments of Allele C0 as shown in Figure 0.8. The matching of the fragment lengths is not perfect. Indeed, the mismatch at other positions is indeed large. The goal of this problem is to maximize this type of matching while minimizing the number and degree of mismatches and keeping the total length of the assembly within reasonable limits. The data for this problem consist of the $M$ cosmid clones with their fragments and the tolerance measure $e$ which is used to determine if two fragments are of the same size. That is, two fragments $F1$ and $F2$ are considered to be of the same size if $|F1 - F2| < e$.

| <u>Clone</u> | | |
|---|---|---|
| C3 | 8131 | 0.92 3.73 19.8 4.43 1.69 1.25 4.68 5.63 |
| C0 | 5154 | 16.55 4.4 1.68 1.07 4.81 8.5 |
| C8 | 21230 | 0.96 1.68 1.08 4.77 8.47 1.44 2.37 6.29 0.62 |

Figure 0.8. An example of fragment assembly.

### 0.4.3 Chromosome Encoding

The encoding for this problem is very simple. Each allele in the chromosome has a label between 0 and $M$ - 1 corresponding to one of the cosmid clones. No two alleles have the same label, and mating and mutation will preserve this constraint. In Figure 0.7, for example, an allele with the label C0 corresponds to the clone with ID 5154 and an allele with the label C9 corresponds to clone ID 10406. The clone sequence (5154, 21230, 10406, 7255, 12282, 27714, 8131, 18993, 7442, 5435), for example, is represented by the chromosome (0 2 9 6 7 8

3 4 1 5). The initial population is generated by picking at random values between 0 and $M$ - 1 without replacement.

| | Number of matches between clones | | | | | | | | | | Total error in the matches | | | | | | | | | |
|---|---|---|---|---|---|---|---|---|---|---|---|---|---|---|---|---|---|---|---|---|
| CLONE | C0 | C1 | C2 | C3 | C4 | C5 | C6 | C7 | C8 | C9 | C0 | C1 | C2 | C3 | C4 | C5 | C6 | C7 | C8 | C9 |
| C0 | 6 | 1 | 4 | 2 | 1 | 0 | 1 | 0 | 2 | 1 | 0 | 5 | 8 | 4 | 4 | 0 | 3 | 0 | 13 | 8 |
| C1 | 1 | 8 | 0 | 1 | 4 | 4 | 4 | 1 | 1 | 0 | 5 | 0 | 0 | 8 | 5 | 2 | 9 | 3 | 9 | 0 |
| C2 | 4 | 0 | 9 | 3 | 2 | 1 | 3 | 2 | 5 | 3 | 8 | 0 | 0 | 14 | 2 | 10 | 16 | 10 | 23 | 5 |
| C3 | 2 | 1 | 3 | 8 | 2 | 0 | 0 | 1 | 2 | 0 | 4 | 8 | 14 | 0 | 11 | 0 | 0 | 9 | 13 | 0 |
| C4 | 1 | 4 | 2 | 2 | 10 | 1 | 3 | 2 | 3 | 4 | 4 | 5 | 2 | 11 | 0 | 10 | 19 | 9 | 6 | 10 |
| C5 | 0 | 4 | 1 | 0 | 1 | 8 | 6 | 3 | 2 | 0 | 0 | 2 | 10 | 0 | 10 | 0 | 10 | 4 | 8 | 0 |
| C6 | 1 | 4 | 3 | 0 | 3 | 6 | 11 | 7 | 7 | 3 | 3 | 9 | 16 | 0 | 19 | 10 | 0 | 7 | 26 | 23 |
| C7 | 0 | 1 | 2 | 1 | 2 | 3 | 7 | 9 | 7 | 4 | 0 | 3 | 10 | 9 | 9 | 4 | 7 | 0 | 20 | 26 |
| C8 | 2 | 1 | 5 | 2 | 3 | 2 | 7 | 7 | 14 | 3 | 13 | 9 | 23 | 13 | 6 | 8 | 26 | 20 | 0 | 5 |
| C9 | 1 | 0 | 3 | 0 | 4 | 0 | 3 | 4 | 3 | 7 | 8 | 0 | 5 | 0 | 10 | 0 | 23 | 26 | 5 | 0 |

Figure 0.9: The M matrix on the left whose entries are the number of fragment matches and the E matrix on the right whose entries are the total errors between any two clones with $e$ = 10.

### 0.4.4 Fitness Function

To calculate the fitness of an individual, the number of fragment matches between all consecutive clones and the error between the fragments is considered The fragment sizes are represented using integer numbers by multiplying the number given in Figure 0.7 by 100, just to avoid dealing with decimals. All values are computed using integers to accelerate computation of the fitness function. Prior to the execution of the GA, two matrices are calculated. One matrix, the match matrix M shown on the left side of Figure 0.9, contains the number of fragments that match between two clones Ci and Cj, within an error tolerance $e$ . The other matrix, the error matrix E shown on the right side of Figure 0.9, contains the total error between the clones being matched. The error between two clones is given by the sum of the errors between all fragments that matched. For example, between clone No. 8131 (C0) and clone No. 5154 (C3) there are two pairs of fragments that match within the specified tolerance of e = 10. The lengths of these fragments are 169 and 168 for one pair and 443 and 440 for the second pair. Thus a 2 appears in (row 2, column 4) of the Match Matrix M. The total error between both pairs of fragments is (169-168) + (443-440) = 4, which is shown in (row 1, col. 4) of the Error Matrix, E.

Our goal is to arrange the cosmid clones as shown in Figure 0.6 so that the lengths of the overlapping fragments match with each other as closely as possible. The necessary matching information is already gathered in the matrix M and the degree of accumulated mismatch per clone is gathered in the matrix E. However, we believe that this information alone is not sufficient to establish which two clones are "adjacent" to each other in the arrangement shown in Figure 0.6. For example, consider how clone C0 (i.e., allele No. 5154) matches with other clones. Inspection of the match matrix M indicates that the degree of match between clone C0 and clone C2 (or, equivalently, allele No. 21230), is 4 matches. Also, fragments in clone C0 match with fragments in clone C3 as well as C8, each with 2 matches. By interpreting this to mean that C2 should be placed nearer to C0 than C3 or C8, we are ignoring information contained in the C0-C3 matches and C0-C8 matches. One more example suffices to make the point. Clone C3 should be placed closer to C2 because they match with each other the maximum number of times, namely 3, although C3 matches with three other clones, each with only 2 matches. This phenomenon makes us think that using only the number of matches between clones is not sufficient to establish the partial order between clones when they possess the same match count. We believe that part of this problem is due to false matches, between fragments of similar sizes, that may occur by chance. We tried to overcome this problem by incorporating the total error in the matches, shown in matrix E, in order to enable our GA to discriminate further between clones. Using the same example, notice that clone C3 has less total error when matched with C0 than clone C8 and therefore indicates that C0 is adjacent to C3. The following equation for fitness captures the essence of the method described so far.

$$fitness = \sum_{i=0}^{M-1} \sum_{\substack{j=-1 \\ j \neq 0}}^{1} \left( \frac{Match[C_i, C_{i+j}]}{Count[C_{i+j}]} \times \left(1 - \frac{Error[C_i, C_{i+j}]}{Match[C_i, C_{i+j}] \times \varepsilon}\right)\right) * 100$$

Here $C_i$ refers to the fact that the cosmid clone Ci has been placed in the ith position of the chromosome, $C_{i+j}$ refers to the clone to the right or left (if any) of the ith position. Match $[C_i, C_{i+j}]$ counts the number of fragments that matched between $C_i$ and $C_{(i+j)}$ within the specified tolerance. That is Match $[C_i, C_{i+j}]$ is the degree of match between $C_i$ and $C_{(i+j)}$. This quantity is divided by Count $[C_{i+j}]$ to get a normalized count for the degree of match. The term Error $[C_i, C_{i+j}]$ refers to the total error accumulated over all fragments that matched between $C_i$ and $C_{(i+j)}$. By dividing this with the number of matches times the error tolerance $\varepsilon$, we are essentially getting the normalized error per fragment. When this normalized error reaches unity, it means that the total error is so large that any apparent matches are worthless. With this interpretation, the second term of the equation essentially tells us the degree of confidence we can place on the normalized matches we are counting in the first term. In the above equation, M is the number of alleles in the chromosome.

By defining the fitness function as above, we are assigning a higher fitness to those clone pairs that match a higher percentage of their regions. For example, Allele 0 with 6 fragments has two matches each with Alleles 3 and 8, each having 8 and 14 fragments respectively. Since 2/8 represents a higher percentage than 2/14, we designed a fitness function that prefers a configuration that places Allele 3 closer to Allele 0 than Allele 8. This is achieved by dividing the number of matches by the number of fragments in the clone.

Before settling on the fitness function described above, others were considered. For example, fitness functions that just counted the number of matches between clones with no regard tp normalization failed to produce the correct answer. A fitness function that just counted the number of matches and then subtracted the total error in those matches also failed to give satisfactory results. It is possible that other fitness functions may give results that are even better than what are reported here. In the future, we plan to include the number of matches as well as the error among groups of three clones as factors and study its effect on performance.

### 0.4.5 Mating and Mutation Operators

The mating operator used in this method is based on a slight modification to the genetic edge recombination operator that was applied successfully to solve the TSP (Traveling Sales Person) problem. As in the TSP problem, the important information here is the adjacency of the alleles, although the order the alleles appear in the chromosome can be derived from the adjacency information. The idea is to recombine the links (pairs of clones) between two parents such that common links are inherited by the offspring. This operator is implemented in two steps as shown in Figure 0.10. First, those links (or traits) that are common to both the parents are identified and passed on to the offspring and the links occupy the same absolute positions in the offspring chromosome. In the example shown in Figure 0.10, the relevant link-pairs are 7-8, 8-1, and 5-0 in the first parent and 1-8, 8-7 and 5-0 in the second parent. Notice that these links are passed on to the two offspring undisturbed. Second, those alleles that are not passed to the offspring (indicated by dashes, in Figure 0.10) are randomly assigned to the available positions while observing the constraint that no link label is repeated.

| Parent 1 | Common links (traits) | Offspring 1 |
|---|---|---|
| (6 7 8 1 2 3 9 4 5 0) | (- 7 8 1 - - - - 5 0) | (3 7 8 1 4 6 2 9 5 0) |
| Parent 2 | | Offspring 2 |
| (1 8 7 9 5 0 2 6 3 4) | (1 8 7 - 5 0 - - - -) | (1 8 7 3 5 0 6 2 4 9) |

Figure 0.10: Modified genetic edge recombination for clone sequencing.

The differences between this operator and the original edge recombination operator are in the number of offspring generated and in the assignment of alleles not transferred from the parent. We generated two offspring instead of one because the location of the links in the clone sequence is important to our problem. In TSP the chromosome is circular, thus the location did not matter. We allow both parents to pass the location of the links to their offspring. To assign the other alleles we select them at random from those clones not passed by their parents. In the original operator the links are assigned from those present in any of the two parents. Alleles with fewer links are assigned first to prevent from running out of links for a given allele.

In the mating operator used here there is excessive exploration of the search space primarily due to the random filling of the unassigned slots, in the second step, while creating the offspring chromosomes. Part of this exploration difficulty is alleviated by the fact that mates are selected using crowding selection and therefore they have common features between them. Exploration is also limited to a smaller region within the entire search space. On the other hand, by allowing unassigned clones to be chosen at random, we are allowing links to re-appear that might not have done so using mutation alone.

Mutation is applied on an individual basis. After an offspring is generated it is mutated if the outcome from the flip of a biased coin is true. When this happens, a link from the offspring is selected at random and all alleles from that link to the last position of the chromosome are reversed. For example, the offspring (1 8 7 3 5 0 6 2 4 9) after mutation can result in (1 8 7 9 4 2 6 0 5 3) if the link between Allele 7 and 3 is selected to mutate.

The mating and mutation operators are compatible with each other in the sense that they both operate on links. The building blocks of this problem are based on the links between clones in the sequence. The GA operates on these links so that the most useful ones are passed from generation to generation.

### 0.4.6 Similarity Function

The similarity function is very simple also. It counts the number of dissimilar links between two individuals. Using the parents from Figure 0.10 once again as an example, notice that there are six dissimilar links, corresponding to the five alleles not assigned to the offspring. For concreteness, these six dissimilar links in Parent 1 chromosome are 6-7, 1-2, 2-3, 3-9, 9-4 and 4-5 and for Parent 2 are 7-9, 9-5, 0-2, 2-6, 6-3 and 3-4. This metric measures the proximity between two clone sequences by counting the different links they have and not the position of the alleles. For example, the sequences (0 1 2 3 4 5 6 7 8 9) and (9 8 7 6 5 4 3 2 1 0) have a distance of zero since all the links are the same. This metric captures the essential aspect of the problem since both solutions are equivalent in our problem.

## 0.5 Results and Discussion

The results presented in this section were obtained on a SGI IRIS 4D computer under IRIX OS running the GA application written in C. The parameters for the GA are the following:

| | |
|---|---|
| Population size: | 200 |
| Mutation probability: | 0.06 |
| Crossover probability: | 1.00 |
| Crowding selection group size ($c_s$) | 10 (5% of population size) |
| Crowding subpopulation size: | 10 (5% of population size) |
| Crowding factor ($c_f$): | 3 |
| Maximum number of generations to execute: | 100 |
| Tolerance *e* | 10 |

These parameters were picked after various trials. In each trial, different values for each of the six parameters were tried, varying one at a time while holding the others constant. All reported results were averaged over five runs. A population size of 200 was found satisfactory. Other sizes (in multiples of 50) were tried, but higher sizes did not provide new information about the problem and lower sizes in some cases did not converge to the best solutions seen before in the allowed number of generations. Mutation was set at 0.06, therefore an average of 12 individuals were mutated every generation. This low value of mutation was selected to avoid eliminating the best of population frequently and allow faster convergence within each niche. On the other hand the mating probability was set to a high value of 1.0 because in our GA all individuals have a high chance of mating with a similar individual. The group size for crowding selection and crowding subpopulation was set at 5% of the population size with a crowding factor of 3. For each individual, at most 5% of the population was examined for selection and at most 15% of the population examined for replacement. These values allowed a diverse population to co-exist during the number of generations allowed and did not restrict competition between individuals from different niches. The tolerance value $E$ was set to 10 to minimize false matches due to chance. Higher values of $E$ increased the false matches more than true matches and therefore more possible clone sequences were found.

Some of the best sequences obtained for two different sets of overlapping clones are shown in Figure 0.11. Data for Set 1 is shown in Figure 0.7 and data for Set 2 is shown in the Appendix. The GA took an average of 50 seconds for each run. The figure shows the actual sequence for the data sets and the clone sequences (with their fitness) obtained by MNC.

Data Set 1 actual sequence and its fitness:

```
(8131 5154 21230 10406 18993 7442 5435 7255 12282 27714) 764
```

The k Best sequences found by the GA and their corresponding fitnesses:

```
(8131  5154 21230 10406 18993  7442  5435  7255 12282 27714) 764
(8131  5154 21230 27714 12282  7255  5435  7442 18993 10406) 749
(8131  5154 21230 10406 27714 12282  7255  5435  7442 18993) 744
(8131 27714 12282  7255  5435  7442 18993 10406 21230  5154) 730
(8131  5154 21230 10406 18993 27714 12282  7255  5435  7442) 725
```

Data Set 2 actual sequence and its fitness:

```
(12595 6722 26999 29626 29064 18301 19811 29035 17755 28828 20235) 750
```

The k Best sequences found by the GA and their corresponding fitnesses:

```
(12595 26999  6722 29626 29064 18301 28828 20235 17755 29035 19811)    757
(20235 28828 17755 29035 19811 18301 29064 29626  6722 26999 12595)    757
(12595 26999  6722 29626 29064 18301 20235 28828 17755 29035 19811)    756
(28828 20235 17755 29035 19811 18301 29064 29626  6722 26999 12595)    755
(12595 26999  6722 29626 29064 18301 28828 20235 17755 19811 29035)    754
(12595 26999  6722 29626 29064 18301 20235 28828 17755 19811 29035)    753
(12595  6722 26999 29626 29064 18301 19811 29035 17755 28828 20235)    750
(20235 28828 17755 29035 19811 18301 29064 29626 12595  6722 26999)    750
(19811 29035 17755 20235 28828 18301 29626 29064 26999  6722 12595)    750
(19811 29035 17755 28828 20235 18301 29064 29626 26999  6722 12595)    749
(12595  6722 26999 29626 29064 18301 19811 29035 17755 20235 28828)    748
(18301 28828 20235 17755 19811 29035 29064 29626  6722 26999 12595)    748
(18301 20235 28828 17755 19811 29035 29064 29626  6722 26999 12595)    747
(20235 28828 17755 29035 19811 18301 29064 29626  6722 12595 26999)    747
(12595 26999  6722 29626 29064 18301 19811 17755 29035 28828 20235)    747
(12595  6722 26999 29626 29064 18301 28828 20235 17755 19811 29035)    747
(12595 26999  6722 29626 29064 18301 28828 20235 29035 17755 19811)    746
```

```
(12595 26999  6722 29626 29064 18301 19811 17755 29035 20235 28828)    744
```

Figure 0.11: Clone sequences obtained by GA and actual sequence.

For Data Set 1 the GA was able to find the actual clone sequence. From the other sequences found, the fitness value of the next best is 15 less than the actual sequence. Similar gaps exist between all the sequences shown in Figure 0.10 for Data Set 1. From the solutions we can see that the last four sequences are a single mutation from the actual sequence.

Data Set 2, shown in the Appendix, presented a more challenging problem for the GA. In this case the best sequence found only had clones 6722 (C7) and 26999 (C2) transposed from the actual sequence. The fitness for the actual sequence is 750, which is the fifth best score when compared with all solutions found. The actual sequence was obtained in some of the runs, but did not survive until the last generation. Another observation is that there is a difference of 13 or less in the fitness between all the sequences found. Some of the sequences are mutations of others, but there is more diversity when compared with the solutions for Data Set 1.

## 0.6 Conclusions

Two points deserve further comment. First, data containing clones with fewer than five fragments were normally sequenced erroneously by the GA. This is due to the lack of opportunity for sufficient fragment matches. Also data pertaining to corner fragments (i.e., fragments lying near the cosmid clone boundaries) are generally more prone to errors. Consequently corner segments will not match well with a high probability with their counterparts in the preceding and succeeding clones. For clones with less than five fragments, this means that on the average, at least half of the data is not useful and in some cases leads to more false matches. Clones with less than five fragments were usually placed first or last in the clone sequence by the GA.

Second, when a large number of overlaps existed between 3 or 4 clones, the GA experienced difficulty deciding the correct sequence. An example of this behavior was observed with Data Set 2. This phenomenon, we believe, is happening because the fitness function is only looking for matches between the clones to the left and right without accounting for the fragments which are common to all three clones. An improved fitness measure is needed to account for fragment matches between three or more clones.

Overall the GA worked well with the data presented to it. Using the correct set of genetic operators was very important to find a GA model that will find good solutions to the problem. Using a multi-modal approach was very useful for this problem also since it prevented premature convergence and at the same time explored the search space in a more efficient manner. Defining the operators for mating, mutation, fitness, and similarity measure to work with adjacency information between the clones rather than clone positions gave the GA the correct set of tools to converge towards the most probable solutions. More information must be incorporated into the fitness evaluation to distinguish even further between the best clone sequences and other similar ones.

## 0.7 Previous Related Work and Scope of Present Work

The DNA restriction fragment map assembly, the subject matter of this paper, resembles and is somewhat related to the *restriction-site mapping,* which deals with the equivalent problem of determining the absolute location of a fragment within a cosmid clone. Here also one uses digestion data from restriction enzymes but the focus is on finding the absolute location of a fragment on a clone. Stefik (1978) used a branch and bound technique with rules to exhaustively eliminate wrong answers from the digest fragment data. This approach is sensitive to error in the data and is computationally intensive. Pearson (1982) exhaustively generated permutations of the single-digest data to compute the error between the generated double-digest and the actual (experimental) double-digest data. This approach is faster but it is limited to a small number of restriction sites also. Krawczak (1988) developed a divide and conquer technique that groups the fragments into compatible clusters and then determines the order of the fragments within each cluster. This approach can process a greater number of restriction sites. Platt and Dix (1993) used Genetic Algorithms (GAs) for restriction-site mapping using double digest data. In their work they did not consider operators suited for multi-modal search spaces and mating which preserve adjacency information.

Other techniques are available to sequence larger DNA regions. Branscomb *et al.* (1990) developed a greedy algorithm to order the most probable clone sequence using overlap probabilities between the clones. The algorithm works well when a large amount of overlap between the clones exists and the fragment data has small errors. This approach is prone to getting stuck in local minima and does not use all the available data gathered at great expense. Techniques using larger clones are also being tried to order, orient, and connect the islands in the original DNA (Olson *et al.*, 1986; Waterman and Griggs, 1986; Stallings *et al.*, 1990; Fickett 1993).

Cuticchia *et al.* (1992) constructed maps using simulated annealing techniques. In their work clones are ordered according to a measure of similarity between them given by the presence or absence of specific sequences. A signature is assigned to each clone and the algorithm uses it to minimize the error between the actual length of the contig and the given length by the hypothetical clone ordering. Matching signatures are used to order the clones. In their work they only considered the relationship between consecutive clones.

**Acknowledgments**

Special thanks to Mr. Tom Slezak and Dr. Elbert Branscomb for describing the problem and providing the test data. This work was supported, in part, by the Applied Mathematics Program of the Office of Energy Research (U.S. Department of Energy) under contract number W-7405-Eng-48 to LLNL Lawrence Livermore National Laboratory and in part by a grant from the Institute of Scientific Computing Research of the Lawrence Livermore National Laboratory.

**References**

D. Beasley, D. R. Bull, and R. R. Martin, A Sequential Technique for Multimodal Function Optimization, To be published in *Evolutionary Computation*, February 1993.

R. K. Belew and L. B. Booker, eds., *Proceedings of the Fourth International Conference on Genetic Algorithms*, Morgan Kaufmann Publishers, San Mateo, CA, July 1991.

E. Branscomb, T. Slezak, R. Pae, D. Galas, A. V. Carrano, and M. Waterman, Optimizing Restriction Fragment Fingerprinting Methods for Ordering Large Genomic Libraries, *Genomics* 8, 351-366, 1990.

D. J. Cavicchio, Adaptive Search Using Simulated Evolution, Doctoral Dissertation, University of Michigan, Ann Arbor, MI, 1970.

W. Cedeño and V. Vemuri, Dynamic multi-modal function optimization using genetic algorithms, *Proc. of the XVIII Latin-American Informatics Conference*, Las Palmas de Gran Canaria, Spain, August 1992.

W. Cedeño and V. Vemuri, Assembly of DNA Restriction-Fragments Using Genetic Algorithms, Submitted to *Evolutionary Computation.*

W. Cedeño, Genetic algorithms in SISAL to solve the file design problem, *Proc. of the Second SISAL User's Conference*, San Diego, CA, December 1992.

A. J. Cuticchia, J. Arnold, and W. E. Timberlake, The use of simulated annealing in chromosome reconstruction experiments based on binary scoring, *Genetics* 132, 591-601, 1992.

L. Davis, (ed.), *Handbook of Genetic Algorithms*, Van Nostrand Reinhold, New York, NY, 1991.

K. A. De Jong, An analysis of the behavior of a class of genetic adaptive systems, Doctoral dissertation, University of Michigan, *Dissertation Abstracts International* 36(10), 5140B, 1975.

K. Deb and D. E. Goldberg, An investigation of niche and species formation in genetic function optimization, *Proceedings of the Third International Conference on Genetic Algorithms*, J. D. Schaffer, ed., 42-50, Morgan Kaufmann Publishers, San Mateo, CA, June 1989.

J. W. Fickett and M. J. Cinkosky, A genetic algorithm for assembling chromosome physical maps, Unpublished, 1993.

S. Forrest, (Ed.) Proc. Fifth International Conference on Genetic Algorithms, Morgan Kaufman, San Mateo, CA, Aug. 1993.

D. E. Goldberg and J. Richardson, Genetic algorithms with sharing for multimodal function optimization, *Proceedings of the Second International Conference on Genetic Algorithms*, J. J. Grefenstette, ed., 41-49, Lawrence Erlbaum Associates, Hillsdale, NJ, June 1987.

D. E. Goldberg, *Genetic Algorithms in Search, Optimization, and Machine Learning*, Addison-Wesley, Reading, MA, 1989.

D. E. Goldberg, K. Deb, and J. Horn, Massive multimodality, deception, and genetic algorithms, *Parallel Problem Solving From Nature 2*, Elsevier Science Publishers, 37-46, 1992.

J. H. Holland, *Adaptation in Natural and Artificial Systems*, University of Michigan Press, Ann Arbor, 1975.

W. Istvanick, A. Kryder, G. Lewandoeski, J. Meidânis, A. Rang, S. Wyman, and D. Joseph, Dynamic methods for fragment assembly in large scale genome sequencing projects, *Proceedings of the Twenty Sixth Annual Hawaii International Conference on System Sciences: Architecture and Biotechnology Computing*, T. N. Mudge, V. Milutinovic, and L. Hunter eds. IEEE Computer Society Press, 534-543, Wailea, Hawaii, 1993.

M. Krawczak, Algorithms for the restriction-site mapping of DNA molecules. *Proc. Natl. Acad. Sci. U.S.A.*, 85, 7298-7301, 1988.

S. W. Mahfoud, Crowding and preselection revisited, *Proceedings of Parallel Problem Solving from Nature 2*, R. Männer and B. Manderick, eds., 27-36, Elsevier Science Publishers B. V., 1992.

M. V. Olson, J. W. Dutchik, M. Y. Graham, G. M. Brodeur, C. Helms, M. Frank, M. MacCollin, R. Scheinman, and T. Frank, Random-clone strategy for genomic restriction mapping yeast, *Proc. Natl Acad Sci. U.S.A.*, 83, 7826-7830, 1986.

J. Opatrny, J., The total ordering problem, *SIAM Journal of Computing*, 8(1): 111-114, 1979.

M. D. Platt and T. I. Dix, Construction of restriction maps using a genetic algorithm, *Proceedings of the Twenty Sixth Annual Hawaii International Conference on System Sciences: Architecture and Biotechnology Computing*, T. N. Mudge, V. Milutinovic, and L. Hunter eds. IEEE Computer Society Press, 756-762, Wailea, Hawaii, 1993.

J. D. Schaffer, ed., *Proceedings of the Third International Conference on Genetic Algorithms*, Morgan Kaufmann Publishers, San Mateo, CA, June 1989.

R. L. Stallings, D. C. Torney, C. E. Hildebrand, J. L. Longmire, L. L. Deaven, J. H. Jett, N. A. Doggett, and R. K. Moyzis, Physical mapping of human chromosomes by repetitive sequence fingerprinting, *Proc. Natl. Acad. Sci. U.S.A.*, 87, 6218-6222, 1990.

M. Stefik, Inferring DNA structures from segmentation data, *Artificial Intelligence* , 11, 85-114, 1978.

G. Syswerda, Uniform crossover in genetic algorithms, *Proceedings of the Third International Conference on Genetic Algorithms*, Morgan Kaufmann Publishers, San Mateo, CA, June 1989.

M. S. Waterman and J. R. Griggs, Internal graphs and maps of DNA, *Bull. Math. Biol.*, 48:189-195, 1986.

D. Whitley, GENITOR: a different genetic algorithm, *Proceedings of the Rocky Mountain Conference on Artificial Intelligence*, Denver Colorado, 1988.

D. Whitley, T. Starkweather and D. Fugway, Scheduling problems and travelling salesmen: The Genetic Edge Recombination operator, *Proc. Third International Conf. on Genetic Algorithms*, Morgan Kaufmann Publishers, San Mateo, CA, June 1989.

## Appendix

Fragment data for set 2 of overlapping cosmid clones:

| CLONE ID | FRAGMENTS (k base-pairs) |
|---|---|
| 29064 | 15.42 3.46 1.50 9.12 4.30 |
| 19811 | 3.13 7.89 7.89 3.02 4.35 14.31 2.65 0.64 |
| 26999 | 4.64 19.69 1.10 1.48 2.82 0.77 9.61 |
| 29626 | 3.46 1.50 9.14 6.47 13.48 |
| 17755 | 1.26 2.62 6.32 2.73 3.54 7.88 7.88 3.02 4.35 1.74 |
| 12595 | 1.48 2.83 0.76 9.68 12.75 1.48 6.06 |
| 20235 | 8.01 12.56 2.62 6.32 2.74 3.54 5.71 |
| 6722 | 2.84 19.72 1.16 1.49 2.84 0.77 9.69 9.20 |
| 28828 | 12.45 2.61 6.27 2.72 3.52 7.89 3.89 |
| 18301 | 2.53 13.99 2.64 16.88 3.44 |
| 29035 | 2.45 2.74 3.53 7.82 7.82 3.02 4.35 7.44 |

| CLONE | C0 | C1 | C2 | C3 | C4 | C5 | C6 | C7 | C8 | C9 | C10 |
|---|---|---|---|---|---|---|---|---|---|---|---|
| C0: | 5 | 1 | 1 | 3 | 2 | 1 | 1 | 2 | 1 | 1 | 2 |
| C1: | 1 | 8 | 0 | 0 | 5 | 0 | 1 | 0 | 2 | 1 | 5 |
| C2: | 1 | 0 | 7 | 1 | 1 | 4 | 1 | 6 | 1 | 0 | 1 |
| C3: | 3 | 0 | 1 | 5 | 1 | 1 | 1 | 2 | 1 | 1 | 1 |
| C4: | 2 | 5 | 1 | 1 | 10 | 1 | 4 | 1 | 5 | 2 | 6 |
| C5: | 1 | 0 | 4 | 1 | 1 | 7 | 1 | 4 | 0 | 0 | 1 |
| C6: | 1 | 1 | 1 | 1 | 4 | 1 | 7 | 1 | 4 | 2 | 2 |
| C7: | 2 | 0 | 6 | 2 | 1 | 4 | 1 | 8 | 0 | 0 | 1 |
| C8: | 1 | 2 | 1 | 1 | 5 | 0 | 4 | 0 | 7 | 2 | 3 |
| C9: | 1 | 1 | 0 | 1 | 2 | 0 | 2 | 0 | 2 | 5 | 3 |
| C10: | 2 | 5 | 1 | 1 | 6 | 1 | 2 | 1 | 3 | 3 | 8 |

Error between the matches

| CLONE | C0 | C1 | C2 | C3 | C4 | C5 | C6 | C7 | C8 | C9 | C10 |
|---|---|---|---|---|---|---|---|---|---|---|---|
| C0: | 0 | 5 | 2 | 2 | 13 | 2 | 8 | 9 | 6 | 2 | 12 |
| C1: | 5 | 0 | 0 | 0 | 5 | 0 | 3 | 0 | 4 | 1 | 23 |
| C2: | 2 | 0 | 0 | 2 | 9 | 9 | 8 | 20 | 10 | 0 | 8 |
| C3: | 2 | 0 | 2 | 0 | 8 | 2 | 8 | 7 | 6 | 2 | 7 |
| C4: | 13 | 5 | 9 | 8 | 0 | 10 | 1 | 10 | 10 | 12 | 14 |
| C5: | 2 | 0 | 9 | 2 | 10 | 0 | 9 | 4 | 0 | 0 | 9 |
| C6: | 8 | 3 | 8 | 8 | 1 | 9 | 0 | 10 | 10 | 12 | 1 |
| C7: | 9 | 0 | 20 | 7 | 10 | 4 | 10 | 0 | 0 | 0 | 10 |
| C8: | 6 | 4 | 10 | 6 | 10 | 0 | 10 | 0 | 0 | 11 | 10 |
| C9: | 2 | 1 | 0 | 2 | 12 | 0 | 12 | 0 | 11 | 0 | 27 |
| C10: | 12 | 23 | 8 | 7 | 14 | 9 | 1 | 10 | 10 | 27 | 0 |

In the *multi-niche crowding* (Cedeño and Vemuri, 1992) both the selection and replacement steps are modified with some type of crowding. The idea is to eliminate the selection pressure caused by FPR and allow the population to maintain some diversity. This objective is achieved in part by encouraging mating and replacement within members of the same niche while allowing some competition for the population slots among the niches. The result is an algorithm that (a) maintains stable subpopulations within different niches, (b) maintains diversity throughout the search, and (c) converges to different local optima. No prior knowledge of the search space is needed and no restrictions are imposed during selection and replacement thus allowing exploration of other areas of the search space while converging to the best individuals in the different niches.

In MNC, the FPR selection is replaced by what we call *crowding selection*. In crowding selection each individual in the population has the same chance for mating in every generation. Application of this selection rule is done in two steps. First, an individual *Ii* is selected for mating. This selection can be either sequential or random. Second, its mate *Im* is selected, not from the entire population, but from a group of individuals of size *Cs* (crowding selection size), picked at random from the population. The mate *Im* thus chosen must be the one who is the most "similar" to *Ii*. The similarity metric used here is not a genotypic metric such as the Hamming distance, but a suitably defined phenotypic distance metric.

Crowding selection promotes mating between members of the same niche while allowing individuals from different niches to mate. Unlike the mating restriction that allows only individuals from the same niche to mate, crowding selection allows some amount of exploration to occur while at the same time looking for the best in each niche.

During the replacement step, MNC uses a replacement policy called *worst among most similar*. The goal of this step is to pick an individual from the population for replacement by an offspring. Implementation of this policy follows these steps. First, *Cf* (crowding factor) groups are created by randomly picking *s* (crowding size) individuals per group from the population. Second, one individual from each group that is most similar to the offspring is identified. This gives *Cf* individuals that are candidates for replacement by virtue of their similarity to the offspring that will replace one of them. From this group of most similar individuals, we pick the one with the lowest fitness to die and that slot is filled with the offspring. Figure 14.8 shows an example of this replacement policy.

Current technological limits are forcing us to limit the sequencing task to small fragments of DNA that are composed of approximately 0.5K base-pairs (Istvanick *et al.,* 1993). In order to divide the DNA into fragments up to this resolution level, techniques using restriction enzymes are used. The restriction enzymes act on the DNA at specific locations which are randomly distributed along the length of the chromosome. Depending on the number of different restriction enzymes used in obtaining the fragments, the data are called single-digest (one enzyme), double-digest (two enzymes), or n-digest (n enzymes) data. Most mappings are done using single- and double-digest data. Scientists use different restriction enzymes to obtain DNA fragments of the appropriate size.

## Chapter 1

**Shumeet Baluja**
School of Computer Science
Carnegie Mellon University

baluja@cs.cmu.edu

# Artificial Neural Network Evolution: Learning to Steer a Land Vehicle



## Abstract

This chapter presents an evolutionary method for creating an artificial neural network based controller for an autonomous land vehicle. Previous studies which have used evolutionary procedures to evolve artificial neural networks have been constrained to small problems by extremely high computational costs. In this chapter, methods for reducing the computational burden are explored. Previous connectionist based approaches to this task are discussed. The evolutionary algorithrm used in this study, Population-Based Incremental Learning (PBIL), is a variant of the traditional genetic algorithm. It is described in detail in this chapter. The results indicate that the evolutionary algorithm is able to generalize to unseen situations better than the standard method of error backpropagation; an improvement of approximately 18% is achieved on this task. The networks evolved are efficient; they use only approximately half of the possible connections. However, the evolutionary algorithm may require considerably more computational resources on large problems.

0-8493-2529-3/95/$0.00 + $.50

## 1.1 Overview

In this chapter, evolutionary optimization methods are used to improve the generalization capabilities of feed-forward artificial neural networks. Many of the previous studies involving evolutionary optimization techniques applied to artificial neural networks (ANNs) have concentrated on relatively small problems. This chapter presents a study of evolutionary optimization on a "real-world" problem, that of autonomous navigation of Carnegie Mellon's NAVLAB system. In contrast to the other problems addressed by similar methods in recently published literature, this problem has a large number of pixel based inputs and also has a large number of outputs to indicate the appropriate steering direction.

The feasibility of using evolutionary algorithms for network topology discovery and weight optimization is discussed throughout the chapter. Methods for avoiding the high computational costs associated with these procedures are presented. Nonetheless, evolutionary algorithms remain more computationally expensive than training by standard error backpropagation. Because of this limitation, the ability to train on-line, which may be important in many realtime robotic environments, is not addressed in this chapter. The benefit of evolutionary algorithms lies in their ability to perform global search; they provide a mechanism which is more resistant to local optima than standard backpropagation. In determining whether an evolutionary approach is appropriate for a particular application, the conflicting needs for accuracy and speed must be taken into careful consideration.

The next section very briefly reviews the fundamental concepts of ANNs. This material will be familiar to the reader who has had an introduction to ANNs. Section 1.3 provides an overview of the currently used artificial neural network based steering controller for the NAVLAB, named ALVINN (Autonomous Land Vehicle in a Neural Network) [16]. Section 1.4 gives the details of the evolutionary algorithm used in this study to evolve a neuro-controller; Population-Based Incremental Learning [4]. Section 1.5 gives the details of the task. Section 1.6 gives the implementation and results. Finally, Sections 1.7 and 1.8 close the chapter with conclusions and suggestions for future research.

## 1.2 Introduction to Artificial Neural Networks

An Artificial Neural Network (ANN) is composed of many small computing units. Each of these units is loosely based upon the design of a single biological neuron. The models most commonly used are far simpler than their biological counterparts. The key features of each of these simulated neurons are the inputs, the activation function, and the outputs. A model of a simple neuron is shown in Figure 1.1. The inputs to each neuron are multiplied by connection weights giving a net total input. This net input is passed through a non-linear activation function, typically the sigmoid or hyperbolic tangent function, which maps the infinitely ranging (in theory) net input to a value between set limits. For the sigmoidal activation function, input values will be mapped to a point in (0,1) and for the hyperbolic tangent activation function, the input will be mapped to a value in (-1,1). Once the resultant value is computed, it can either be interpreted as the output of the network, or used as input to another neuron. In the study presented in this chapter, hyperbolic tangent activations were used.

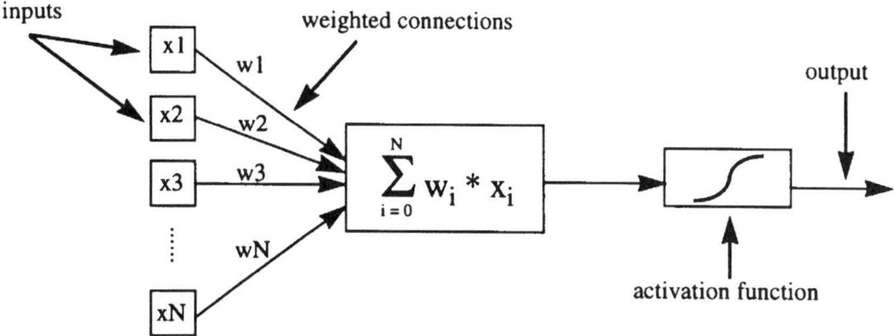


Figure 1.1: The artificial neuron works as follows: the summation of the incoming (weights * activation) values is put through the activation function in the neuron. In the above shown case, this is a sigmoid. The output of the neuron, which can be fed to other neurons, is the value returned from the activation function. The **x**'s can either be other neurons or inputs from the outside world.

Artificial neural networks are generally composed of many of the units shown in Figure 1.1, as shown in Figure 1.2. For a neuron to return a particular response for a given set of inputs, the weights of the connections can be modified. "Training" a neural network refers to modifying the weights of the connections to produce the individual output vector associated with each input vector.

A simple ANN is composed of three layers, the input layer, the hidden layer and the output layer. Between the layers of units are connections containing weights. These weights serve to propagate signals through the network. (See Figure 1.2.) Typically, the network is trained using a technique which can be thought of as gradient descent in the connection weight space. Once the network has been trained, given any set of inputs and outputs which are sufficiently similar to those on which it was trained, it will be able to reproduce the associated outputs by propagating the input signal forward through each connection until the output layer is reached.

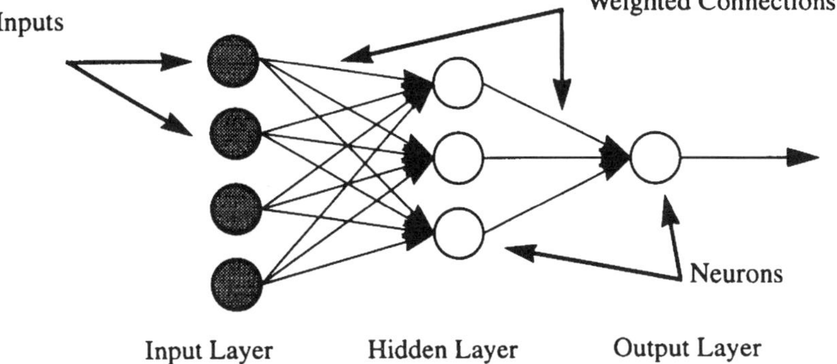


Figure 1.2: A fully connected three layer ANN is shown. Each of the connections can change its weight independently during training.

In order to find the weights which produce correct outputs for given inputs, the most commonly used method for weight modification is error backpropagation. Backpropagation is simply explained in Abu-Mostafa's paper "Information Theory, Complexity and Neural Networks"[1]:

> ...the algorithm [backpropagation] operates on a network with a fixed architecture by changing the weights, in small amounts, each time an example $y_i = f(x_i)$ [where $y$ is the desired output pattern, and $x$ is the input pattern] is received. The changes are made to make the response of the network to $x_i$ closer to the desired output, $y_i$. This is done by gradient descent, and each iteration is simply an error signal propagating backwards in the network in a way similar to the input that propagates forward to the output. This fortunate property simplifies the computation significantly. However, the algorithm suffers from the typical problems of gradient descent, it is often slow, and gets stuck in local minima.

If ANNs are not overtrained, after training, they should be able to generalize to sufficiently similar input patterns which have not yet been encountered. Although the output may not be exactly what is desired, it should not be a catastrophic failure either, as would be the case with many non-learning techniques. Therefore, in training the ANN, it is important to get a diverse sample group which gives a good representation of the input data which might be seen by the network during simulation. A much more comprehensive tutorial of artificial neural networks can be found in [12].

### 1.3. Introduction to ALVINN

ALVINN is an artificial neural network based perception system which learns to control Carnegie Mellon's NAVLAB vehicles by watching a person drive, see Figure 1.3. ALVINN's architecture consists of a single hidden layer backpropagation network. The input layer of the network is a 30x32 unit two dimensional "retina" which receives input from the vehicle's video camera, see Figure 1.4. Each input unit is fully connected to a layer of four hidden units which are in turn fully connected to a layer of 30 output units. In the simplest interpretation, each of the network's output units can be considered to represent the network's vote for a particular steering direction. After presenting an image to the input retina, and passing activation forward through the network, the output unit with the highest activation represents the steering arc the network believes to be best for staying on the road.

To teach the network to steer, ALVINN is shown video images from the onboard camera as a person drives and is trained to output the steering direction in which the person is currently steering. The backpropagation algorithm alters the strengths of connections between the units so that the network produces the appropriate steering response when presented with a video image of the road ahead of the vehicle.

Because ALVINN is able to learn which image features are important for particular driving situations, it has been successfully trained to drive in a wider variety of situations than other autonomous navigation systems which require fixed, predefined features (e.g., the road's center line) for accurate driving. The situations ALVINN networks have been trained to handle include single lane dirt roads, single lane paved bike paths, two lane suburban neighborhood streets, and lined divided highways. In this last domain, ALVINN has successfully driven autonomously at speeds of up to 55 m.p.h., and for distances of over 90 miles on a highway north of Pittsburgh, Pennsylvania.

Figure 1.3: The Carnegie Mellon NAVLAB Autonomous Navigation testbed.

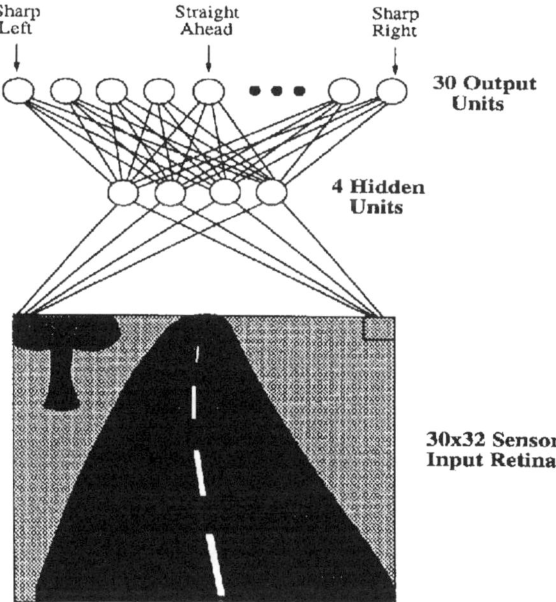


Figure 1.4: The ALVINN neural network architecture.

The performance of the ALVINN system has been extensively analyzed by Pomerleau [16][17][18]. Throughout testing, various architectures have been examined, including architectures with more hidden units and different output representations. Although the output representation was found to have a large impact on the effectiveness of the network, other features of the network architecture were found to yield approximately equivalent results [15][16]. In the study presented here, the output representation examined is the one currently used in the ALVINN system, a distributed representation of 30 units.

### 1.3.1 Training ALVINN

To train ALVINN, the network is presented with road images as input and the corresponding correct steering direction as the desired output. The correct steering direction is the steering direction the human driver of the NAVLAB has chosen. The weights in the network are altered using the backpropagation algorithm so that the network's output more closely corresponds to the target output. Training is currently done on-line with an onboard Sun SPARC-10 workstation.

Several modifications to the standard backpropagation algorithm are used to train ALVINN. First, the weight change "momentum" factor is steadily increased during training. Second, the learning rate constant for each weight is scaled by the fan-in of the unit to which the weight projects. Third, a large amount of neighbor weight smoothing is used between the input and hidden layers. Neighbor weight smoothing is a technique to constrain weights which are spatially close to each other, in terms of their connections to the units in the input retina, to similar values. This is a method of preserving spatial information in the context of the backpropagation algorithm.

In its current implementation, ALVINN is trained to produce a Gaussian distribution of activation centered around the appropriate steering direction. However, this steering direction may fall between the directions represented by two output units. A Gaussian approximation is used to interpolate the correct output activation levels of each output unit. Using the Gaussian approximations, the desired output activation levels for the units successively farther to the left and the right of the correct steering direction will fall off rapidly on either side of the two most active units. A representative training example is shown below, in Figure 1.5. The 15x16 input retina displays a typical road input scene for the network. The target output is also shown. This corresponds to the steering direction the driver of the NAVLAB chose during the test drive made to gather the training images. Also shown is the output of an untrained network. Later in the chapter, trained outputs will be shown for comparison.

One of the problems associated with this training is that the human driver will normally steer the vehicle correctly down the center of the road (or lane). Therefore, the network will never be presented with situations in which it must recover from errors, such as being slightly off the correct portion of the road. In order to compensate for this lack of real training data, the images are shifted by various amounts relative to the road's center. The shifting mechanism maintains the correct perspective, to ensure that the shifted images are realistic. The correct steering direction is determined by the amount of shift introduced into the images. The network is trained on the original and shifted images.

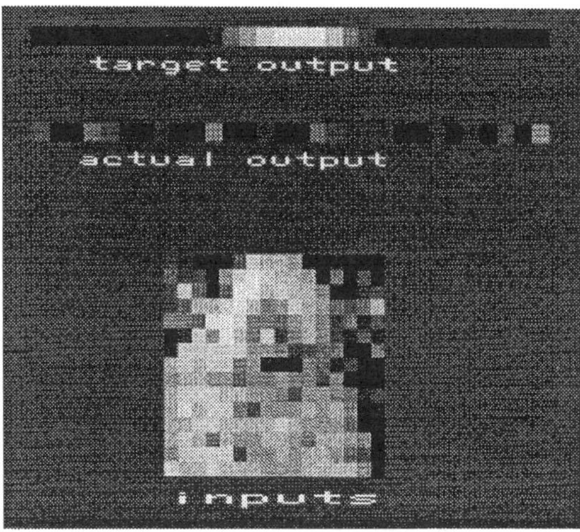


Figure 1.5: Input image, target and actual outputs before training.

## 1.4 The Evolutionary Approach

The majority of approaches in which evolutionary principles are used in conjunction with neural network training can be broadly subdivided into two groups. The first concentrates on formulating the problem of finding the connection weights of a predefined artificial neural network architecture as a search problem. Traditionally backpropagation, or one of its many variants, has been used to train the weights of the connections. However, backpropagation is a method of gradient descent through the weights space, and can therefore get stuck in local minima. Evolutionary algorithms (EAs) are methods of global search, and are less susceptible to local minima. Finding the appropriate set of weights in a neural network can be formulated as parameter optimization problem to which EAs can be applied in a straightforward manner. A much more comprehensive overview of evolutionary algorithms, and their applications to parameter optimization tasks, can be found in [3] [9].

The second method for applying EAs endeavors to find the appropriate structure of the network for the particular task; the number of layers, the connectivity, etc., are defined through the search process. The weights can either be determined using backpropagation to train the networks specified by the search, or can simultaneously be found while searching for the network topology. The method explored in this chapter is a variant of the latter approach, and will be described in much greater detail in the following sections. The advantage to this method is that if there is very little knowledge of the structure of the problem, and therefore no knowledge, other than the number of inputs and outputs needs that need to be incorporated into the network, the structure of the network does not need to be predefined in detail.

Given the possibility of backpropagation falling into a local minima, and the potential lack of knowledge regarding the appropriate neural network architecture to use, using EAs appears to be a good alternative. However, the largest drawback of EAs, and the one which has made them prohibitive for many "real world" learning applications, is their enormous computational burden. As EAs do not explicitly use gradient information (as backpropagation does), large amounts of time may be spent searching before an acceptable solution is found.

Previous work has been done to measure the feasibility of evolutionary approaches on standard neural network benchmark problems, such as the encoder problem, and exclusive-or (XOR) problems, etc. More complicated problems have also been attempted, such as the control of an animat which learns to play soccer, given a small set of features about the environment such as the ball position, the status of the ball, etc. with good results [14]. Other work, which has concentrated on solving a "search and collection task" of simulated ants has used the evolution of recurrent neural networks, with evolutionary programming, again with successful results [2].

Many of the studies which have used evolution as the principle learning paradigm of training artificial neural networks have often modelled evolution through genetic algorithms [6][10][14]. However, genetic algorithms are very computationally expensive for large problems. In order to reduce the search times, a novel evolutionary search algorithm is used in this study. The algorithm, Population Based Incremental Learning (PBIL), is based upon the mechanisms of a generational genetic algorithm and the weight update rule of supervised competitive learning [12]. Although a complete description of its derivation and its performance compared with other evolutionary algorithms is beyond the scope of this chapter, a description of its fundamental mechanisms can be found below. More detailed descriptions of the algorithm and results obtained in comparisons with genetic algorithms and hillclimbing can be found in [4].

### 1.4.1 Population-Based Incremental Learning

PBIL is an evolutionary search algorithm based upon the mechanisms of a generational genetic algorithm and supervised competitive learning. The PBIL algorithm, like standard genetic algorithms, does not use derivative information; rather, it relies on discrete evaluations of potential solutions. In this study, each potential solution is a fully specified network; both the topology and the connection weights can be encoded in the potential solution and evolved in the search process. The PBIL algorithm described in this chapter operates on potential solutions defined in a binary alphabet. The exact encodings of the networks will be described in the next section.

The fundamental goal of the PBIL algorithm is to create a real-valued probability vector which specifies the probability of having a '1' in each bit position of the potential solution. The probabilities are created to ensure that potential solutions, from which the individual bits are drawn with the probabilities specified in the probability vector, have good evaluations with a high probability. The probability vector can be considered a "prototype" for high evaluation vectors for the function space being explored.

A very basic observance of genetic algorithm behavior provides the fundamental guidelines for the performance of PBIL. One of the key features in the early portions of genetic optimization is the parallelism inherent in the search; many diverse points are represented in the population of a single generation. In representing the population of a GA in terms of a probability vector, the most diversity will be found in setting the probabilities of each bit position to 0.5. This specifies that generating a 0 or 1 in each bit position is equally likely. In a manner similar to the training of a competitive learning network, the values in the probability vector are gradually shifted towards the bit values of high evaluation vectors. A simple procedure to accomplish this is described below. The probability update rule, which is based upon the update rule of standard competitive learning, is shown below.

$$probability_i = (probability_i \times (1.0 - LR)) + (LR \times solutionVector_i)$$

$probability_i$ is the probability of generating a 1 in bit position i.
$solutionVector_i$ is the value of the ith position in the high evaluation vector.
*LR* is the learning rate (defined by the user).

The *probability vector and the solutionVector* are both the length of the encoded solution.

The step which remains to be defined is determining which solution vectors to move towards. The vectors are chosen as follows: a number of potential solution vectors are generated by sampling from the probabilities specified in the current probability vector. Each of these potential solution vectors is evaluated with respect to the goal function. For this task, the goal function is how well the encoded ANN performs on the training set. This is determined by decoding the solution vector into the topology and weights of the ANN, performing a forward pass through the training samples, and measuring the sum squared error of the outputs. The probability vector is pushed towards the generated solution vector with the best evaluation: the network with the lowest sum squared error. After the probability vector is updated, a new set of potential solution vectors is produced; these are based upon the updated probability vector, and the cycle is continued.

During the probability vector update, the probability vector is also moved towards the complement of the vector with the lowest evaluation. However, this move is not made in all of the bit positions. The probability vector is moved towards the complement of the vector with the lowest evaluation only in the bit positions in which the highest evaluation vector and the lowest evaluation vector differ.

In addition to the update rule shown above, a "mutation" operator is used. This is analogous to the mutation operator used in standard genetic algorithms. Mutation is used to prevent the probability vector from converging to extreme values without performing extensive search. In standard genetic algorithms the mutation operator is implemented as a small probability of randomly changing a value in a member of the population. In the PBIL algorithm, the mutation operator affects the probability vector directly; each vector position is shifted in a random direction with a small probability in each iteration. The magnitude of the shift is small in comparison to the learning rate.

The probability vector is adjusted to represent the current highest evaluation vector. As values in the bit positions become more consistent between highest evaluation vectors produced in subsequent generations, the probabilities of generating the value in the bit position increases. The probability vector has two functions, the first is to be a prototype for high evaluations vectors, and the second is to guide the search from which it is further refined.

In the implementation used in this study, the population size is kept constant at 30; the population size refers to the number of potential solution vectors which are generated before the probability vector is updated. This is a very small population size in comparison to those often used in other forms of evolutionary search. Because of the small size and the probabilistic generation of solution vectors, it is possible that a good vector may not be created in each generation. Therefore, in order to avoid moving towards unproductive areas of the search space, the best vector from the previous population is also kept in the current population. This solution vector is only used in case a better evaluation vector is not produced in the current generation. In genetic algorithm literature, this technique of preserving the best solution vector from one generation to the next is termed "elitist selection," and is often used in parameter optimization problems to avoid losing good solutions, by random chance, once they are found.

```
P <-- initialize probability vector. (Each position = 0.5)
loop # GENERATIONS
   ***** Generate Samples *****
   i ← loop #SAMPLES
      sample_i ←  generate  sample  vector  according  to
                                        probabilities in P
      evaluation_i ← Decode_Network_and_Perform_Forward_Pass
                                        (sample_i)
   best ← find vector corresponding to best evaluation
   worst ← find vector corresponding to worst evaluation
   ***** Update Probability Vector towards best network *****
   I ←loop #LENGTH
      P_I ← P_I * (1.0 - LR) + best_I * (LR)
   ***** Update Probability Vector away from worst network *****
   I ←loop #LENGTH
      if (best_I ≠ worst_I) P_I ← P_I * (1.0 - NEGATIVE-LR) + best_I
      * (NEGATIVE-LR)
   ***** Mutate Probability Vector *****
   I ←loop #LENGTH
      if (random (0,1) < MUT_PROBABILITY)
         P_I ← P_I (1.0 - MUT_SHIFT) + random (0.0 or 1.0) *
         (MUT_SHIFT)

USER DEFINED CONSTANTS:
```

```
GENERATIONS: how many iterations the algorithm is allowed to
   continue.
SAMPLES: the number of vectors generated before update of the
   probability vector
LENGTH: the number of bits in a generated vector
MUT_PROBABILITY: the probability for a mutation occurring in each
   position
MUT_SHIFT: the amount a mutation alters the value in the bit
   position
LR: the learning rate, how fast to exploit the search performed
   so far.
NEGATIVE-LR: the negative learning rate, how much to learn from
   negative examples.
```

Figure 1.6: The general PBIL algorithm for a binary alphabet. The "elitist" selection mechanism is not shown.

A complete discussion of the merits and drawbacks of this algorithm, as compared to a standard genetic algorithm, is beyond the scope of this chapter. This algorithm, which is far less complex than even a simple genetic algorithm, very quickly optimizes many of the functions which are used to gauge genetic algorithm performance. It does not, however, use the crossover (recombination) operators, or define operations directly on the members of the population, both of which are common to genetic algorithms. The basic algorithm is shown in Figure 1.6. In the set of standard GA benchmark problems on which PBIL has been compared, the resulting solutions found by PBIL are better than those found with the genetic algorithm, and are discovered with far less computational cost [4].

## 1.5 Task Specifics

The central task explored in this chapter is to develop ANNs for control of an autonomous land vehicle. The specific goal is to develop ANNs which are able to generalize beyond their training set. The motivation for this task is the current project at Carnegie Mellon University to create a pool of pre-trained ALVINN networks, each of which is trained on different road types, under different conditions, etc., from which the appropriate network can be chosen to achieve the most accurate steering direction. Special purpose hardware is currently being designed to allow pre-trained neural networks to control the steering direction. The hardware design does not support modification of the weights; therefore on-line training of the networks is not possible. The goal of this project is to create pre-trained networks which perform well in the encountered situations. This task differs in several ways from the standard ALVINN task. The first difference is that a large degree of good generalization, while important, is not crucial in standard ALVINN tests, as ALVINN is frequently trained to adapt to changing conditions. Secondly, this task does not have to be done on-line. In the standard ALVINN task, training speed is crucial, as it must be able to adapt "on-the-fly" to changing lighting conditions, changing road-types, and changing weather conditions, etc. In this task, the on-line training is not required, as the network pool is trained before any of the networks are used.

For the experiments presented here, four sets of data were collected. Two sets of data were obtained by driving back and forth on a partially shaded single lane paved bicycle path. The other two sets of data were obtained by driving back and forth on a a two lane suburban neighborhood street. The training for all of the experiments in this chapter was done on a total of 1000 images. In the training set, 500 images from the first road type were used, traveling only in one direction, and 500 images were used of the second road type, again only traveling in a single direction. Each of these four sets of images are composed of images of typical road scenes, and images which contain shifts and rotations of the road in the original images. More details of how these shifts and rotations are performed, and their use in training standard ALVINN networks, can be found in [16].

As mentioned before, the PBIL algorithm used in this study operates on binary strings. One of the drawbacks of using a binary alphabet is that the values of the weights of the encoded network must be discretized to a specified precision. As the solution string lengths used in this study are also of fixed size, the number of bits allocated to represent each weight are pre-specified before the algorithm is started. The translation of these bits to weights assumes that the bits encode a base-2 number which specifies the value of the weight within a pre-specified range of possible values. It has been found through empirical testing that overestimating the number of required bits did not hinder performance, although it did increase the search time.

The encodings of the networks into binary strings was as follows: each connection in the network is determined to be either present or absent by a single bit. The weight of the connection, if it is present, is determined by a pre-specified number of bits, and is encoded as a base-2 number. See Figure 1.7 for an example. Although in these experiments the number of bits to represent a connection weight were prespecified, this is not a requirement for evolutionary procedures. An alternative method which avoids this limitation, in which the granularity of detail is also evolved in the search process, is described in [14].

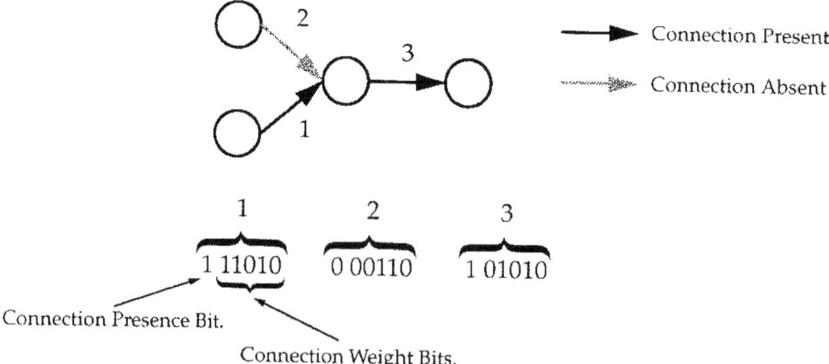


Figure 1.7: Network Encoding into binary form. In this example, the weight of each connection is represented with 5 bits. The presence of each connection is determined by the value of an additional bit.

In the experiments attempted in this chapter, the *maximal network,* a network which specifies the maximum number of feed forward connections allowed (this is user defined), is used as the basis of the solution encodings used in PBIL. This network employed for this study is similar to the one shown in Figure 1.4, with 30 output units, 5 hidden units, and a 15x16 input retina. This architecture is the maximal network, the number of connections ultimately used by the final architecture is determined through the search process.

## 1.6 Implementation and Results

In an evolutionary approach, the need to evaluate each ANN is the source of the largest time penalties. Each ANN must be evaluated to determine which network in the current population has the smallest sum squared error and the largest sum squared error on the training set, as these two examples are used for adjusting the probability vector in the PBIL algorithm. As mentioned before, the evaluation of each network is proportional to the sum squared error between the target and predicted outputs for each image in the training set. In the experiments presented in this chapter, the training set size was 1000 images. With this size training set, evaluating each network is very computationally expensive. A training method designed to reduce the computational burden is presented below.

Rather than evaluating each network on the entire training set, for each network evaluation a small, randomly selected, portion of the training set is used to measure the network's performance. Although this does not provide an exact indication of the performance of the network on the entire training set, it provides an estimate. The larger the sample size, the more accurate the estimate. The drawback of using only a subset of the training set for each evaluation is the potential noise in each evaluation. However, the consequences of this drawback are reduced as the "survival" of networks throughout a number of consecutive generations will most likely be an indication of their ability to work well on large portions of the training set. In practice, this provides an effective method of reducing the computational burden, with little loss in generalization ability. However, this method should not be used if generalization is not crucial. If the network's capability of memorizing the training set is important, this method does not perform as well [5].

Tests were performed with 1000 training, 100 validation and 800 image test sets. Training was only guided by the errors the network accrued on the training set. The validation set was used to gauge which network should be chosen at the end of the search to test the generalization ability of the network. The generalization ability is gauged on the 800 image test set. The test set was only used once per run, to gauge the performance of the single network which had the lowest average error on the validation set. All of the results are reported in terms of the error on the test set.

For these experiments, 7 bits were used to represent each weight, with the ranges of weights between -1.0 and 1.0. The search was allowed to progress 3000 generations with 30 networks evaluated per generation. To avoid problems with noise in the network's evaluation, one fifth to one third of the entire training set (200-333 images) were used to evaluate each network.

Other tests have been performed with networks which have only a single output unit [5]. The steering direction is determined by the activation of the output unit. In those tests, the same training, validation, and testing sets were used. Results revealed that the use of samples sizes as small as 50 images for each network's evaluation led to approximately equivalent performance, in terms of generalization ability, with using the entire 1000 image training set for each evaluation [5]. However, larger sample sizes are needed for networks which use the 30 output distributed steering direction representation described in this chapter.

The encoding of the network used a bit string which is longer than is used in most evolutionary search procedures. For each connection, its absence or presence required 1 bit, the encoding of the weight required 7 bits. The 7 bits were interpreted as a base-2 number; the value was mapped to a number between [-1, +1]. In the maximal network, there were a total of 1114 possible connections: (240+1) * 4 (input to hidden), and (4 +1) * 30 (hidden to output). The (+1) factors are used for connections to a *bias* input unit, this is a unit whose value is permanently clamped to a value of (+1.0), to which all units are connected [12]. With 1114 connections, the total size of the bit string was 8912 bits.

In order to give a base-line performance from which to compare the performance of the evolutionary algorithms, results using the backpropagation algorithm are also given. The backpropagation algorithm incorporated all of the modifications which are used in standard ALVINN training [15]. All of the results presented in this chapter are the average of at least 10 training sessions. Backpropagation was able to achieve an average sum squared error of 9.40, measured on the 800 image test set. The PBIL algorithm was able to achieve an average error of 8.19, an error reduction of approximately 13%. The final networks successfully pruned away, on average, approximately half of the possible connections. The drawback of the PBIL approach, however, is the training time; the backpropagation algorithm took only several minutes, on average, to train the networks to its lowest error. The evolutionary approach took over an hour. As mentioned before, the need for generalizability must be carefully weighed with the desire for efficiency when choosing the appropriate training method. In Figure 1.8, sample input and output are given after training by PBIL. Samples attained by backpropagation are also shown.

A second test was conducted to determine whether PBIL could perform better if the architecture was pre-specified, and only the weights of the connections were evolved. For this experiment, the maximal network previously described was used. PBIL was only allowed to modify the weights, not the architecture of the ANN. The hope was that if PBIL was constrained to a pre-specified architecture, it may do better because the search space is more constrained. Further, the connections can be effectively eliminated by setting the connection weight to 0.0. Using only the evolution of weights, PBIL was able to find networks which, on average, had an average sum squared error of 7.96; this reflects approximately a 15% improvement over standard backpropagation. This network was encoded in 7798 bits. This is smaller than the encoding used in the previous experiment as each connections was assumed to be present.

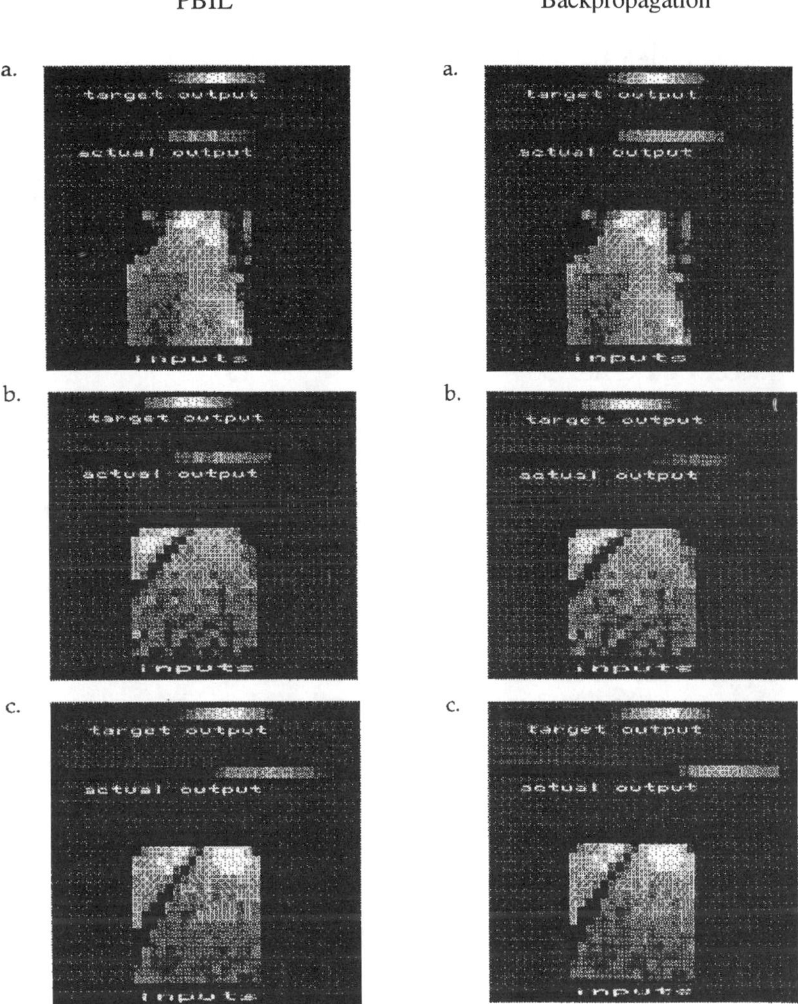


Figure 1.8: (Left) The PBIL derived network's output. The actual output for this image has the correct general location as the target output. However, the output is not as smooth as the target. (Right) Typical outputs of a network architecture trained with backpropagation. Figures are taken from the test set. Sum squared errors are as follows (PBIL) a: 4.2, b: 8.2, c: 9.9. and (Backpropagation) a: 2.0, b: 15.0, c: 23.7. Images chosen from the test set to show a wide range of errors.

To this point, the effectiveness of evolutionary search techniques has been compared to that of backpropagation using the sum squared error metric. As both algorithms were trained only with this error metric, it is correct to measure their general relative effectiveness by comparing the performances based upon this error. However, in terms of absolute performance, domain specific error metrics are often more indicative of real performance improvements. In this domain, such an error metric is *Gaussian peak position error* (GPPE). This is a measure of the distance (in output units) between the correct peak in the Gaussian, and the predicted peak in the Gaussian. This is a linear translation of the steering error. The performances of the training methods gauged by this error metric are also provided in Table 1.1.

| 30 Output | TRAINING METHOD | | |
|---|---|---|---|
| | PBIL topology & weights | PBIL weights | Back-Propagation |
| SSE | 8.19 | 7.96 | 9.40 |
| % ERROR DECREASE 30 output BP (SSE) | 12.9 | 15.3 | n/a |
| GAUSSIAN PEAK POSITION ERROR (GPPE) | 2.96 | 2.90 | 3.35 |
| % ERROR DECREASE 30 output BP (GPPE) | 11.6 | 13.4 | n/a |

Table 1.1: Results with training using the SSE error metric.

The errors are shown in Table 1.1 in terms of both average sum squared error and Gaussian peak position error. Since the evolutionary technique only used network evaluations based upon sum squared error (SSE), larger improvements are seen when performance is gauged using this error metric than when gauged with the GPPE error metric. Nonetheless, the GPPE error metric is more indicative of the actual performance of the system. In the next section, network evaluations which are based upon the GPPE metric are explored. It should be emphasized that although simple backpropagation could not have easily used the GPPE error metric, evolutionary algorithms can very easily incorporate such information to guide the search.

### 1.6.1 Using a Task Specific Error Metric

To this point, all of the experiments have used the sum squared error metric to guide the evolutionary search. Nonetheless, for many problems, the output which is produced by the ANN must be translated into a different form to be used by the specific task. For example, in order to use the 30 output representation, a gaussian is fit to the outputs, and the peak of the Gaussian is used to determine the predicted steering direction. It is the distance between the predicted and actual peaks of the gaussian which is crucial to good performance, since this determines the error in the network's steering direction. However, the peak difference error measure does not provide an easy mapping of credit or blame to each specific output unit, as is needed for backpropagation. Therefore, sum squared error is often used as the guiding error metric.

In other domains, alternate error metrics have been proposed for backpropagation to better capture the underlying requirements of specific tasks. One such error metric, the Classification Figure of Merit (CFM) error metric has been used for problems such as the 1-of-N classifier. The standard SSE error metric attempts to minimize the difference between each output node and its target activation. The CFM error metric attempts to maximize the difference between the activation of the output node representing the correct classification and the output of all the other nodes which represent incorrect classifications [11]. The CFM error metric concentrates changes on ensuring that the correct classification is made rather than ensuring that the target output is matched exactly.

The CFM error metric focuses effort towards performing the underlying task of classification rather than reproducing the exact target vector. Similarly, using the GPPE error metric focuses the training towards yielding accurate peak position interpretation of the output vector. Unlike the CFM error metric, however, error cannot easily be assigned to individual output units with the GPPE error metric. Nonetheless, because most evolutionary techniques do not use explicit credit assignment, the GPPE error metric can still be used to guide the evolutionary search. In this section, networks are evolved which explicitly reduce the GPPE rather than the sum squared error; each network is evaluated by its ability to reduce the GPPE error on each image in the training set, without regard to the sum squared error.

Using the GPPE error metric changes the goal of the search algorithm. Using the SSE error metric, the goal is to reproduce the entire target output vector exactly. Using the GPPE error metric, the goal is to place a larger output activation on the portions of the output vector which correspond to the correct steering direction than those which do not. In determining where the peak of the Gaussian lies in the output vector, many of the small amounts of noise can be ignored. This gives the search procedure the flexibility to not be as precise in large portions of the output vector and still achieve a high score; this is clearly displayed in Figure 1.9.

The average error using the GPPE error metric was 2.76 (GPPE); this is approximately an 18% improvement over backpropagation. The error measured in terms of SSE, however, was much higher than in the previous experiments: 19.3. The large difference in SSE error, in comparison to the other experiments presented before, indicates that doing well in terms of SSE is not a prerequisite for good performance on the error metric of interest, GPPE.

## 1.7 Conclusions

Various parameter settings were used for all of the training methods. In addition, the backpropagation algorithm used in this study maintained spatial information about the input retina, through neighbor weight smoothing, which the evolutionary algorithm was not given. Nonetheless, the evolutionary approach performed better, on average, than backpropagation. Although the evolutionary approach provides performance improvements, it also incurs severe computation penalties. For example, the backpropagation method was able to achieve its minimum in several minutes. The evolutionary approach took over an hour. In deciding whether to use an evolutionary approach or backpropagation, it is necessary to carefully weigh the need for accuracy with the desire for speed.

This chapter has presented several techniques for increasing the efficiency of evolutionary procedures for training artificial neural networks. The first is the evolutionary procedure used — the population based learning algorithm. Although it is far less complex than a simple genetic algorithm, it is effective in the optimization problems to which genetic algorithms are often applied [4]. The second is the evaluation of each network on a subset of the original training set.

The majority of the time in the evolutionary search procedure is spent evaluating the effectiveness of each network. The time to create a new potential solution vector, including the generation of a new solution vector, and the translation of it to the structure and weights of the ANN, is very small in comparison. Therefore, a reduction in the time spent in the network evaluation portion of the algorithm has a tremendous impact on the overall speed of learning.

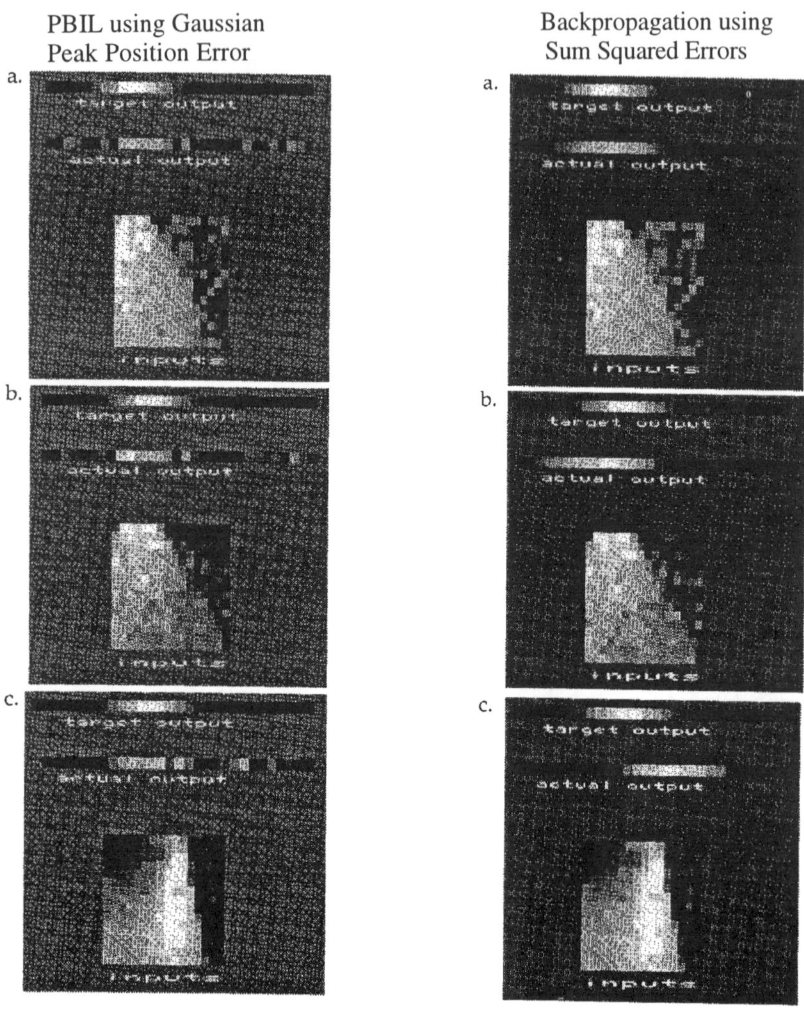


Figure 1.9: Sample input and output using the GPPE error metric. Images taken from test set. Images were chosen to show different amounts of GPPE errors: image A: 0.53, B: 1.16, C: 1.71. Backpropagation, trained with the standard SSE error metric, achieves the following GPPE errors: 0.49, B: 1.70, C: 5.12.

Evolutionary search procedures have the ability to use direct information of error metrics other than those which can easily be used by backpropagation. This can lead to improved performance on the specific task. In this chapter, either the

GPPE or SSE error metric was used to guide the evolutionary search. Although it was not tried in this chapter, perhaps using a combination of both the error metrics may give more information from which to guide the search, and therefore to find better solutions. This direction is open for future research.

In many of the previous studies in which artificial neural networks were evolved, either the entire network structure was evolved from only the inputs and outputs specified, or only the weights were evolved once the network was entirely prespecified. The method presented in this chapter, which is similar to the one chosen by [14], is a hybrid between these two extremes. In this method, a maximal network is specified. The evolutionary algorithm searches for the network topology by either maintaining or throwing away possible connections. The network topologies which are evolved in this study are efficient; they use approximately only half of the total number of possible connections. This hybrid approach allows any *a priori* information regarding the network topology to be easily incorporated into the search. Although a more automatic specification of the network would be desirable, it is suspected that as the need for more automatic specification of nctworks is increased, using genetic or PBIL search methods will increase the already large search times.

## 1.8 Future Directions

The version of population based incremental learning which was used in this study was very simple. More advanced versions of the algorithm may yield better results. For example, although the cardinality of the networks encoding was 2, the PBIL algorithm can also work on alphabets of larger cardinality. Either using a larger cardinality alphabet or using real valued features may improve the performance of the algorithm. Fogel *et al.,* have studied evolving neural networks with real values rather than binary encoded values, and have achieved good results [8].

In the experiments presented here, a fraction of the entire training set was used. However, no method for determining the correct fraction of the training set to use was presented. If too large a portion is used, search can be slowed down tremendously. If too small a portion is used, the noise in the evaluation of each network may lead the search algorithm away from finding networks which perform well on the entire training set. As using samples from the training set has the potential to greatly reduce the amount of time used for developing networks with good generalization capabilities, more exploration should be done to determine how much of the training set should be used for individual evaluations.

A promising area for future research is the integration of backpropagation with evolutionary search procedures. Evolutionary search has the ability to perform global search while backpropagation provides the ability to perform local optimizations. Therefore, backpropagation could be used as a "post processing" step after the EA is completed. Another benefit for the integration of these two procedures is potential reduction in the time needed for search by the EA. For example, backpropagation can be periodically used to locally optimize the best networks found through search. This makes the goal of the evolutionary search to find a good basin of attraction, and let backpropagation optimize the neural network once a good basin is found. Such an approach is taken by [13].

The long-term future goal, of which this project is a part, is to collect a pool of the evolved networks which can be installed into the NAVLAB. However, before this can be done, several issues need to be resolved, one of which is that of training networks with more than a single road type. The desire to use a small number of "road specific" networks must be weighed against the potential performance degradation of networks trained on more than a single road type. The transition to using evolved networks, whether they are eventually evolved with single or multiple road types, will not be difficult. The evolved networks use the same inputs as the ALVINN networks, and their outputs are translated into steering commands in exactly the same manner as the ALVINN networks already in use. The results shown here appear promising in their error reduction; nonetheless, only actual use will determine their true efficacy.

**Acknowledgements**
I am grateful to Dean Pomerleau for his extended discussions of the ideas contained within this report. Thanks are also due to Charles Thorpe for his discussions and encouragement. Todd Jochem kindly provided the sample images from which all training and testing were conducted. Finally, thanks are also due to the members of the CMU-Vision and Autonomous Systems Center for enduring the computational costs associated with this project.

The author is supported by a National Science Foundation Graduate Fellowship. The views and conclusions contained in this document are those of the author and should not be interpreted as representing official policies, either expressed or implied, of the National Science Foundation, ARPA, or the U.S. Government.

**References**
[1] Abu-Mostafa, Y., Information Theory, Complexity and Neural Networks, *IEEE Communications Magazine.* Vol. 27, No. 11, 1989.

[2] Angeline, P., Saunders, G. & Pollack, J., An Evolutionary Algorithm that Constructs Recurrent Neural Networks. *IEEE Transaction on Neural Networks.* Vol. 5, No. l, January, 1994.

[3] Baeck, T. & Schwefel, H.P. (1994) An Overview of Evolutionary Algorithms for Parameter Optimization, in *Evolutionary Computation.* Vol. 1, No. l, pp. l-24, 1993.

[4] Baluja, S. (1994) Population Based Incremental Learning: A Method for Integrating Genetic Search Based Function Optimization and Competitive Learning. Carnegie-Mellon University Technical Report, CMU-CS-94-163.

[5] Baluja, S. (1994) Evolution of an Artificial Neural Network Based Autonomous Vehicle Controller. Paper in Progress.

[6] Belew, R. (1993) Interposing an Ontogenic Model Between Genetic Algorithms and Neural Networks. In Hanson, Cowan, Giles (ed). *Advances in Neural Information Processing Systems* 5. Morgan Kaufmann Publishers, San Mateo, CA 99-106.

[7] Fogel, D.B. (1994) An Introduction to Simulated Evolutionary Optimization. *IEEE Transaction on Neural Networks.* Vol. 5, No. 1, January, 1994.

[8] Fogel, D.B., Fogel, L.J. & Porto, W. (1990) Evolving Neural Networks. *Biological Cybernetics* 63. 487-493.

[9] Goldberg, D. (1989) *Genetic Algorithms in Search, Optimization and Machine Learning.* Reading, MA: Addison-Wesley Publishing Company.

[10] Gruau, F. (1993) "Genetic Synthesis of Modular Neural Networks". In Forrest, S. (ed). *Proceedings of the Fifth International Conference on Genetic Algorithms.* 318-325. Morgan Kaufmann Publishers, San Mateo, CA.

[11] Hampshire, J.B. & Waibel, A.H., (1989) "A Novel Objective Function for Improved Phoneme Recognition Using Time-Delay Neural Networks". Carnegie Mellon University Technical Report. CMU-CS-89-118.

[12] Hertz, J., Krogh, A., & Palmer, R.G. (1993) *Introduction to the Theory of Neural Computation,* Addison-Wesley, Reading, MA.

[13] Keesing, R. & Stork, D. (1991) "Evolution and Learning in Neural Networks: The Number and Distribution of Learning Trials Affect the Rate of Evolution". In Lippman, Moody, Touretzky (ed). *Advances in Neural Information Processing Systems 3.* Morgan Kaufmann Publishers, San Mateo, CA.

[14] Maniezzo, V. (1994) Genetic Evolution of the Topology and Weight Distribution of Neural Networks. *IEEE Transaction on Neural Networks.* Vol. 5, No. 1, January, 1994.

[15] Pomerleau, D.A. (1994) Personal Communication. Carnegie-Mellon University.

[16] Pomerleau, D.A. (1993) *Neural Network Perception for Mobile Robot Guidance,* Kluwer Academic Publishing.

[17] Pomerleau, D.A. (1992) Progress in Neural Network-based Vision for Autonomous Robot Driving. In the *Proceedings of the 1992 Intelligent Vehicles Conference,* I. Masaki (ed.), pp. 391-396.

[18] Pomerleau, D.A. (1991) Efficient Training of Artificial Neural Networks for Autonomous Navigation. In *Neural Computation* 3:1 pp. 88-97.

# Chapter 2

**Michael Levin**

Cell and Developmental Biology Dept.
Harvard Medical School
Boston, MA

mlevin@husc8.harvard.edu

## Locating Putative Protein Signal Sequences



### Abstract

This chapter presents an application of genetic algorithms to a problem in molecular biology. Many proteins occurring in cells participate in biochemical events such as degradation, chemical modification, directional transport, etc. It has been shown that in certain cases, a string of amino acids serves as a specific signal; thus proteins which carry this sequence within their primary structures participate in some molecular event, while proteins lacking this sequence do not (the endoplasmic reticulum retention signal "KDEL" is a good example). Finding the sequence of a specific possible signal based only on the primary structures of a group of proteins thought to carry it is a very difficult task. No good algorithm currently exists for locating brand new signals. A genetic algorithm is described here which is able to discover such sequences. This algorithm is able to search the enormous state space of all possible signals in reasonable time, and locate likely signal sequences (which can then be tested empirically). The algorithm can also be used to find signature sequences in related proteins. Because genetic algorithms are domain independent, a parametrization study is also presented, which shows optimal values of certain constants for this specific task.

0-8493-2529-3/95/$0.00 + $.50

## 2.1 Introduction

Many proteins important in cell function participate in various processes (retention in or targeting to specific organelles, chemical modification, degradation, secretion, etc.). In certain cases, the signal which determines exactly which proteins participate in a given process is a short string of amino acids within the primary structure of the proteins. The endoplasmic retention signal, KDEL, is a good example of this (Pelham, 1990).

So many examples of signals have been found (Bairoch, 1991) that when one has several proteins (called the "in" group), all of which undergo some particular event, it becomes tempting to search for a sequence which might serve as the recognition signal. Once a potential sequence has been found (one that occurs only in that group of proteins), the hypothesis can be tested by artificially grafting the signal onto a protein which doesn't normally participate in the event. If the protein is seen to then undergo the event, the hypothesis is confirmed.

One problem with this process is that given the primary structures of several proteins, it is a very difficult task to come up with a potential signal sequence; if the proteins are of significant length, it is very hard to identify a common (but unique to the group) region by eye, especially since certain amino acid homology rules and groupings may apply. If one has a pretty good idea what this signal might be, simple pattern matching, weight matrix analysis, or discriminant analysis can be used. However, there are no good methods for easily finding a completely new signal.

This problem reduces to the task of finding the longest string which matches optimally somewhere within all members of the functionally-defined "in" group, and does not match a random set of proteins not belonging to the group (the "out" group). A closely related problem is to find a signature sequence which can be used to tell certain proteins apart from similar ones (such as the "A-G-L-x-F-P-V" signature for histone H2B, Wells, 1989). This task thus touches on features of machine learning, pattern recognition, classification systems, and feature abstraction.

Considering the fact that the data is noisy (i.e., one or more of the "in" group proteins may not carry the signal, but participate in the event of interest for other reasons), and the fact that the signal sequence may not be 100% conserved among all proteins, the search space of all possible signals of a given length (usually 3-10 amino acids long) is a very difficult one. If a very fast computer is available, and one is willing to restrict the search to signals less than about six amino acids long, an exhaustive search of all possible short strings may be feasible. However, as the length of the proteins involved grows, and one wants to look at signals which may be somewhat longer, this quickly becomes impractical with respect to the time involved to perform the search (the time required is proportional to $30^N$ where N is the maximum number of characters in the signal).

A set of algorithms which has recently been shown to be able to find solutions in difficult search spaces are known as "genetic algorithms" (Goldberg, 1989, Davis, 1991, Holland, 1992, Koza, 1992). These domain-independent algorithms simulate evolution by retaining the best of a population of potential solutions and mutating these to arrive at the next generation's population. This process is repeated until a solution of sufficient quality is found (or computational resources are exhausted). The algorithms have proven to be robust and effective for a wide variety of problems, such as symbolic regression, process control, generation of emergent behavior, classification, and pattern recognition (see Koza, 1992 and references therein). GAs have also been used in molecular biology (Dandekar, 1992).

This approach can be used to locate likely candidates for functional protein signals (De La Maza and Tidor, 1992, used a very similar problem to study the effects of Boltzmann selective pressure). This is done by performing random mutation and fitness selection over a population of candidate signal sequences. Each individual in this population is a string of amino acids of some length. Its fitness is proportional to how well this sequence matches the members of the "in" group, and inversely proportional to its match with the "out" group. Genetic algorithms are used, rather than the more general technique of genetic programming because in this case the map from discrete character set genome to the possible solution space is a very natural one. This algorithm is shown to effectively and easily locate potential signal sequences with no initial data other than the functional grouping of proteins and their primary structure. Since the genetic algorithm approach is domain-independent, a parameterization study is performed, to determine the optimal parameters for locating such sequences.

### 2.2 Implementation

The algorithm is fairly simple, and is easily coded in C. Each member of the population is a string (called a "schema") of some length (minimum length is usually set to 3, maximum length to 10) over the alphabet consisting of the single letter codes for amino acids, plus the symbols _* (the wildcard, or "any amino acid" symbol), a ("an acidic amino acid"), b ("a basic amino acid"), n ("a neutral amino acid"), h ("a hydrophobic amino acid"), p ("a polar amino acid"), and c ("a charged amino acid"). This string can be directly compared to any protein's primary structure. Thus for example, the hypothetical string "TY*Sa" would match a protein which contained a threonine, followed by a tyrosine, followed by any amino acid, followed by a serine, followed by any acidic amino acid. Additional symbols can be added (such as the numbers 0 through 9) which can stand for certain other homologies (for example, 0 may stand for "either S or A here"). The string is the member's genetic material — one chromosome.

Several operators are used. First, a fitness function has to be defined, which can evaluate the worth (i.e., success at differentiating the "in" proteins from the "out" proteins) of any individual. This is returned as a scalar floating point number, which allows unambiguous ordered ranking of all individuals. This number is based on parameters designed to take into account several desirable qualities of a candidate solution; higher numbers indicate better schemata. The fitness function used in this implementation, when applied to a schema S, returns a number which is equal to:

$$k_1.\textit{knowledge}(S) + k_2.\textit{size}(S) - k_3.\textit{vagueness}(S) \qquad (2.1)$$

where knowledge(S) is equal to:

$$match\_ins(S) - k_4.match\_outs(S) \tag{2.2}$$

*match_ins*(S) determines how well schema S matches the proteins in the "in" group. *match_outs*(S) does the same for the "outs" group. The degree of match is computed as the sum of all matches to proteins in a group, divided by the number of members in the group. The degree of match to a given protein is defined as the number of matching characters (at the best-matching position within the protein) divided by the length of the schema. The match to the "out" group is subtracted to ensure that the best individuals are those which match the "out" group least.

*size*(S) determines the effects of the schema size. In general, the value of *size*(S) should be proportional to the length of the schema, because it is better to have the complete signal sequence than a part of it. It may, however, include special nonlinear terms to punish schemata that are too long.

*vagueness*(S) determines how specific the schema is. It is proportional to the number of non-specific symbols occurring in S (such as * etc.). This term is also subtracted in equation 1 because it is best to have as specific a sequence as possible (while still matching optimally).

The constants $k_1$ through $k_4$ are parameters that the user can adjust for specific effects. Normally $k_1 >> k_2, k_3$ because the most important thing is for the schema to differentiate between the "in" group and the "out" group. However, the other terms ensure that if two individuals have similar matching ability, the more specific and longer ones will be considered more fit. The constant $k_4$ can be changed to control how specific the signal is to the "in" group. It is usually less than 1.0 because even a signal that occurs somewhat in non-belonging proteins can be useful if it always occurs in belonging ones.

Once the most fit members of a population are identified, their genotypes are used to construct the next generation. Two possible operators are mutation, and crossover. A single mutation event performed on a schema S (asexual reproduction), as used in this algorithm, consists of choosing at random among: deleting a symbol at a random position within S, adding a random symbol somewhere within S, or changing a random symbol within S to some other random symbol. Appropriate safeguards are used to ensure that schemata don't become too small or too large. Other than that, the mutation is completely random, with no foresight as to the effects on its performance. Crossover consists of picking two individuals, and producing two new ones by swapping random parts of the parents' genome.

Crossover was used in the initial trials of these experiments, but resulted in premature convergence of the population on suboptimal solutions (data not shown). Thus, all results shown in this chapter utilize simple mutation only. These results are consistent with those of Fogel and Atmar, 1990, who conclude that complex genetic operators such as crossover and regional inversion do not compare favorably with simple mutation (unlike Holland, 1975 and Koza, 1992, who claim that crossover produces better results than mutation).

The control flow of the algorithm is shown in Figure 2.1. After the parameters are set, the "in" and "out" groups are read in from disk. The "out" group should ideally consist of proteins which are related to the proteins in the "in" group, but known (from empirical evidence) not to participate in whatever event functionally defines the "in" group. Alternatively, the "out" group can consist of randomly chosen proteins, or even of random sequences of amino acid symbols.

An initial random population is then created. The population size is the parameter $P_1$ — this is what determines how many solutions the algorithm is working with at any time. The bigger the value of $P_1$, the longer it takes to evaluate each generation; however, higher values of $P_1$ make it more likely that a good solution will be found. Typical values of $P_1$ can be from 300 to 1000. The user can, at this point, seed the initial population with several initial guesses. This can be used to improve a guess obtained by other means, or to help speed up the search when some of the signal is known, but it is not a good practice in general because it can cause the search to prematurely converge on some solution and ignore one which may turn out to be better.

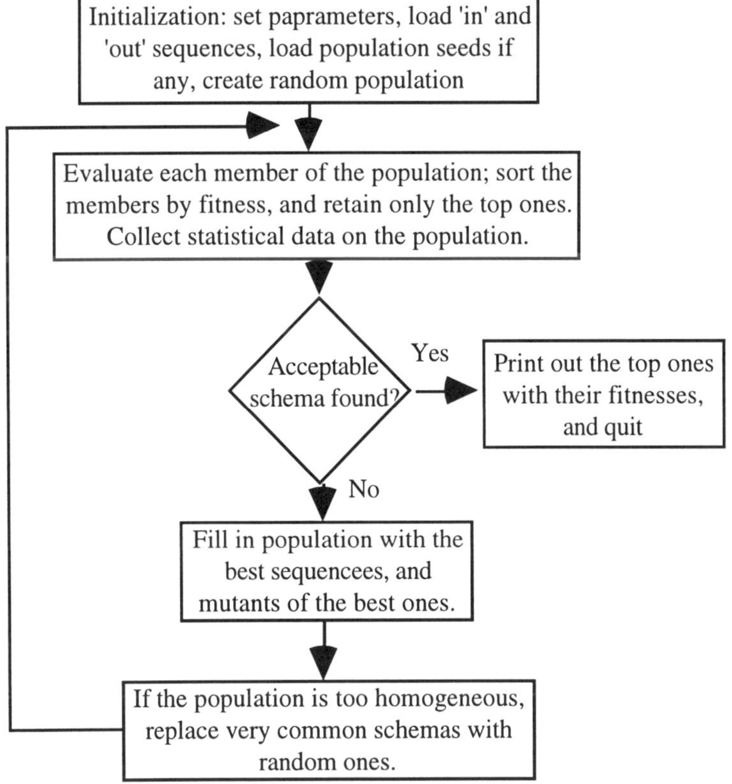

Figure 2.1: The genetic algorithm flow of control.

Then, each individual is evaluated according to the fitness function, and the top $P_2$ schemas are chosen. $P_2$ is usually between 10% and 70% of $P_1$. Too high a value of $P_2$ results in slow convergence, while too low a value may cause premature convergence due to early elimination of potentially good schemata. It is important to note that evaluating the fitness of a given individual is the most computation-intensive step in this algorithm. As the population homogeneity begins to rise, a simple trick can be used to cut down the evaluation time (which can be critical, when the "in" group is large). This method takes advantage of the fact that if more than one schema in the population have identical sequences, only one has to be evaluated, and its fitness can be assigned to all of them. Thus, previous to evaluating fitnesses, the population is sorted by alphabetical order. For each schema $S_n$ (n>l) if it is identical to $S_{n-1}$, then the fitness assigned to $S_n$ is simply copied from $S_{n-1}$; otherwise, the fitness of $S_n$ is calculated explicitly.

The population is then rebuilt, to consist of mutated copies of the best individuals, as well as unchanged versions of these individuals. This is "elitist" selection, and ensures that good schemas are never lost from the population. This process continues until either an acceptable solution is found, or the time limit expires. This process can contain several additional features. For example, if the population homogeneity becomes too high, some copies of the most frequent individuals can be replaced with random schemas, or mutated heavily in an attempt to inject variety into the system (the elitist selection ensures that this cannot decrease the maximum fitness found in the population). The whole algorithm is summarized by the following pseudocode:

1. Read initial data — in and out groups, parameters N, P, Q, R, S, etc. Place protein sequence in two-dimensional string array

2. Build up a random population of schemas, or read them in from a file. Place sequences into two-dimensional string array. Crossover and mutation are accomplished as string operations (i.e., character and substring substitutions, deletions, inversions, etc.) on the members of these arrays.

3. Until top fitness is acceptable, or allotted time has expired, do:

    A. Computer fitness for each member of the population, by matching its sequence to each member of the in and out groups. Fitness is calculated as in Eq. 1 above, using simple string matching.

    B. Sort the population. Leave the top *n* members unchanged. Set the next *P* members to strings which arise from crossovers between randomly chosen members of the *N* best. The choice is biased to favor crossovers between dissimilar schemata.

    C. Introduce *q* mutations into the members resulting from crossover, and set the remainder of the population to consist of mutated versions of the top *N* members.

D. Compute and plot the top fitness, average fitness, and homogeneity of population as a function of generation number.

E. Compute total homogeneity of population. If this is higher than an acceptable level *R*, then eliminate all but one copy of each individual, and fill in the rest of the population with crossovers between the remaining individuals and random schemata.

4. Print out the top *S* non-identical schemata, their fitnesses, and their locations within each member of the in group.

In this algorithm, the computational complexity (as measured in the number of string comparisons per generation) as a function of total protein lengths is O(n). That is, it increases only linearly with increases in the number of total amino acids in the in and out groups. However, the total time spent on the search is not necessarily O(n) because different numbers of total generations are required to find adequate solutions for different sets of proteins, and because of the stochastic nature of the algorithm.

### 2.3 Results of Sample Applications

This algorithm was tested on many different kinds of signals (data not shown). Two examples are illustrated here in detail. Figures 2.2 and 2.3 show the progress of the search over time (in generations on the abscissa). Three quantities (explained below) are monitored; their magnitude is normalized between 0 and 1 (on the ordinate).

The first sample application illustrates how the algorithm finds the KDEL signal (given the sequences of the following proteins found in the GenEMBL database): *H. vulgare* GRP94 homologue, rat immunoglobulin heavy chain binding protein (BiP), rat calreticulin, and rat protein disulfide isomerase (accession numbers X67960, M14050, X53363, X02918, respectively). For this run, the parameters are set as shown in column 1 of Table 2.1. The results of the run are seen in Figure 2.2. The KDEL sequence is found in 64 generations, which represents about 2.5 hours of real time on a lightly loaded (average system load during run = 1.05) DecStation 5000 workstation. Interestingly (perhaps), it initially found other sequences common to these proteins (with 100% fit to each): "EED" and "EEEa". Once a schema has been found, and determined (empirically) not to be of interest, others can be searched for by entering this sequence into the "out" group (to ensure that the search disregards it).

| Parameter | KDEL | Histone signture | Default |
|---|---|---|---|
| Population size ($P_1$) | 800 | 800 | 800 |
| Survival size ($P_2$) | 300 | 300 | 30% |
| $K_1$ | 15 | 15 | 15 |
| $K_2$ | 2 | 2 | 2 |
| $K_3$ | 3 | 3 | 3 |
| $K_4$ | 0.1 | 0.8 | 0.1 |

Table 2.1: Parameter settings for various runs.

The second sample application illustrates how this method can be used to find signature sequences. In this case, the histone H2A signature *A-G-L-x-F-P-V* (Wells, 1989) can be found by running H2A variants in the "in" group and the H2B, H3, and H4 proteins in the "out" group. In this experiment, the "in" group consisted of sea urchin *(P. miliaris)* late histone H2A-2, human histone H2A gene (lambda-HHG55), *P. miliaris* histone H2A-2.1 gene, and the murine H2A gene (accession numbers M11085, K01889, M14140, X16495, respectively). The "out" group consisted of sea urchin *(P. miliaris)* late histone H2B-2, *P. miliaris* histone H2B-2.2 gene, *P. miliaris* gene for histone H3, chicken histone H3 gene, *A. thaliana* histone H3 gene, *X. laevis* histone H4-I gene, and the newt histone H4 gene (accession numbers M11088, M14143, VO1140, J00869, M35387, M23776, M23777, J00954, respectively). For this, the specificity constant needs to be higher than usual. The constants in this experiment are given in column 2 of Table 2.1. Figure 2.3 shows that the H2A signature is found at generation 206. Interestingly, another one is found, which is considered by the algorithm to be even better (because it doesn't contain any non-specific characters): LQFPVGR at generation 84.

Several interesting things can be noted from these sample runs. The solid line shows the fitness of the best individual at each generation. This curve is monotonic, since the elitist selection ensures the best individuals are never lost. In these and some other runs (data not shown) the maximum fitness curve is reminiscent of the punctuated equilibrium hypothesis (Eldredge, 1985) — long stretches of little change interrupted by sharp improvements. This may be due to the fact that the mutation rate used here is too low to cause changes in top fitness over small time periods.

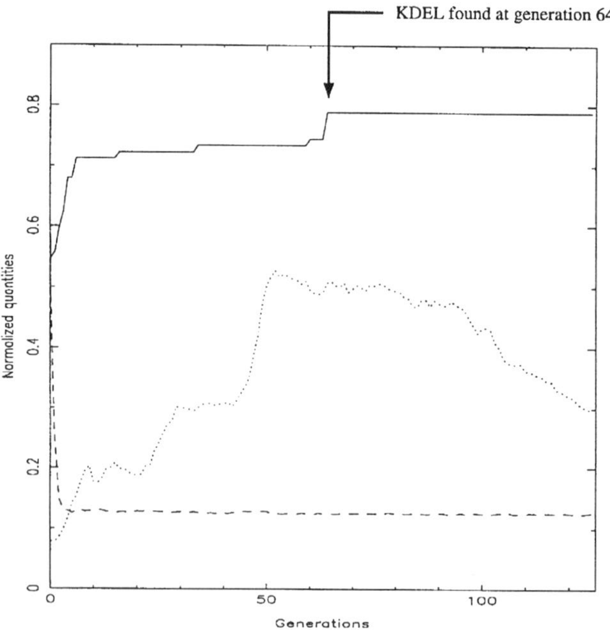


Figure 2.2: Locating the KDEL sequence.

A genetic algorithm search was performed with the parameters given in column 1 of Table 2.1. The solid line represents the fitness of the most fit schema at any generation. The dashed line represents the average length of the schemata at a given generation. The dotted line represents the homogeneity of the population.

The maximum fitness of the second plot starts out much lower than that of plot 2, since the target string of experiment 2 is more complex, and the average fitness of a random individual is likely to be lower. The dashed line representing the average length of all schemata drops quickly to the optimal length. This is somewhat surprising since the length constant in the fitness function is low, and it might be expected that the length not be important (and thus not be selected for) until the fitness becomes quite high and the population converges. The dotted line represents the population homogeneity (as computed by taking the sum of the average similarities of each individual to all others in the population). Interestingly, it is non-monotonic and complex; this is an emergent phenomenon — there is nothing in the fitness function to directly cause such a curve. Note that these curves are very different between Figures 2.2 and 2.3, suggesting that the large-scale population dynamics are different for different instances of this search problem.

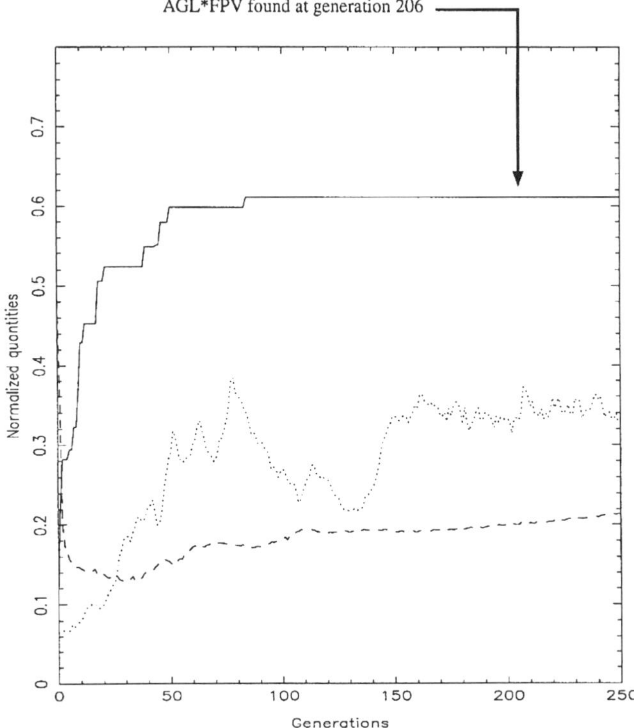


Figure 2.3: Locating the Histone H2A Signature Sequence.
A genetic algorithm search was performed with the parameters given in column 1 of Table 2.1. The solid line represents the fitness of the most fit schema at any generation. The dashed line represents the average length of the schemata at a given generation. The dotted line represents the homogeneity of the population.

## 2.4 Parametrization study

With so many variables in this domain-independent algorithm, it becomes interesting to: 1) determine what combination of settings are optimal for the protein signal problem, and 2) examine the properties of the algorithm as they vary with the parameters. For these purposes, a parametrization study was performed. In all of these studies, the dependent variable was the generation number in which the desired answer first appeared ("generation of discovery"). The problem set in all cases was to locate the KDEL sequence (using half-lengths of the proteins given above). All parameters except the one being changed are set to the values in column 3 of Table 2.1. A study of variation (since the algorithm is a non-deterministic one) was performed; 20 repetitions of exactly the same problem and parameters showed that differences in generation of discovery were of the range ±13 (data not shown). This is to be considered as the significant difference level for the experiments described below. In all of the figures, the value shown is the average of 10 repeat runs.

The first part of this study examined the dependence of the algorithm's efficiency in finding the KDEL sequence on the size of the population used. A population size of 400 found the solution in 41 generations, while a population size of 1400 found the solution in only 17 generations. Intermediate values of population size produced intermediate values of generation of discovery. Populations of sizes 300 and smaller did not tend to locate the solution at all (within 2000 generations). Figure 2.4 summarizes the dependence of generation of discovery on the size of the population. Clearly it is better to use larger generation sizes. However, since larger generation sizes also take longer to evaluate, it is interesting to examine how the time of discovery relates to the generation size. It is important to note that these times are relative (because they depend on what kind of computer the tests are run on).

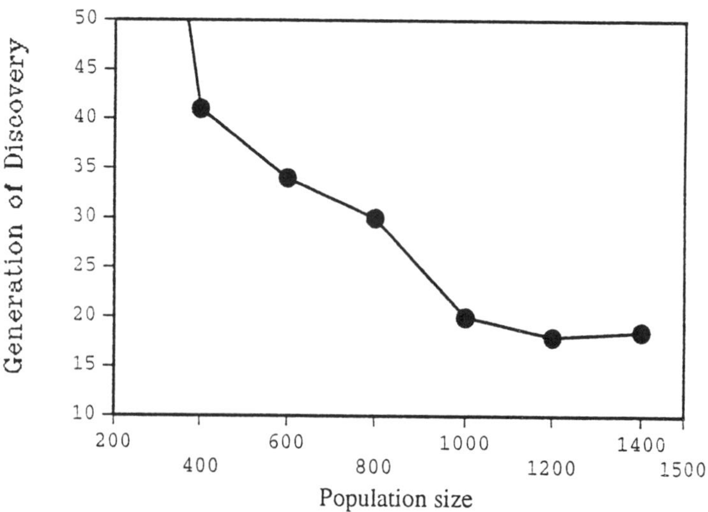


Figure 2.4. Dependence of solution rate on population size.

A series of genetic search algorithms was performed on the KDEL problem, each using a different population size (given on the X axis). The other parameters are set as in the third column of Table 2.1. The generation number at which the KDEL sequence was found is plotted on the Y axis.

A population of size 400 found the solution in 58 minutes, while a population of size 1400 found it in 80 minutes. Intermediate population sizes produced intermediate results. Figure 2.5 summarizes this data, showing a U-shaped relationship. For small generation sizes, it takes longer to find the solution because of the large number of generations necessary. For large population sizes, it also takes longer, because of the computational cost of evaluating large populations. The optimal value seems to be about 800, which allows the solution to be found in just 31 minutes.

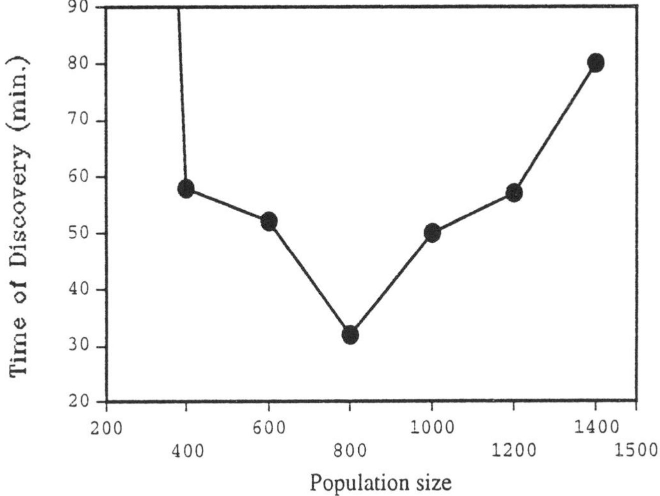


Figure 2.5: Dependence of Time to Discovery on Population Size.
A series of genetic search algorithms was performed on the KDEL problem, each using a different population size (given on the X axis). The other parameters are set as in the third column of Table 2.1. The time (in minutes) at which the KDEL sequence was found (relative to start time) is plotted on the Y axis.

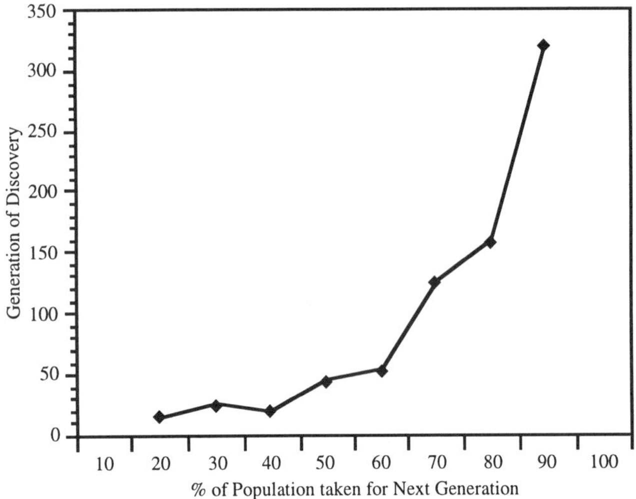


Figure 2.6: Dependence of Generation of Discovery on Survival Size
A series of genetic search algorithms was performed on the KDEL problem, each allowing a different percentage of the top individuals to contribute genetic material to the next generation (given on the X axis). The other parameters are set as in the third column of Table 2.1. The generation number at which the KDEL sequence was found is plotted on the Y axis.

The second part of this study examined the role of the number of survivors at each generation. Since this value in itself does not alter the computation time, only generation of discovery (not absolute time of discovery) was studied. When 20% of the best individuals are allowed to reproduce at each generation, the solution can be found in 20 generations. When 90% are allowed, the average is 320 generations. Reproduction values of less than 20% tended not to find the solution at all. Too few reproducers lead to premature convergence on local maxima, while too many lead to very slow convergence to the global maximum. Figure 2.6 summarizes this data, and shows that the optimal tradeoff seems to occur at a survival size of about 20%.

The final part of this study looked at the role of mutation. In these experiments, the same problem as above was examined, with varying numbers of mutations in each offspring. Figure 2.7 summarizes the data, which shows that when each offspring is subject to between 1 and 256 mutations, the solution is found on average at the same generation number (around 35). The differences between these values are not significant, showing (surprisingly) that the algorithm's efficiency is tolerant to a wide range of mutation incidences.

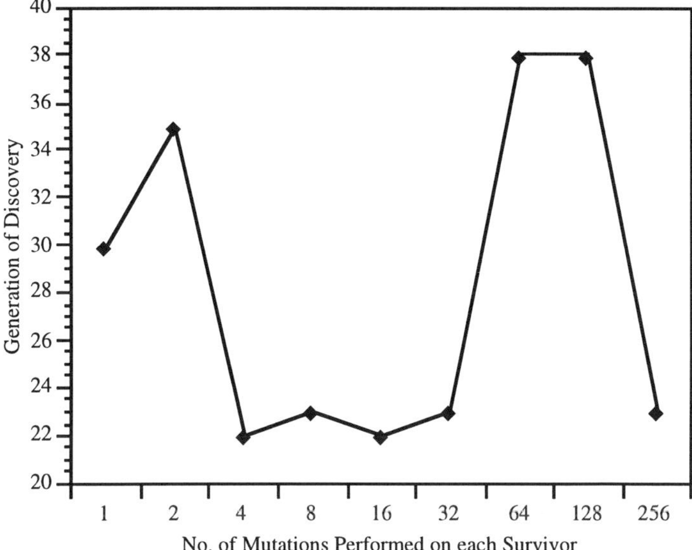


Figure 2.7: Dependence of Generation of Discovery on Mutation Incidence. A series of genetic search algorithms was performed on the KDEL problem, each allowing a different number of mutation events to occur when a top individual contributes genetic material to the next generation (given on the X axis). The other parameters are set as in the third column of Table 2.1. The generation number at which the KDEL sequence was found is plotted on the Y axis.

## 2.5 Future directions

There are several ways in which this algorithm could be improved. First, it would easily lend itself to parallelization on a computer such as the Connection Machine. Immense savings in time would be accomplished by running the fitness evaluations of each individual in parallel. The algorithm could also be made to deal better with noise in the experimental data by choosing to disregard a member of the "in" group if a schema is found which matches all the other members very well, but does not match it. Other varieties of genetic algorithms (steady-state populations, demes, etc.) may also produce better results.


## Acknowledgments

I would like to acknowledge several helpful discussions with David Fogel.


## References


Bairoch A., (1991), PROSITE: a dictionary of sites and patterns in proteins, *Nucleic Acids Res.,* 19:2241-2245.

Dandekar, T., (1992), Potential of genetic algorithms in protein folding, *Protein Engineering,* 5(7): 637-645.

Davis, Lawrence, Handbook of Genetic Algorithms, Van Nostrand Reinhold, NY: 1991.

De La Maza, Michael, Tidor, Bruce, (1992), Increased flexibility in genetic algorithms, in *Proceedings of the ORCA CSTS Conference: Computer Science and Operations Research: New Developments in Their Interfaces,* pp. 425-440.

Eldredge, Niles, *Time Frames,* Simon and Schuster, New York: 1985.

Fogel, D. B., (1990), Comparing genetic operators with Gaussian mutations in simulated evolutionary processes using linear systems, *Biological Cybernetics,* 63:111-114.

Goldberg, David E., *Genetic Algorithms in Search, Optimization, and Machine Learning,* Addison-Wesley, MA: 1989.

Holland, John H., *Adaptation in Natural and Artificial Systems,* Univ. of Michigan Pr., Ann Arbor: 1975.

Holland, John H., *Adaptation in Natural and Artificial Systems,* MIT Press, MA: 1992.

Koza, John R., *Genetic Programming,* MIT Press, MA: 1992.

Michalewicz, Z., *Genetic Algorithms + Data Structures = Evolution Programs,* Springer-Verlag, NY: 1992.

Pelham, H.R.B., (1990), The retention signal for soluble proteins of the endoplasmic reticulum, *Trends Biochem. Sci.,* 15:483-486.

Wells D.E., McBride C., (1989), A comprehensive compilation and alignment of histones and histone genes, *Nucleic Acids Res.,* 17:r311-r346.

## Chapter 3

**Peter J.B. Hancock**
Department of Psychology
University of Stirling,
Scotland, FK9 4LA

pjh@compsci.stirling.ac.uk

# Selection Methods for Evolutionary Algorithms



## Abstract

Selection pressure can have a decisive effect on the outcome of an evolutionary search. Try too hard, and you will end up converging prematurely, perhaps on a local maximum, perhaps not even that. Conversely, too little selection pressure, apart from wasting time, may allow the effects of genetic drift to dominate, again leading to a suboptimal result. In nature, there are two aspects to breeding success: surviving long enough to reach reproductive maturity, and then persuading a mate to be your partner. In simulations, such subtleties are mostly the province of artificial life experiments where, for example, an animal that fails to find enough food may die. In such systems it is possible for the whole population to die out, which may be realistic but does rather terminate the search. In most Evolutionary Algorithms (EA), therefore, a more interventionist approach is taken, with reproductive opportunities being allocated on the basis of relative fitness. There are a variety of selection strategies in common use, not all of which use the fitness values directly. Some order the population, and allocate trials by rank, others conduct tournaments, giving something of the flavour of the natural competition for mates. Each of the schools of EA has its own methods of selection, though GA practitioners in particular have experimented with several

0-8493-2529-3/95/$0.00 + $.50

algorithms. The aim of this chapter is to explain the differences between them, and give some indication of their relative merits.

At its simplest, selection may involve just picking the better of two individuals. Rechenberg's earliest Evolution Strategy proceeded by producing a child by mutation of the current position and keeping whichever was better. Genetic algorithms require a substantial population, typically of the order of a hundred, in order to maintain diversity that will allow crossover to make progress. Holland's original scheme for GAs assigned each individual a number of offspring in proportion to its fitness, relative to the population average. This strategy has been likened to playing a two-armed bandit, with uncertain payoffs. How should one best allocate trials to each arm, given knowledge of the current payoff from each? The best strategy turns out to be to give an exponentially increasing number of trials to the apparently better arm, which is exactly what fitness proportional selection does for a GA. However, the approach suffers from a variety of problems, which will be illustrated, along with possible solutions, below.

A full comparison of selection methods might involve their use on a range of tasks and the presentation of large tables of results. These would probably be ambiguous, since it seems unlikely that there is any one best method for all problems. Instead, this chapter follows the lead of Goldberg and Deb, who compared a number of the common GA selection methods in terms of their theoretical growth rate and time complexity. They considered an extremely simple problem, where there are just two string values, arbitrarily 1 and 1.5. The initial population contains one copy of 1.5. They then looked at how quickly different selection methods would cause this string to take over the population, without any mutation or other genetic operators. Three similar simple problems are used here. In all cases, results reported are the average of 100 runs.

1. Take-over. A population of $N$=100 individuals are initialised with random values between 0 and 1, except for one, which is set to 1. This population is acted on by selection alone, to give take-over curves analogous to those produced by Goldberg and Deb. However, the range of values in the initial population allows observation of the worst values, as well as the best. If poor individuals are removed too quickly, genetic diversity needed for the final solution may be lost.

2. Growth. Some of the selection schemes considered produce exponential take-over rates. To allow comparisons under slightly more realistic conditions, mutation was added. The whole population is initialised with random values in the range 0-0.1. When an individual is reproduced, the copy has added to it a Gaussian random variable, with standard deviation of 0.02, subject to staying within the range 0-1. The population gradually converges towards 1, at a rate mostly determined by the selection pressure, though clearly limited by the size of the mutation.

3. Noise. Many target objective functions are noisy, and one of the claims made about Genetic Algorithms is that they are relatively immune to its effects. As will be seen, the degree of immunity depends on which selection method is used.

The task is the same as the previous one, except that another Gaussian random variable is added to each individual's value. The noisy score is used to determine the number of offspring allocated, the true value is then passed on to any children, subject to small mutation as before.

The time complexity of the different algorithms is not considered here, because it is rarely an issue in serious applications, where the time taken to do an evaluation usually dominates the rest of the algorithm. If this is not the case, then the whole run will probably only take a few seconds: one or two more shouldn't hurt!

On the other hand, stochastic effects, ignored by Goldberg and Deb, are considered here. A selection algorithm might specify 1.6 offspring for a given individual. In practice, it will have to get a whole number, and there are different ways to do the required sampling. Some methods are prone to errors, such that even the best individual may not get any offspring. Where this happened during the take-over simulations, the best value was replaced, arbitrarily overwriting the first member of the population. If this were not done, the graphs would be more affected by the particular number of runs that lost the best value than by real differences in the take-over rate in the absence of such losses. Suppose two sets of 100 runs of an algorithm are conducted, where the best string is lost with probability 0.5 on any one run. One set of runs might lose the best, say, 48 times, the other 55. The latter will appear to grow more slowly, simply because more zero values are being averaged in. The number of occasions such replacement was needed will be reported.

The results of the simulations require interpretation — it is certainly not simply the case, for example, that faster growth rates are "better". A working assumption behind the interpretation offered below is that, other things being equal, greater diversity in the population is beneficial.

This chapter is only concerned with single "panmitic" population models, where all individuals compete for selection in a single pool. There are a variety of interesting parallel models, including multiple small populations that occasionally exchange individuals, and spatial populations, where each individual sees only its immediate neighbours. Such models would be difficult to compare meaningfully by the simple methods employed here.

Also not considered here are a variety of methods used to influence selection, usually to encourage diversity in the population. This might simply be to improve the search by preventing premature convergence or perhaps to allow multiple solutions to be found. Techniques such as niching (Deb and Goldberg 1989), sharing (Goldberg and Richardson, 1987), crowding (De Jong, 1975), mate-selection (Todd and Miller, 1991) and incest prevention (Eshelman, 1991) all find their place in the literature.

### 3.1 Fitness proportionate selection (FPS)

The traditional GA model selects strings in proportion to their fitness on the evaluation function, relative to the average of the whole population. Holland's original scheme actually suggested picking only one parent according to fitness. If a second is required for crossover, this is picked at random. This produces rather lower selection pressure, but results that are qualitatively similar to the now more common practice of picking both according to fitness (Schaffer, 1987). FPS unfortunately suffers from well-known problems to do with scaling. Suppose you have two strings, with fitness 1 and 2, respectively. The second string will get twice as many reproductive opportunities as the first. Now suppose that the underlying function is altered simply by adding 10 to all the values. Our two strings will now score 11 and 12, a ratio of only 1.09. It might be hoped that such a simple translation of the target function would have no effect on the optimisation process. In practice the selection pressure would be significantly reduced.

This scaling effect causes another problem. Suppose we are optimising a function with a range of 0-10. Initially, the random population might score mostly in the range 0-1. A lucky individual with a score of 3 will then be given a large selective advantage. It will take over the population, reducing, and eventually removing, the genetic diversity. If this potential hazard is avoided, the fitness of the population might improve, say to the range 9-10. Now, for the reason described in the previous paragraph, there will be very little selection pressure, and the search will stagnate. In summary: if there is little variation in the fitness of the strings, there will be little selective pressure. The problem of stagnation has been addressed by using a moving baseline: windowing and sigma scaling.

### 3.2 Windowing

One way to ameliorate the problem is to use the worst observed score as a baseline, and subtract that value from all the other fitnesses. This then converts our stagnating population in the range 9-10 back to the range 0-1. However, it will give the worst string a fitness of zero, and, as noted above, it is not generally wise to exclude weaker strings completely. The selection pressure is therefore usually reduced by using the worst value observed in the $w$ most recent generations as a baseline, where $w$ is known as the window size, and is typically of the order of 2-10. The dramatic effect of this moving baseline is shown in Figure 3.1a, which shows the increase in the number of copies of the optimal value under selection only. FPS initially converges rapidly, but then tails off as all of the population approaches a score of 1. Moving the baseline maintains the selection pressure, more strongly for smaller window size. Subtraction of the worst value also solves the problem of what to do about negative values. A negative number of expected offspring is not meaningful. Simply declaring negative values to be zero is not sufficient, since with some evaluation functions the whole population might then have a score of zero.

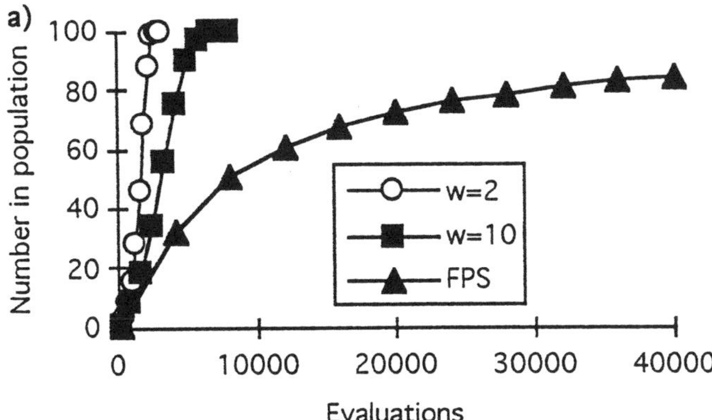


Figure 3.1a) Take-over rates for fitness proportionate selection, with and without baseline windowing.

**3.3 Sigma scaling**

As noted above, the selection pressure is related to the scatter of the fitness values in the population. Sigma scaling exploits this observation, setting the baseline $s$ standard deviations (sd) below the mean, where $s$ is the scaling factor. Strings below this score are assigned a fitness of zero, with a consequent potential for the loss of diversity. This method helps to overcome a potential problem with particularly poor individuals ("lethals") which with windowing would put the baseline very low, thus reducing selection pressure. Sigma scaling keeps the baseline near the average. It also allows the user to adjust the selection pressure, which is inversely related to the value of $s$. By definition, the average fitness of the scaled population will be $s$ times sd. Thus an individual that has an evaluation one standard deviation above the average will get $(s+1)/s$ expected offspring. Typical values of $s$ are in the range 2-5, with stronger selection again given by smaller values. The effect on take-over rate is shown in Figure 3.1b, for $s$ values of 2 and 4: selection pressure is rather greater than with a window of size 2.

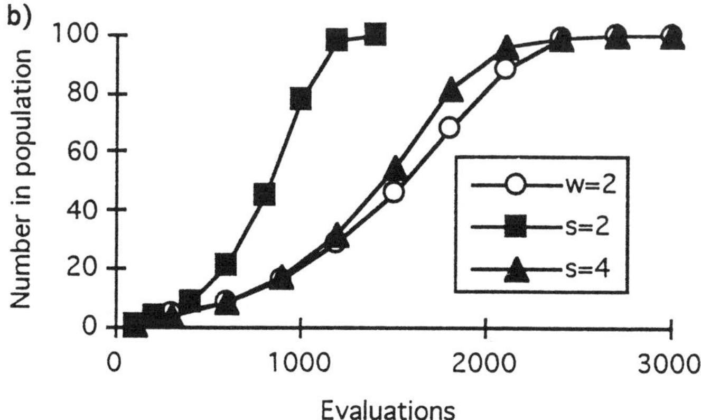

Figure 3.1b) Take-over rates for window, and sigma-scaling baseline methods. Note the x scale change.

These moving baseline techniques help to prevent the search from stagnating, but may exacerbate the problem of premature convergence caused by a particularly fit individual because they increase its advantage relative to the average. The sigma scaling method is slightly better, in that good individuals will increase the standard deviation, thereby reducing their selective advantage somewhat. However, a better method is desirable.

### 3.4 Linear scaling

Linear scaling adjusts the fitness values of all the strings such that the best individual gets a specified number of expected offspring. The other values are altered so as to ensure that the correct total number of new strings are produced: an average individual still expects one offspring. Exceptionally fit individuals are thus prevented from reproducing too quickly.

The scaling factor $s$ specifies the number of offspring expected for the best string and is typically in the range 1.2 to 2, again giving some control on the selection pressure. The expected number of offspring for a given string is given by:

$$1 + \frac{(s-1)(fitness - avg)}{(best - avg)}$$

It may be seen that this returns $s$ for the best, and 1 for an average string. There is still a problem for low-scoring strings, which may be assigned a negative number of offspring. It can be addressed by assigning them zero, but this would require that all the other fitness values be changed again to maintain the correct average. It also risks loss of diversity. An alternative is to reduce the scaling factor such that just the worst individual gets a score of zero:

$$s = 1 + \frac{(best - avg)}{(avg - worst)}$$

The algorithm may be summarised in the following C-code, which adds another variable *ms*, set to 1 less than the modified $s$ value to save a subtraction in the for loop:

```
if (s > 1 + (best-avg)/(avg-worst) )
        ms = (best-avg)/(avg-worst);
else
        ms = s - 1;
for (i = 0; i< N; i++)
        fitness(i) = 1 + ms * (fitness(i) - avg)/(best - avg);
```

The effects on convergence rate are shown in Figure 3.2a. As expected, increasing the scaling factor increases the convergence rate. With a linear scaling factor of 2, the convergence is between that obtained from a window size of 2, and a sigma scaling factor of 2. At low selection pressures, the convergence rate is proportional to $s$. Thus in this simulation, the best value takes over the population in 4000 evaluations for $s$=1.2. With $s$=1.1, it takes 8000 evaluations.

This would suggest convergence in less than 1000 evaluations when $s$=2, where in fact it takes 2000. The reason is the automatic reduction in selection pressure caused by the need to prevent negative fitness values. In this application the convergence produced with $s$=2 is very similar to that produced with $s$=1.5. The effective selection pressure is therefore still determined to some extent by the spread of fitness values in the population. A very poor individual will effectively terminate the search, so it is worth monitoring the actual value of $s$ during the run and if necessary discarding such lethals.

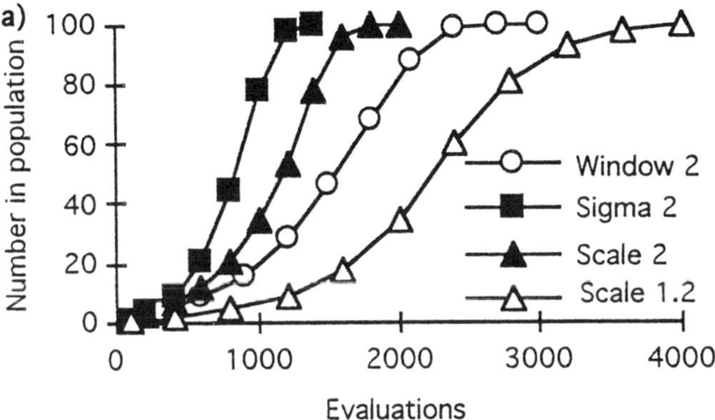


Figure 3.2a) Take-over rates for baseline window, sigma and linear scaling.

The growth rates in the presence of mutation for these scaling methods are shown in Figure 3.2b. All are quite similar, simple FPS being able to maintain selection pressure because of the range of fitness values caused by the mutation. Windowing and sigma scaling come out ahead precisely because they fail to limit particularly fit individuals. Fortuitous mutations are therefore able to reproduce rapidly.

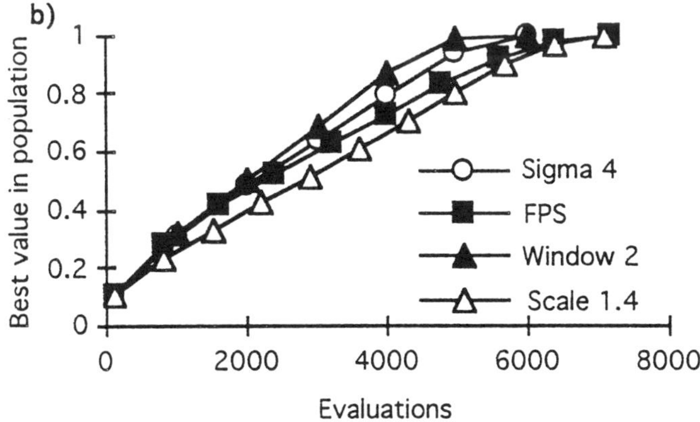


Figure 3.2b) Growth rates for FPS and three scaling methods.

### 3.5 Sampling algorithms

The various methods just described all deliver a value for the expected number of offspring for each string. Thus with direct fitness measurements, a string with twice the average score should be chosen twice. That is straightforward to implement, but there are obvious problems with non-integer expected values. The best that can be done for an individual with half the average fitness score, that expects 0.5 offspring, is to give it a 50% probability of being chosen in any one generation. Baker, who considered these problems in some detail, refers to the first process as selection, and the second as sampling (Baker, 1987).

A simple, and lamentably still common way to perform sampling may be visualised as spinning a roulette wheel, the sectors of which are set equal to the fitness values of each string. The wheel is spun once for each string selected. The wheel is more likely to stop on bigger sectors, so fitter strings are more likely to be chosen on each occasion. Unfortunately this simple method is unsatisfactory. Because each parent is chosen individually, there is no guarantee that any particular string, not even the best in the population, will actually be chosen in any given generation. This sampling error can act as a significant source of noise. The problem is well-known: De Jong suggested ways to overcome it in his 1975 thesis. The neatest solution is Baker's Stochastic Universal Sampling (SUS) algorithm (Baker, 1987), which produced the results of Figures 3.1 and 3.2. Figure 3.3 shows the difference in results for the two methods with fitness proportional selection. The rate of take-over of the best value is reduced, a reflection of the fact that the roulette wheel simulation lost the best value from the population an average of 9.1 times per run. Conversely, the worst value current in the population increases more rapidly, because it is quite likely for poor strings to be missed by the random selection. Both effects are likely to be deleterious to performance.

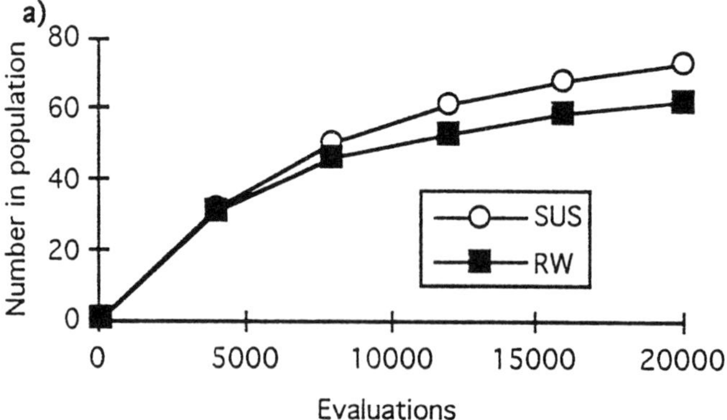


Figure 3.3a) Take-over rates for simple FPS, using roulette wheel (RW) and Baker's Stochastic Universal Sampling algorithm (SUS).

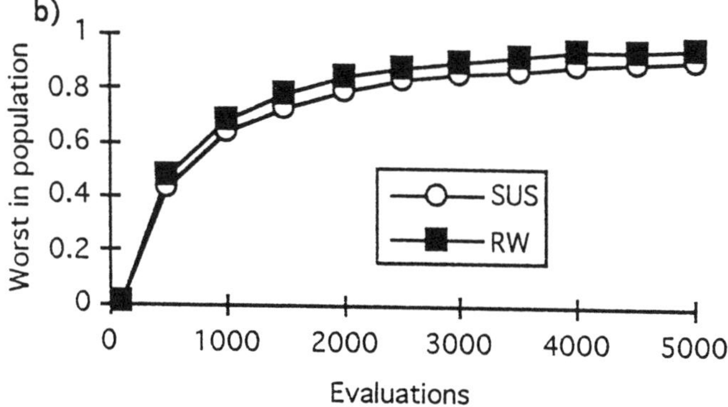


Figure 3.3b) Rise in the worst value in the population.

Baker's algorithm does the whole sampling in a single pass, and requires only one random number. The essence is to sum up the expected values, crediting the current string with an offspring every time the total goes past an integer. Thus if the initial random number is 0.7, and the first string expects 1.4 offspring, it will get two, since the total will be 2.1. If the random number is less than 0.6, it will get only one, since the total will be less than 2.

```
num = rand();
picked =1;
for (i=0; i<N; i++)
{
        num+= expected_off(i);
        while (picked < num);
        {
                parent[picked - 1] = i;
                  picked++;
        }
```

The only catch is that there will be sampling errors if the population is sorted, as it will be, for instance, for rank selection (see below). If two adjacent individuals each expect between 0.5 and 1 (exclusive) offspring, you can guarantee that one will get a child, while the other will not. The population should therefore be shuffled, adding to the complexity of the process. Many of the more obvious ways to do a shuffle produce a surprisingly biased result. An unbiased shuffle of the strings is given by the following code, where random_int returns an integer in the range i to N-1 inclusive:

```
for (i=0;i<N;i++)
{
        j=random_int(i ,N-1);
        temp = sorted_pop[j];
        sorted_pop[j]=sorted_pop[i];
        sorted_pop[i]=temp;
}
```

Alternatively, since copies of the strings will be required at some stage in the reproduction process, it may be better to shuffle them during the copying process. Baker's sampling algorithm can then be used to copy the winners back into the original array, ready for the application of reproduction operators:

```
for (i=0;i<N;i++)
{
        j = random_int(i ,N-1);
        new_pop[i] =sorted_pop[j];
        sorted_pop[j] =sorted_pop[i];
}
```

### 3.6 Ranking

Baker (1985) suggested rank selection in an attempt to overcome the deficiencies of the direct fitness based approach. First the population is ordered according to the measured fitness values. A new fitness value is then ascribed, inversely related to the string's rank. Two methods are in common use.

### 3.7 Linear ranking

The best string is given a fitness $s$, between 1 and 2. The worst string is given 2-$s$. Intermediate strings' fitness values are given by interpolation, assuming a *C* type array that starts at 0 for rank 1:

$$f(i) = s - \frac{(2i(s-1))}{(N-1)}$$

Since this prescription automatically gives an average fitness of 1, the fitness values translate directly as the expected number of reproductive opportunities. If $s$ is set to 2, the worst string gets no chance of reproduction. In principle, $s$ could be increased beyond 2 to achieve higher selection pressures, but then some of the worst strings would be given negative expected offspring. These could be truncated to zero, but then the remaining fitness values would need rescaling to give the correct total number of offspring. A simpler method of achieving higher selection pressures, which also gives some chance to the worst members of the population, is to use a non-linear ranking, such as that described in the next section. However, such high pressures are rarely needed. The selection pressure generated by linear ranking is proportional to $s$-1. Thus with $s$=1.1, convergence takes about 12,000 evaluations, with $s$=1.2 it takes 6000, and with $s$=2 it takes 1000.

In his 1985 paper, Baker discusses methods of preventing premature convergence. He proposes a measure of percentage involvement, being the proportion of the population that gets to produce an offspring in any generation. He suggests a target of at least 94%, which is given by linear ranking with $s$ of about 1.1. Using $s$=2 gives only 75% involvement.

### 3.8 Exponential ranking

The best string is given a fitness of 1. The second best is given a fitness of $s$, typically about 0.99. The third best is then given $s^2$ and so on down to the last, which receives $s^{(N-1)}$. The ascribed fitness values need to be divided by their average to give the expected number of offspring for each string.

```
s=0.99 ## or whatever ##
tot_fitness=0;
fitness=1;
for (i =0; i<N; i++)
{
        f(i) = fitness;
        tot_fitness += fitness;
         fitness *= s;
}
avg_fitness = tot_fitness/N;
for (i=0; i<N; i++)
        f(i) /= avg_fitness;
```

Those who are particularly conscious of cpu cycles may wish to trade storage for time and construct an array containing the expected number of offspring for each rank. Depending on the program's internal data structures, such an approach might do away with the need for any assignment of fitness at this stage of the selection procedure. However, some workers recommend varying the selection pressure during the run, in which case the array would need updating whenever $s$ is changed.

The selection pressure is proportional to $1-s$, thus $s=0.994$ gives twice the convergence rate of $s=0.998$. With $s=0.999$, convergence takes about 25,000 evaluations, with $s=0.968$, it takes about 700. Figure 3.4 shows a family of take-over curves, confirming that doubling the selection pressure halves the take-over time. For $s$ below 0.98, the expression $N(1-s)$ gives a good approximation to the expected number of offspring for the best individual. In fact $s=0.98$ gives about 2.3 offspring to the best of a population of 100, but $s=0.95$ gives almost exactly 5, should that much pressure be wanted for some reason.

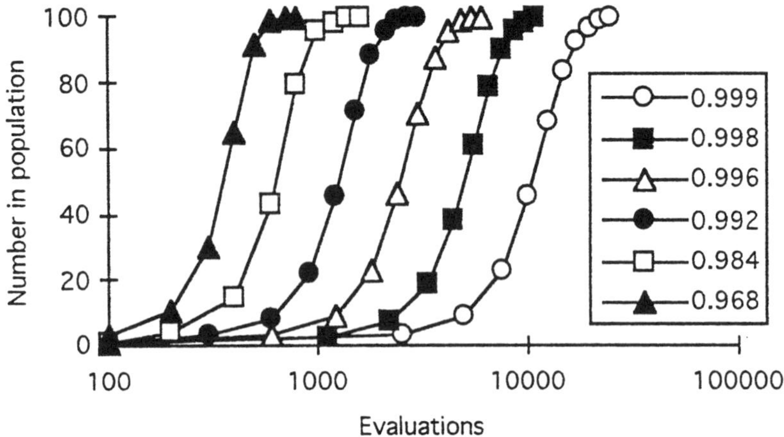


Figure 3.4. A family of take-over curves for exponential rank-based selection, at varying values of s.

The difference between the two ranking methods is illustrated in Figure 3.5. Figure 3.5a shows the expected number of offspring against rank for linear ranking with $s$=1.8 and exponential ranking with $s$=0.986. The expectation for the best individual is very similar, and as a consequence, so is the take-over time and the growth rate in the presence of mutation. However, the rate of loss of the worst value is considerably less for exponential ranking, as shown in Figure 3.5b, because of the increased chance given to weaker strings. For equivalent growth rates, exponential ranking ought to give a more diverse population, albeit at the expense of more average individuals.

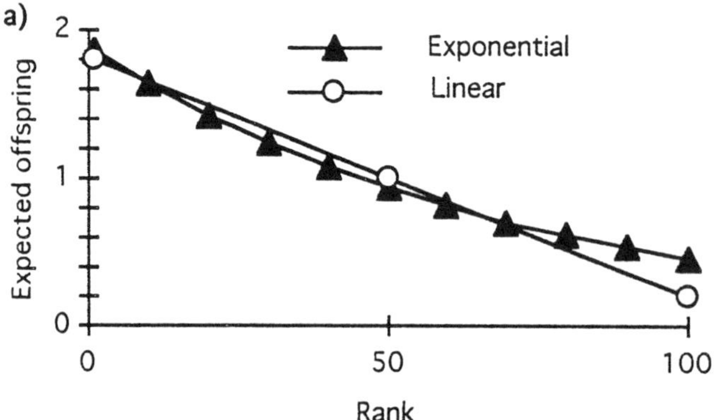


Figure 3.5a) Expected number of offspring for individual of given rank, for exponential ranking with s=0.986, and linear ranking with s=1.8.

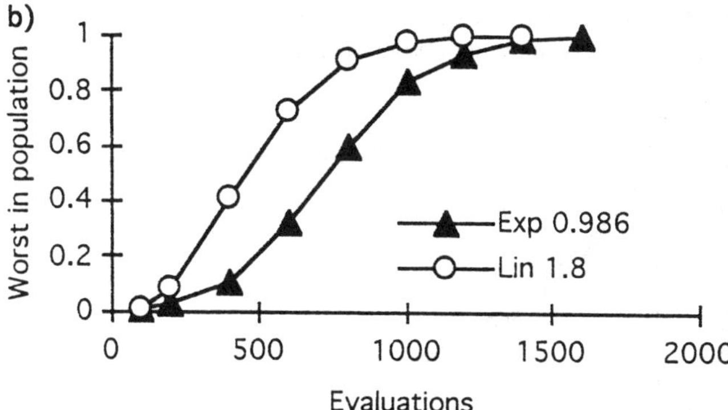


Figure 3.5b) Increase in worst value in the population, during take-over simulation.

Other kinds of non-linearity are possible, and possibly desirable. For instance, inverting the curve given by exponential ranking would improve the chances of above-average individuals at the expense of the worst. This would implement a softer version of the Top-n selection described below. Kuo and Hwang (1993) implement what they call disruptive selection, which uses a non-monotonic fitness function that gives more trials to good and bad strings, at the expense of intermediate values.

**3.9 Tournament selection**

In tournament selection, $n$ individuals are chosen at random from the population, with the best being selected for reproduction. A fresh tournament is held for each parent-to-be. This method is quite popular, perhaps partly because of the echoes of the mating battles often seen in nature. Goldberg and Deb show that the expected result for a tournament with $n$=2 is exactly the same as linear ranking with $s$=2. That this is the case may be seen intuitively by considering various cases. Every string should expect to be picked twice. The best string in the population will win both its tournaments, while the worst will never win, and thus never be selected. A string with performance equal to the average will expect to win half its tournaments, and so be picked half as often as the best string, exactly as linear ranking specifies.

Tournament selection can be made to emulate linear ranking with $s$<2 by making it only probable that the better string will win. If the better string wins with probability of 0.5, the process reduces to random selection, with no bias in favour of better strings. This is equivalent to linear ranking with $s$=1. The conversion between probability in tournament selection and $s$ in linear ranking is to double the probability, thus a probability of 0.8 is equivalent to $s$=1.6. The selection pressure generated by the tournaments may be increased by using $n$>2. With $n$=3, an average string can expect to win only a quarter of its tournaments. Since it should be selected for 3 tournaments, it can expect 0.75 offspring. The graph of expected offspring against rank becomes non-linear, resembling that produced by exponential ranking. Figure 3.6a shows the comparison: when the best string gets four offspring, exponential ranking still favours the worst strings at the expense of those somewhat above average.

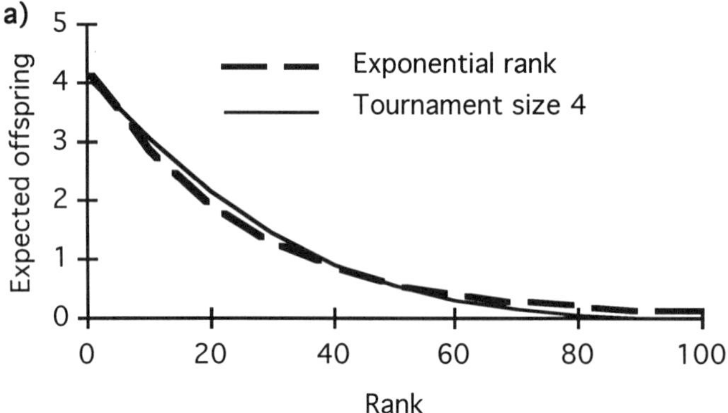


Figure 3.6a) Expected number of offspring for individual of given rank, for exponential ranking with s=0.96 and tournament selection with n=4.

Figure 3.6b shows the convergence of the population under tournament and linear ranking selection methods. The difference is caused by stochastic errors in tournament selection. Because each tournament is carried out individually, the method suffers from exactly the same sampling errors as roulette wheel selection and linear ranking, with Baker's selection procedure, should usually be used instead. It is interesting that Goldberg and Deb, who omitted consideration of stochastic effects, concluded that tournament selection was preferable to linear

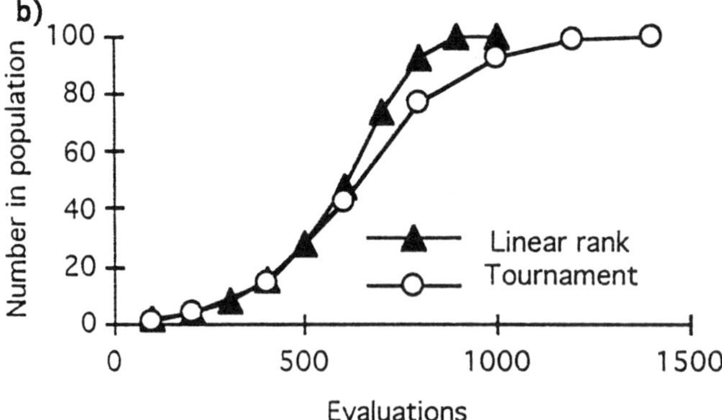


Figure 3.6b) Take-over rates for linear ranking, s=2 and tournament selection, n=2.

ranking because of the lower time complexity. This can be particularly significant in parallel models where a local tournament may be appropriate. Each processor can carry out the tournaments needed to generate the strings it will evaluate, whereas linear ranking selection needs to be carried out centrally, with the remaining processors waiting for the results to be distributed. Some models such as ASPARAGOS (Gorges-Schleuter, 1989) use a spatially distributed population, with tournaments being held only amongst nearby members of the population. However, those using tournament selection should be aware of the implied sampling errors.

Occasionally, holding tournaments may be the only way to do the evaluation, for instance when evolving game-playing programs. Generation of a rank ordering by playing every program against every other may be too time-consuming. In this case also, tournament selection is the natural way to proceed.

**3.10 Genitor or steady state models**

A major argument in the GA camp centres on whether to replace the whole population at a go (generational model), or some subset, often one or two (incremental model). Whitley, with his Genitor system, is one of the major proponents of one-at-a-time, or steady-state reproduction (Whitley and Knuth, 1988). Any of the above methods of selection could be used to pick the parents of the single offspring, but Whitley uses linear ranking (Whitley, 1989), and provides a neat algorithm for picking an individual from the population with the desired probability.

$$n = N \times \frac{s - \sqrt{(s \times s - 4 \times (s-1) \times rand())}}{2 \times (s-1)}$$

Steady-state reproduction inevitably carries the same kind of sampling errors as roulette wheel selection. Some users therefore employ stochastic tournament selection to pick the two parents, since it can't add any extra error and it obviates the need to sort the population.

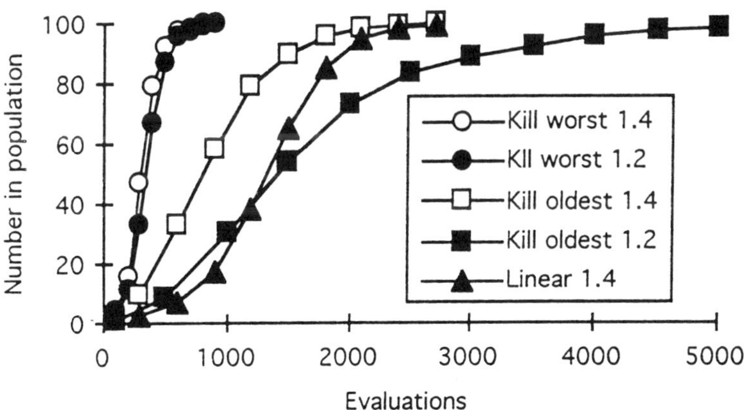


Figure 3.7. Take-over rates for incremental algorithms, with kill-worst and kill-oldest, and generational model using linear ranking.

Goldberg and Deb show that the Genitor model produces very high selection pressure. Most of this comes from always replacing the worst member of the population. Changing the linear ranking scale factor has very little effect. There are various ways to reduce the selection pressure. One is to pick an individual from the population to be replaced at random. This produces growth curves almost identical to a generational model with the same selection pressure (Syswerda, 1991).

An alternative is to replace the oldest member of the population. Figure 3.7 compares kill-oldest with kill-worst. In all the graphs, the x-axis units are evaluations, to allow direct comparison between the different population models. Kill-oldest is comparable with a generational model of the same selection pressure. However, it is faster at the start of run, and slower to finish off, probably the opposite of what is desirable. As might be expected, loss of the worst is much more rapid as well (not shown). Slow finishing is the consequence of the sampling errors, like those of roulette wheel selection, that inevitably result from breeding one at once. With $s=1.4$, kill-oldest lost the best value an average of 2.5 times per run.

Figure 3.8 shows that, for growth in the presence of mutation, kill-oldest is closer to kill-worst than to a generational model with the same nominal selection pressure. One reason may be that the effects of sampling error are less significant in this task — in the take-over task, it is rather crucial if the best individual is lost. Another may be a manifestation of the claim sometimes made for steady-state reproduction that they can exploit good new individuals more rapidly, because they are immediately available for reproduction. In the simple take-over example, this has no effect, because the additional copies of better strings are exactly balanced by the increase in average fitness (De Jong and Sama, 1991). Even where progress is obtained because of genetic operators, the effects may not be that big, because although a new string *may* be selected immediately, it is more likely to have to wait a significant fraction of $N$ evaluations at normal selection pressures. Because of the stochastic sampling effects, it may have to wait more than $N$ evaluations, if it is allowed to live that long, or even fail to be selected at all. Clearly, the stronger the selection pressure, the more significant the effect becomes.

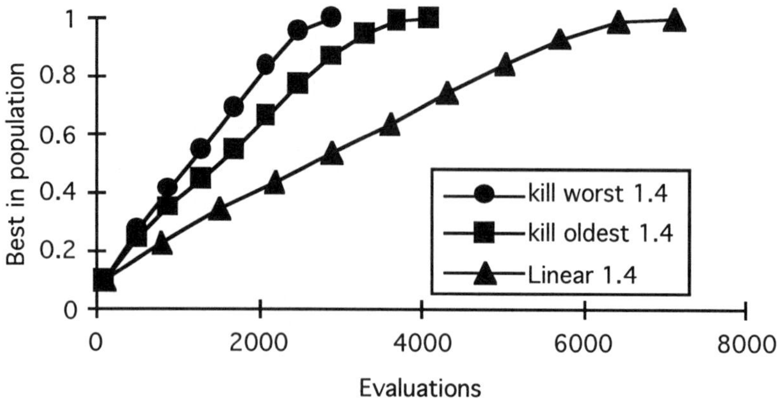


Figure 3.8. Growth rates in the presence of mutation for incremental and generational models, all using linear rank selection with s=1.4.

A third alternative is to delete worse members of the population preferentially, for instance by inverse ranking. Syswerda typically uses this method, and shows that it produces growth curves similar to kill-worst (Syswerda, 1991). However, he is using inverse exponential ranking with $s$=0.9, which is a very high selection pressure. A more typical usage might be to select for reproduction from the top with linear ranking, $s$=1.2, and kill from the bottom with the same selection rate. If we ignore for a moment the possibility that it will be superseded during its lifetime, the best string can expect $s/(2-s)$ offspring before being deleted. However, part of this comes from its increased life-expectancy, and during any period of N evaluations, equivalent to one cycle of a generational model, it will only expect $1/(2-s)$ offspring: 1.25 for $s$=1.2. Added to this will be a factor caused by the preferential deletion of worse individuals, which will cause the average fitness to rise. The net result is a growth rate that rises faster than $s$. Figure 3.9 shows that a steady-state model, with selection from top and bottom with $s$=1.2, produces a near identical growth rate to a generational model with linear ranking, $s$=1.4. However, the steady-state model with $s$=1.4 is faster than generational with $s$=1.8, while it is slower for smaller values of s. The approximate equivalence at $s$=1.2/1.4 appears to be genuine, and is not affected by things such as the size of the mutation. For those wishing to compare the techniques on their own problems, equivalent values for lower selection pressures are 1.13/1.2 and 1.085/1.1.

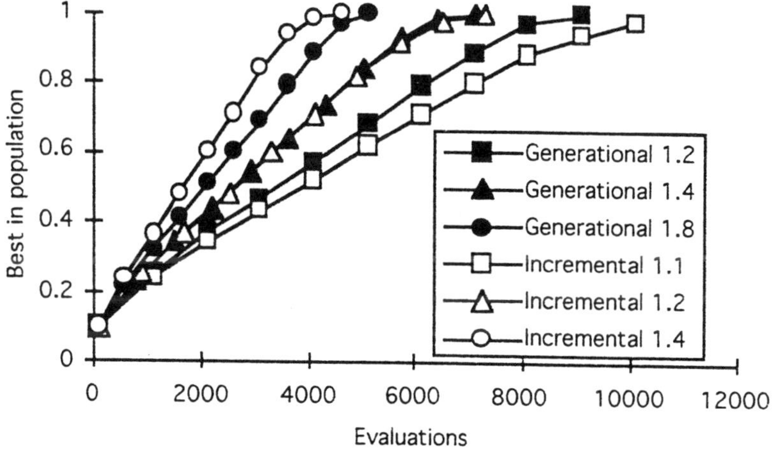


Figure 3.9. Growth rates for generational model, with linear ranking, and incremental model, with rank-based selection for reproduction, and inverse rank-based selection for deletion.

GA practitioners repeatedly claim significantly faster convergence using steady-state reproduction. In practice, this may often be the result of inadvertently high selection pressure acting on a fairly easy problem. However, another potentially crucial difference is that steady-state models are often run with the elimination of duplicates. Any new string is only admitted to the population if different from those already present. This will have a radical effect on the behaviour of the algorithm, which cannot be illustrated with the simple simulations used here, but is certainly worth trying on real problems.

A serious drawback with steady-state reproduction is the effect of noise in the evaluation function. This will be considered below, suffice it now to say that it does not do well.

### 3.11 Evolution strategy and evolutionary programming methods

The selection methods used in ES and EP algorithms are rather similar, producing high take-over rates. The approach differs from that typical of GAs, being more like selection of the fittest than fitness proportional reproduction. However, in some guises the effects can be very similar.

### 3.12 Evolution strategy approaches

There are two main methods of selection used in ESs, known as (m+l) and (m,l), where m is the number of parents and l is the number of offspring (for a review, see Hoffmeister and Bäck, 1992). The top m individuals form the next generation, selection being from parents and children in the (m+l) case, children only for (m,l). Typically, l is one to five times m (it obviously must be bigger than m for the (m,l) model). With these schemes, take-over by the best value is exponential. Thus for a (100,200) ES, it takes $\log_2(100) = 7$ generations, for (100+200), it takes $\log_3(100) = 5$ generations. The CHC genetic algorithm (Eshelman, 1991) uses an (m+m) selection procedure, combined with recombination operators designed to promote search.

### 3.13 Top-n selection

Some workers select the n best individuals, and give them each $N/n$ offspring (Nolfi, 1990). This clearly has the potential for extremely rapid take-over — with $n$=10, the best value will take over in two generations. It differs from the ES (m,l) approach only in what is called the population, thus Top-n with $n$=50 and $N$=100 is equivalent to a (50,100) ES. Bäck and Hoffmeister (1991) term such selection algorithms as extinctive, in that some individuals are guaranteed to get no offspring. They show, as might be expected, that such harsh selection is appropriate for unimodal objective functions.

### 3.14 Evolutionary programming methods

The canonical EP selection algorithm is a stochastic version of a (m+l) ES. Each individual produces one offspring, each of the $2N$ individuals plays $c$ others chosen at random (with replacement) in a tournament. The N with most wins forms the next generation. If the tournaments are deterministic, the result will converge to that of an $(N+N)$ ES as $c$ increases. The size of $c$ has little effect on the simple task used here: the best value always wins its competitions, and takes over the population in 7 generations.

As before, the selection may be softened by making the tournaments stochastic. Note a problem with terminology here: what is referred to above as a deterministic tournament, because the better individual always wins, Fogel refers to as a stochastic tournament (Fogel, 1994), because the opponents are picked at random. Here, a stochastic tournament is one in which the fitter individual may lose. One approach to this is to make the probability of the better string winning depend on the relative fitness of the two strings: $p_i = f_i/(f_i + f_j)$ (Fogel, 1988). This has the effect of reducing selection pressure to zero as the population becomes uniform, and produces a take-over curve remarkably similar to simple proportionate selection, Figure 3.10a. However, loss of the worst is more rapid, since poor strings initially get little chance to reproduce (not shown). Note, however, that since EP does not use recombination, there may be less requirement for diversity than in a GA.

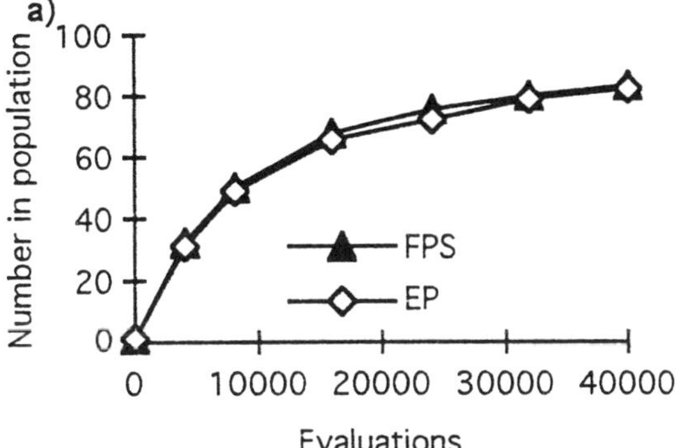


Figure 3.10a) Take-over rates for fitness proportional selection and stochastic evolutionary programming selection, with tournament size.

Figure 3.10b shows growth rates under mutation of a number of the selection methods described. Genitor produces the highest growth rate, but note that exponential ranking can be made to converge similarly fast, or much the most slowly, by varying the selection pressure. Linear ranking and linear scaling can also be set to give similarly slow growth rates (not shown). Despite the very rapid take-over given by the ES methods, growth rate is less spectacular, and actually very similar to that provided by simple fitness proportionate selection. This does relatively well because of the high number of offspring allocated to particularly fit individuals, not in general a good idea.

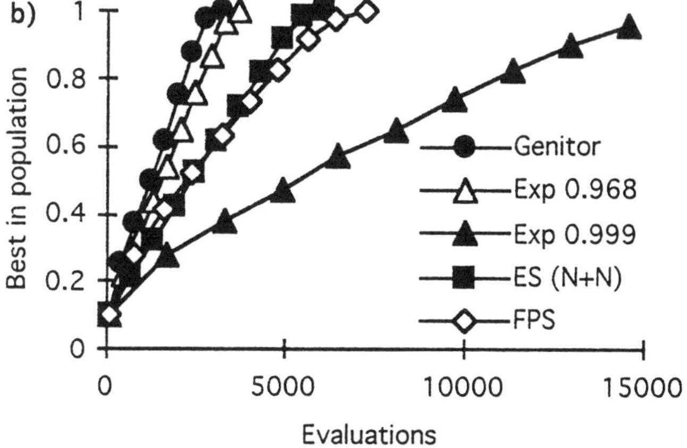

Figure 3.10b) Growth rates for Genitor (incremental, kill-worst), with s=1.2, exponential ranking, (100+100) ES and FPS.

### 3.15 The effects of noise

It is often claimed that Genetic Algorithms are relatively immune to the effects of noise in the evaluation function. However, it is to be expected that the various selection schemes would differ in their susceptibility. This was assessed by adding noise to the evaluation, and observing the effect on the growth in the presence of mutation. Another Gaussian random variable was added to the true value of each individual and used to allocate its number of offspring. The true value was then passed to any offspring, subject to the small mutation as before. In order to have a significant effect on the rate of convergence, it was found to be necessary to add noise with a standard deviation of 0.2: 10 times that of the mutation. This is not so much a signal-to-noise ratio as a noise-to-signal ratio, and gives some credence to the claimed noise immunity.

Figure 3.11a shows the effects of adding noise on two traditional GA methods, linear ranking and sigma scaling. Even with this level of noise, the time to convergence is less than halved. However, note that sigma scaling deteriorates rather less than the ranking method. The obvious conclusion is that sigma scaling is more noise tolerant, but this would be a mistake. The real reason for the result is quite subtle, but similar to that responsible for the growth rates in Figure 3.2b. The effect of the noise is to reduce the accuracy of the selection procedure. In the limit, if noise swamps the evaluation entirely, all individuals would expect one offspring. Here, the best individual can expect somewhat more than one, but less than it should get in the absence of noise. With ranking and $s$=1.8, it will therefore get somewhere between 1 and 1.8 offspring. Figure 3.11b shows the actual number of offspring allocated to the best individual, averaged over the 100 runs. Linear ranking scores about 1.2 at the start of run. Sigma scaling sets the baseline according to the spread of values in the population, which will be determined mostly by the noise. Lucky individuals can appear 2 or 3 standard deviations above the mean. Because there is not the upper limit imposed by ranking, the best individual averages higher, about 1.5 initially in this case. The faster growth rate is therefore effectively a case of premature convergence, but since the problem is so simple, there is only the correct solution for it to converge to.

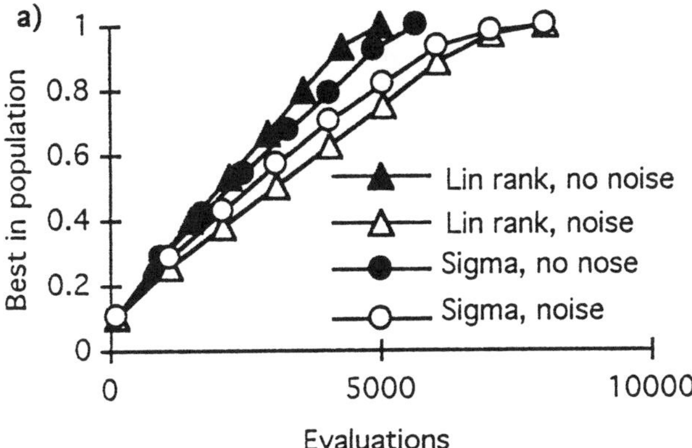


Figure 3.11a) Growth rates with and without noise for linear ranking with s=1.8 and sigma scaling with s=4.

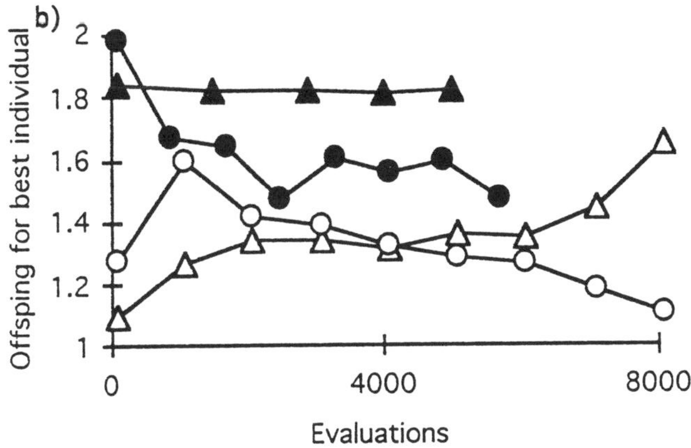


Figure 3.11b) Measured number of offspring for the best individual in the population during the same simulations.

Figure 3.12 shows the effects on steady-state reproduction. Because Genitor kills the apparent worst in the population, lucky individuals that got a much better evaluation than they merited will remain in the population. The effects on convergence rate are disastrous. Killing the oldest performs much better, echoing the findings of Fogarty (1993). Not shown is the effect on a model using inverse rank based deletion of worse individuals. As expected, it also deteriorates, so that steady state with $s$=1.2 from both ends converges in around 14,000 evaluations, compared with 10,000 for generational linear ranking, with $s$=1.4. As noted above, these two are almost identical in the absence of noise. Compared with kill-oldest, the deterioration is somewhat worse.

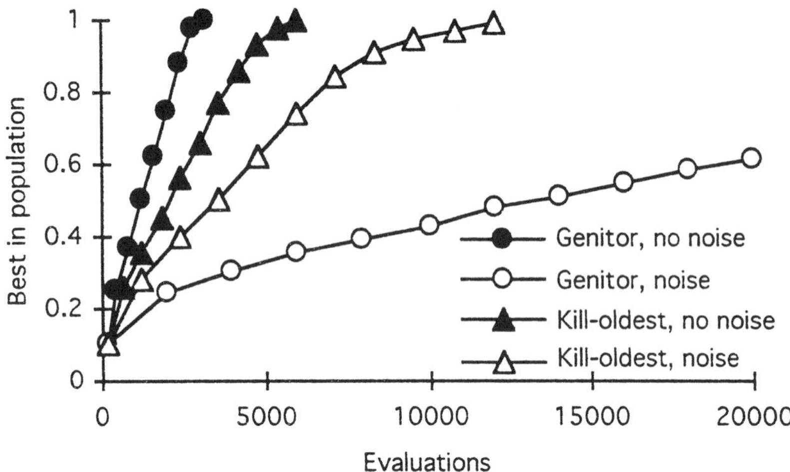


Figure 3.12. Effect of adding noise to growth rate of Genitor (kill-worst) and kill-oldest incremental models.

In theory, the noise sensitivity of steady-state models may be reduced by re-evaluating existing members of the population. One approach is to pick an individual at random for each new offspring produced. Since most of the problem appears to come from fortuitous individuals that are ranked higher than they should be, it might seem better to select high ranks preferentially, as for reproduction. The old and new evaluations should be averaged, to improve the estimate of the true value. When tried, the latter approach did perform slightly better, but the improvement given by either method was not nearly sufficient to discount the extra evaluations. They just took slightly less than twice as many. The most robust way of handling noise in incremental algorithms appears to be to kill the oldest.

Figure 3.13 tells a similar story for the ES selection methods. Without noise, (100+100) converges rapidly. Allowing the parents to pass to the next generation ensures that nothing is lost, giving rapid gains on the simple task. A (100+200) ES converges more rapidly in terms of generations, but requires more evaluations in total. Comparison with the (100,200) model illustrates the advantage of conserving the best parents. In the presence of noise, however, the conservative approach again fails. The (100,200) ES deteriorates by a similar amount to other generational techniques, (100+100) deteriorates dramatically. The deterministic version of EP selection performs very similarly to a (100+100) ES, depending on the number of tournaments held (not shown).

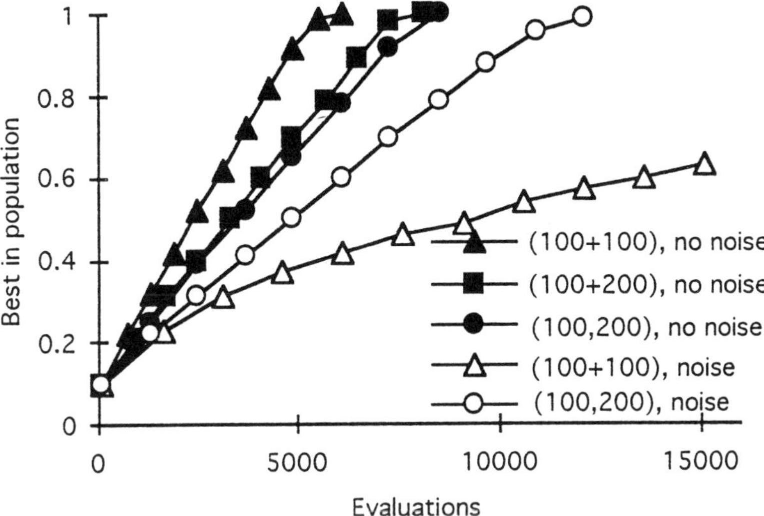


Figure 3.13. Effect of noise on growth rate of ES selection methods.

**Conclusions**

By now it should be apparent that there are fewer significant differences between the various selection schemes than might be thought. The decisions to be made include the following:

1. Whether to use direct fitness measures, with appropriate scaling, or rank-based selection. The latter provides better control over selection pressure, at the expense of the link between fitness and reproductive success.

2. Whether to use a generational or incremental model. The latter suffers in the presence of noise and also from the same kind of sampling errors as the roulette wheel. Its benefit is the ability to exclude duplicates, the advantage of which could not be illustrated here.

3. Whether to introduce a deliberate non-linearity between fitness and allocated offspring. Exponential rank selection benefits the worst members, at the cost of above-average individuals. Top-n selection goes to the opposite extreme, giving all the offspring to the top few. It may be appropriate to alter the balance during a run, keeping worse individuals initially, and moving towards harsher selection.

4. Whether to use fitness or rank proportional reproduction (GA) or selection of the fittest (ES/EP). It is probably relevant that the latter stress the importance of mutation as a search operator, while GAs rely on recombination.

A common technique that has not been mentioned yet is the elitist strategy (De Jong, 1975), which simply ensures that the best individual survives into the next generation. This is not simply to ward off errors in sampling, since the best string may be correctly selected, but be disrupted by recombination or mutation. It is a conservative strategy, often found to give an improvement in performance, but again suspect with noisy evaluations. It can also hinder progress, by anchoring the population to a local maximum. A softer alternative is to lose the best if no progress has been made for some number of generations (e.g. 5).

Selection is only one part of the algorithm, and decisions about which to use need to be made in parallel with decisions about recombination operators. For instance, Eshelman (1991) deliberately combines a conservative "survival of the fittest" selection method with disruptive recombination operators. It isn't possible to say which is best without defining evaluation criteria. Thus Goldberg and Deb preferred tournament selection to linear ranking because they considered time complexity, but not stochastic effects. Freisleben and Härtfelder (1993) used a meta-level GA to tune the parameters of another GA. It was able to alter parameters such as population size, mutation rates and the selection method. The choice was tournament, rank and two forms of FPS. Tournament was the clear winner, which is odd, because it is just rank with added noise.

How could adding noise help? Their task was learning the weights for a neural net simulation. This is plagued with symmetry problems — many different genetic string produce identical nets, because the order of hidden units is immaterial, while that of the genetic string is not. Crossover therefore does not work successfully. The GA has to decide which of the permutations to use. The most likely explanation for Freisleben and Härtfelder's result is that the noise of tournament selection allowed one permutation to get its nose ahead of the rest and take over the population, thus resolving the problem. Had the meta-GA had available recombination operators able to handle the permutations, it might well have chosen a different selection procedure. Hopefully this chapter willbe of some assistance in making such decisions.

**Acknowledgements**
This work was partly supported by grant no. GR/H93828 from the UK Science and Engineering Research Council.

**References**

Bäck, T. and Hoffmeister, F. Extended selection mechanisms in genetic algorithms. Pages 92-99 of Proceedings of the fourth international conference on Genetic Algorithms, Belew, R.K. and Booker, L. (Eds), Morgan Kaufmann. 1991.

Baker, J.E. Adaptive selection methods for Genetic Algorithms. Pages 101-111 of Proceedings of an international conference on Genetic Algorithms, Grefenstette, J.J. (ed), Lawrence Earlbaum. 1985.

Baker, J.E. Reducing bias and inefficiency in the selection algorithm. Pages 14-21 of Proceedings of the second international conference on Genetic Algorithms, Grefenstette, J.J. (ed), Lawrence Earlbaum. 1987.

De Jong K.A. An analysis of the behavior of a class of genetic adaptive systems. Ph.D. thesis, Department of Computer and Communication Studies, University of Michigan, 1975.

De Jong K.A. and Sarma J. Generation gaps revisited. In Foundations of Genetic Algorithms 2, Whitley, D. (Ed). Morgan Kaufmann. 1991.

Eshelman, L.J. The CHC adaptive search algorithm: how to have safe search when engaging in nontraditional genetic recombination. In Foundations of Genetic Algorithms, Rawlins, G.J.E. (ed), Morgan Kaufmann. 1991.

Fogel, D.B. An evolutionary approach to the travelling salesman problem. Biological Cybernetics **60**, 139-144, 1988.

Fogel D.B. An introduction to simulated evolutionary optimization. IEEE Transactions on Neural Networks **5**, 3-14, 1994.

Fogarty, T.C. Reproduction, ranking, replacement and noisy evaluations: experimental results. Technical report, Faculty of Computer Studies and Mathematics, University of the West of England, 1993.

Freisleben, B. and Härtfelder, M. Optimization of genetic algorithms by genetic algorithms. In Artificial Neural Networks and Genetic Algorithms, Albrecht, R.F. and Reeves, C.R. and Steele, N.C. (Eds). 1993.

Goldberg, D.E. and Richardson, J. Genetic algorithms with sharing for multimodal function optimization. Pages 41-49 in Genetic algorithms and their applications: proceedings of the second international conference on Genetic Algorithms, Lawrence Earlbaum. 1987.

Goldberg, D.E, and Deb, K. A comparative analysis of selection schemes used in Genetic Algorithms. Foundations of Genetic Algorithms, Rawlins, G.J.E. (Ed), Morgan Kaufmann, 1991.

Gorges-Schleuter, M. ASPARAGOS: An asynchronous parallel genetic optimization strategy. Pages 422-427 of Proceedings of the third international conference on Genetic Algorithms, Schaffer, J.D. (Ed), Morgan Kaufmann, 1989.

Hoffmeister, F. and Bäck, T. Genetic algorithms and evolution strategies: similarities and differences. Technical report SYS-1/92, University of Dortmund, 1992.

Kuo, T. and Hwang, S.-Y. A genetic algorithm with disruptive selection. In proceedings of the fifth international conference on Genetic Algorithms, Forrest, S. (Ed), 1993.

Nolfi, S., Elman, J.L. and Parisi, D. Learning and Evolution in Neural Networks, Technical report CRL TR 9019, UCSD, July 1990.

Schaffer, J.D., Some effects of selection procedures on hyperplane sampling by genetic algorithms. Pages 89-103 of Genetic Algorithms and Simulated Annealing, Davis, L. (Ed), Pitman, London. 1987.

Syswerda, G. A study of reproduction in generational and steady-state genetic algorithms. In Foundations of Genetic Algorithms, Rawlins, G.J.E. (ed), Morgan Kaufmann. 1991.

Todd P.M. and Miller. G.F. On the sympatric origin of species: mercurial mating in the quicksilver model. Pages 547-554 of Proceedings of the fourth international conference on Genetic Algorithms, Belew, R.K. and Booker, L. (Eds), Morgan Kaufmann. 1991.

Whitley, D., and Knuth, J. GENITOR: a different genetic algorithm. Pages 118-130 of Proceedings of the Rocky Mountain Conference on Artificial Intelligence, Denver, Colorado. 1988.

Whitley, D. The Genitor algorithm and selection pressure: why rank-based allocation of trials is best. Pages 116-121 of Proceedings of the third international conference on Genetic Algorithms, Schaffer, J.D. (ed), Morgan Kaufmann. 1989.

<u>Chapter 4</u>

**El-Ghazali Talbi**
LGI/IMAG
BP53 38041
Grenoble France

**Pierre Bessiere,**
**Juan-Manuel Ahuactzin,**
**& Emmanuel Mazer**
LIFIA/IMAG
46, Av. Felix Viallet 38031
Grenoble Frence

# Parallel Cooperating Genetic Algorithms: An Application to Robot Motion Planning



## Abstract
The goal of the work described in this paper is to build a path planner able to drive a robot in a dynamic environment where the obstacles are moving.

In order to do so, we propose a method, called "ARIADNE'S CLEW algorithm", to build a global path planner based on the combination of two parallel genetic algorithms: an EXPLORE algorithm and a SEARCH algorithm. The purpose of the EXPLORE algorithm is to collect information about the environment with an increasingly fine resolution by placing landmarks in the searched space. The goal of the SEARCH algorithm is to opportunistically check if the target can be easily reached from any given placed landmark.

The ARIADNE'S CLEW algorithm is shown to be very fast in most cases allowing planning in dynamic environments. Hence, it is shown to be complete, which means that it is sure to find a path when one exists. Finally, we describe a massively parallel implementation of this algorithm.

0-8493-2529-3/95/$0.00 + $.50

## 4.1 Introduction

The goal of this work is to build a path planner able to drive a robot in a dynamic environment where the obstacles are moving.

Designing a path planner is a central question in robotics research. A review of the existing approaches can be found in Latombe's book [1]. There are two main ways to deal with this problem: the global and the local approaches. The global approaches suppose that a complete representation of the configuration space has been computed before looking for a path. The global approaches are complete in the sense that if a path exists it will be found. Unfortunately, computing the complete configuration space is very time consuming, worse, the complexity of this task grows exponentially as the number of degrees of freedom increases. Consequently, today most of the robot path planners are used off-line: the planner is invoked with a model of the environment, it produces a plan which is passed to the robot controller which, in turn, executes it. In general, the time necessary to achieve this is not short enough to allow the robot to move in a dynamic environment. The local approaches need only partial knowledge of the configuration space. The decisions to move the robot are taken using local criteria and heuristics to choose the most promising directions. Consequently, the local methods are much faster. Unfortunately, they are not complete, it may happen that a solution exists and is not found. The local approaches consider planning as an optimisation problem, where finding a path to the target corresponds to the optimisation of some given function. As any optimisation technique, the local approaches are subject to get trapped in some local optima, where a path to the goal has not been found and from which it is impossible or, at least, very difficult to escape.

The ultimate goal of a planner is to find a path in the configuration space from the initial position to the target. However, while searching for this path, an interesting sub-goal to consider may be to try to collect information about the free space and about the possible paths to go about that space. The ARIADNE'S CLEW algorithm tries to do both at the same time. An EXPLORE algorithm collects information about the free space with an increasingly fine resolution, while, in parallel, a SEARCH algorithm opportunistically checks if the target can be reached. The EXPLORE algorithm works by placing landmarks in the searched space in such a way that a path from the initial position to any landmark is known. In order to learn as much as possible about the free space the EXPLORE algorithm tries to spread the landmarks all over the space. To do so, it tries to put the landmarks as far as possible from one another. For each new landmark produced by the EXPLORE algorithm the SEARCH algorithm checks with a local method if the target may be reached from that landmark. The ARIADNE'S CLEW algorithm is very fast, however, we will show that it is a complete planner which will find a path if one exits. The resolution at which the space is scanned and the time spent to do so, automatically adapts to the difficulty of the problem.

Both the EXPLORE and the SEARCH algorithms may be seen as solving optimisation problems. We first introduce the optimisation technique we are using, namely, genetic algorithms. We then describe successively in some details, the SEARCH algorithm, the EXPLORE algorithm and the concatenation of both. We finally explain a massively parallel implementation of our method and present some results proving that using this method we are able to drive a robot in a dynamic environment. We conclude with a discussion and some perspectives for future work.

## 4.2 Principles of Genetic Algorithms

Genetic algorithms are programs used to deal with optimisation problems. They have first been introduced by Holland [2]. Their goal is to find optimum of a given function F on a given search space S. For instance, the search space S may be $2^N$, a point of S is then described by a vector of N bits and F is a function able to compute a real value for each of the $2^N$ vectors.

In an initialisation step a set of points of the search space S (called a "population" of "individuals"), is drawn at random (the "genotype" of each individual is a vector of N bits). Then, the genetic algorithm iterates over the following 4 steps until a satisfying optimum is reached (see Figure 4.1 below):

**1 Evaluation**: The function F is computed for each individual, ordering the population from the worst to the best.

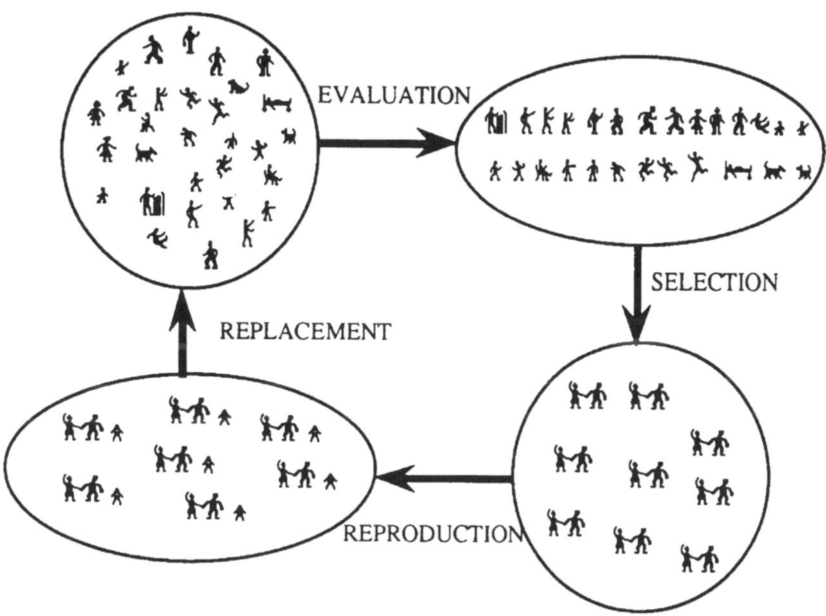


Figure 4.1. The basic principles of genetic algorithms.

**2 Selection**: Pairs of individuals are selected, best individuals having more chance to be selected than poor ones (one individual may appear in different pairs).

**3 Reproduction**: New individuals (called "offspring") are produced from these pairs.

**4 Replacement**: A new population is generated by replacing some of the individuals of the old population by the new ones.

Reproduction is done using some "genetic operators". Number of them may be used but the two most common are mutation and crossing-over. The mutation operator picks at random some mutation locations among the N possible sites in the vector and flip the value of the bits at these locations as represented in Figure 4.2.

Figure 4.2: Mutation operator.

The cross-over operator selects at random a cut point among the N possible sites in the binary genotype and exchanges the last parts of the two parents vectors as shown in Figure 4.3.

Figure 4.3: Cross-over operator.

Genetic algorithms have many applications and exhibit very impressive optimisation capabilities compared to other optimisation techniques especially when the search space is big ($\approx 2^{300}$) and F quite irregular (see [3] for a recent survey).

Besides their scientific interest as a model of biological evolution, genetic algorithms have two main technological interests:

1 They are very robust techniques able to deal with a very large class of optimisation problems.

2 They are very easy to program in parallel and the acceleration obtained by doing so is considerable (see [4]).

We proposed a parallel genetic algorithm and developed an implementation on a massively parallel machine based on Transputers (see [5]). This algorithm and the performances obtained by the parallel implementation have been an essential achievement for the success of the work described in this paper.

The principle of this parallel genetic algorithm is described by Figure 4.4. It consists in one parallel process running for each individual in the population. The processes are organised in a torus structure where each process has 4 neighbours. At each generation all the individuals, in parallel, choose among their 4 neighbours with whom they want to breed and reproduce with the chosen bride. The parallel genetic algorithm iterates over the following 4 steps until a satisfying optimum is reached:

1 **Evaluation**: Evaluate in parallel all the individuals.

2 **Selection**: Select in parallel, among the four neighbours, the bride with the best evaluation.

3 **Reproduction**: Reproduce in parallel with the chosen bride.

4 **Replacement**: Replace in parallel the parents by the offspring.

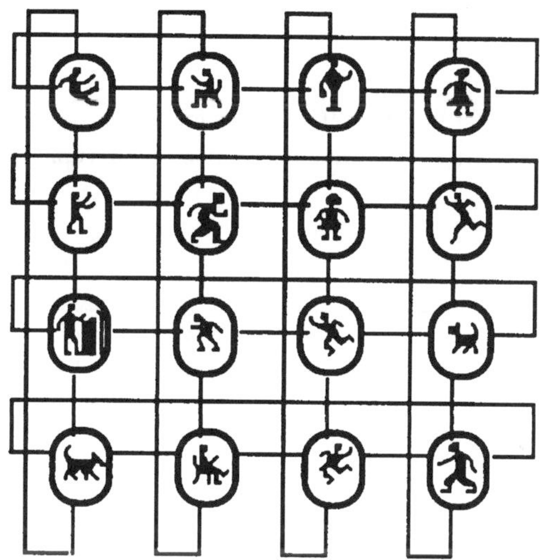

Figure 4.4: The principle of the parallel genetic algorithm.

## 4.3 The Search Algorithm

The purpose of the SEARCH algorithm is to determine if the target $\tau$ may be reached "simply" from a given point $\pi$. In order to do so, it looks for fixed length Manhattan motions in the configuration space starting at $\pi$ and ending at $\tau$.

Given a system with N degrees of freedom $\{\theta_1, \theta_2, ..., \theta_N\}$, a Manhattan motion of length 1 consists in moving each degree of freedom $\theta_i$ successively once by $\Delta\theta_i$. A Manhattan motion of length L is a succession of L Manhattan motions of length 1 or of LxN elementary motions of a single degree of freedom. Such a Manhattan motion M is denoted as:

$$M = \left(\Delta\theta_1^1, \Delta\theta_2^1, ..., \Delta\theta_i^1, ..., \Delta\theta_N^1, \Delta\theta_1^2, \Delta\theta_2^2, ..., \Delta\theta_N^L\right)$$

Let us call $\tau_i^j$ the point reached in the configuration space after ixj elementary motions. Let us call $\tau_a^b$ the furthest point reached along M before a collision occurred We are looking for a collision-free Manhattan motion such that $\tau_a^b = \tau_N^L = \tau$.

The SEARCH algorithm may be expressed as an optimisation problem for the parallel genetic algorithm where:

> The search space $S_S$ is the set of all Manhattan motions of length L starting at $\pi$.
>
> The evaluation function $F_S$ applied to a Manhattan motion M given a target $\tau$ is defined as follow:
>
> $F_S(M,\tau)=0$ if any $\tau_i^j$ of M preceding $\tau_a^b$ is in the BACKPROJECTION$_S$ of $\tau$. (The BACKPROJECTION$_S$ of $\tau$ is the set of all points of the searched space from which $\tau$ may be reached by a Manhattan motion of length 1).
>
> Otherwise $F_S(M,\tau)=\|\tau - \tau_a^b\|$

The SEARCH algorithm tries to minimise the evaluation function $F_S(M,\tau)$ over the search space $S_S$.

Manhattan motions have been chosen because for the NxL elementary motions of M, it is possible to compute simply in parallel, both the corresponding $\tau_i^j$ and the collision-free test on the path from $\pi$ to $\tau_i^j$ (see [6]). Furthermore, in a 3 dimensional physical space, the collision-free test itself consists of three processes running in parallel checking, respectively, that there is no vertex-to-plan collisions, plan-to-vertex collisions and edge-to-edge collisions. Finally, each of these three processes may be expressed as the parallel evaluation of AxB processes where A is the number of elements in the first set of the test (A is the number of vertices, plans or edges) and where B is the number of elements in the second set (B is the number of plan, vertices or edges).

The SEARCH algorithm may be used as a planner by itself. It has been used as such for several applications. Let us describe briefly two of them (a more detailed presentation may be found in [7]).

The first application is a planner for a planar arm with two degrees of freedom. By restricting ourselves to two dimensions we can graphically represent the configuration space and give the reader a better feeling of the method. However, the proposed method does not make any hypothesis about the number of degrees of freedom and can be used without modification for arms with a much larger number of degrees of freedom. Figure 4.5a shows a Manhattan motion in the configuration space and the associated "individual" of the genetic algorithm. Figure 4.5b shows the initial and final configuration of the arm in the operational space. Figure 4.5c shows the path found in the operational space and Figure 4.5.d the path found in the configuration space. Finally, Figure 4.5e shows the portion of the configuration space which has been evaluated. It should be noticed that only a very restricted part of the configuration is really computed, this is one of the main explanation of the efficacy of the algorithm and this is why this algorithm is able to handle planning in dynamic environments.

Figure 4.5a

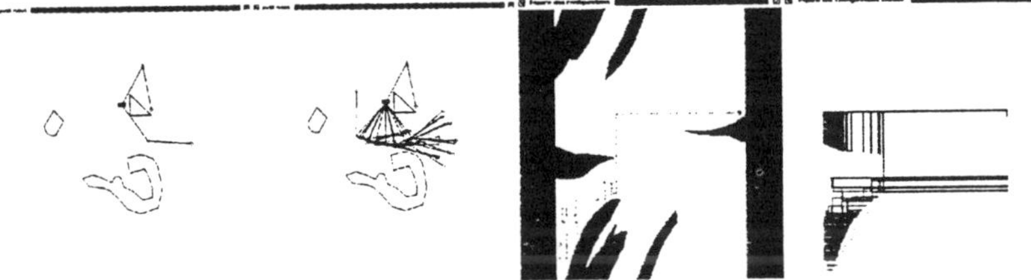

Figure 4.5b, c, d, and e

The second application is a planner for a holomatic mobile robot. Figure 4.6 shows how the planner behaves in a dynamic environment. Figure 4.6a shows the initial found path. Figure 4.6b shows the path re-planned after the closing of the door. The used version of SEARCH has been implemented on a massively parallel Transputers machine. The planning time for a given path was less than 1 second on a machine of 64 Transputers.

As shown by the two previous examples, the use of SEARCH as a planner is very interesting. However, it may happen that the genetic algorithm gets trapped in some local minima. In that case the planner does not find a solution even if one exists. The SEARCH algorithm is not complete, this is its main drawback.

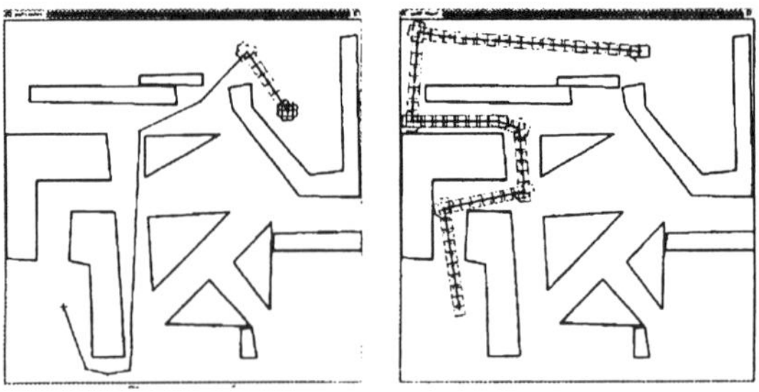

Figure 4.6a and b.

## 4.4 The Explore Algorithm

The purpose of the EXPLORE algorithm is to collect information about the free space. The EXPLORE algorithm works by placing landmarks in the searched space in such a way that a collision-free Manhattan path from the initial position x to any landmark $\lambda_k$ is known. In order to learn as much as possible about the free space the EXPLORE algorithm tries to spread the landmarks all over the space. To do so, it tries to put the landmarks as far as possible from one another. Let us call $\Lambda = \{\lambda_1,\lambda_2,...\lambda_k,...\}$ the set of already placed landmarks at a given step of the program. It is possible to define the distance between a point $\alpha$ of the searched space and the set $\Lambda$ by $D(\alpha,\Lambda) = \text{Min} \; \|\lambda_k - \alpha\|$ on all landmarks $\lambda_k \in \Lambda$

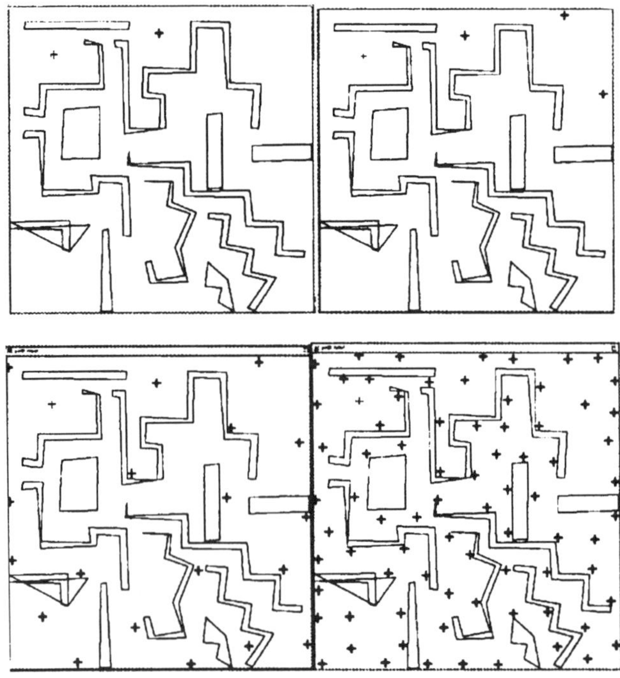

Figure 4.7 shows how the landmarks spread in the environment.

The EXPLORE algorithm may be expressed as an optimisation problem for the parallel genetic algorithm where:

> The search space $S_e$ is the set of all Manhattan motions of length L starting from any landmark $\lambda_k$ of $\Lambda$.
>
> The evaluation function $F_e$ applied to a Manhattan motion M of $S_e$ is defined as follows:
>
> $F_e(M)=D(\tau_a^b,\Lambda)$ where, $\tau_a^b$ is still the furthest point reached along M baefore a collision occurred.

The EXPLORE algorithm tries to maximise over the search space $S_e$ the evaluation function $F_e(M)$.

Figure 4.8 shows "the ARIADNE'S CLEW": a tree of landmarks allowing to go about the free space.

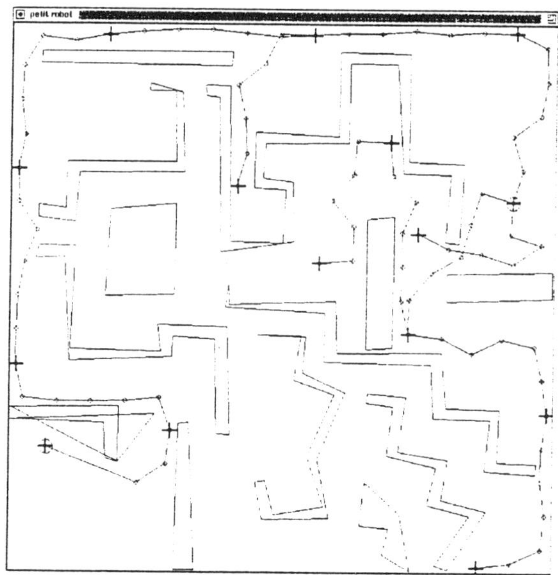


Figure 4.8: The ARIADNE'S CLEW

## 4.5 The ARIADNE's CLEW Algorithm

The purpose of the ARIADNE'S CLEW algorithm is to find a path from a given point $\pi$ to a target $\tau$. The ARIADNE'S CLEW algorithm is the following:

1 - Use the SEARCH algorithm to find if a "simple" path exists between $\pi$ and $\tau$.

2- If no "simple" path found by step 1, then do until a path is found
 2.1 * Use EXPLORE to generate a new landmark $\lambda$.
 2.2 * Use SEARCH to look for a "simple" path from $\pi$ to '$\tau$.

It is interesting to notice that SEARCH may be seen as a backprojection function for EXPLORE. SEARCH could be called $\text{BACKPROJECTION}_e$ because it plays relatively to EXPLORE the exact same role than $\text{BACKPROJECTION}_s$ relatively to SEARCH. A quite complicated backprojection function indeed, which usually produces very big backprojection allowing EXPLORE to stop after placing just a few landmarks.

The ARIADNE'S CLEW algorithm has three very important qualities:

- It reduces to the very fast SEARCH algorithm for most of the cases.

- It is complete, in the sense that if a path exists it will be found (see proposition 2 below).

- It automatically adapts the resolution at which it scans the space to the complexity of the problem (see proposition 3 below).

Figures 4.9a and b show two complex paths found by the ARIADNE'S CLEW algorithm.

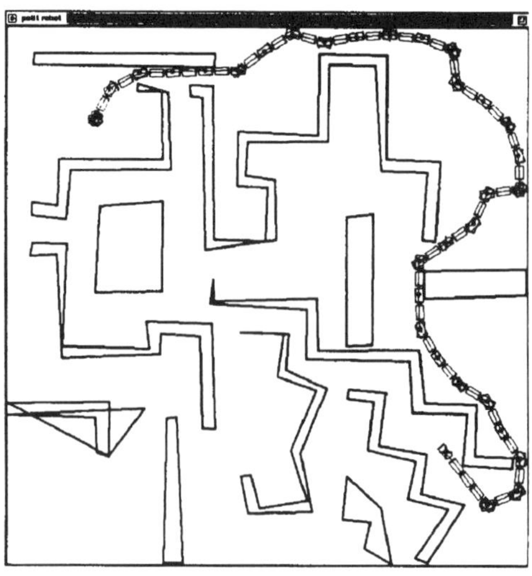

Figure 4.9a

**Definition 1**: a PATH P from an initial point π to a target τ in an N dimension metric space is defined as an N-uplet (F1(t), F2(t), ..., FN(t)) of N continuous functions from [0,1] -> $\Re$ such that (F1(0), F2(0), ..., FN(0)) are the coordinates of ~ and (FI(1), F2(1), ..., FN(1)) are the coordinates of τ.

**Definition 2**: a MANHATTAN MOTION OF LENGTH 1 in an N dimensions space is defined as an N-uplet $\mathbf{M}_1 = \left(\Delta\theta_1^1, \Delta\theta_2^1, ..., \Delta\theta_i^1, ..., \Delta\theta_N^1\right)$ where each $\Delta\theta_i^1$ is an integer corresponding to the length of the move along dimension i expressed in some given elementary length unit υ.

**Definition 3**: a MANHATTAN MOTION OF LENGTH L in an N dimension space is defined as an

NxL-uplet: $\mathbf{M}_L = \left(\Delta\theta_1^1, \Delta\theta_2^1, ..., \Delta\theta_i^1, ..., \Delta\theta_N^1, \Delta\theta_1^2, \Delta\theta_2^2, ..., \Delta\theta_N^L\right)$ where each $\Delta\theta_i^j$ is an integer corresponding to the length of .the jth move along dimension i expressed in some given elementary length unit υ.

**Proposition 1**: for any ε > 0, for any path P, it is possible to find υ, L and a Manhattan motion $M_L$ of length L, such that the path P is approximated by $M_L$ with an error less than ε.

*Sketch of proof:*
- Direct application of the Stone-Weirstrass theorem.

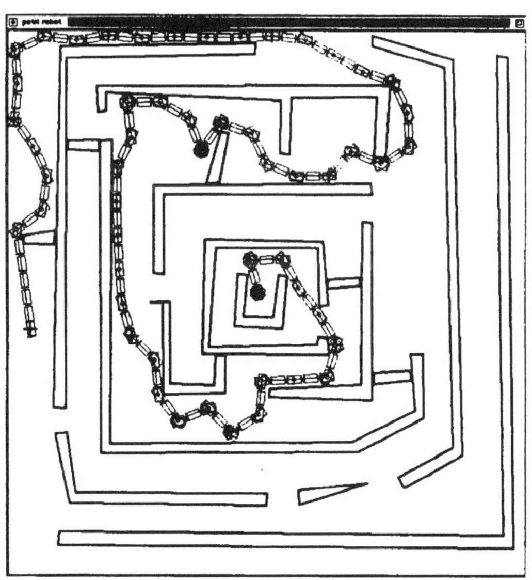

Figure 4.9b

**Proposition 2**: the ARIADNE'S CLEW algorithm is complete, which means that, for any given ε > 0, if a path exists from the initial point π to the target τ it will find (in a finite time) L and a Manhattan motion of length L $M_L$ starting at $\pi$ and ending at $\tau$ with an error less than $\varepsilon$.

*Sketch of proof:*
- Proposition 1 insures that such a Manhattan motion $M_L$ exists.

- The ARIADNE'S CLEW algorithm searches a discrete finite space.

- The ARIADNE'S CLEW algorithm insures that all the produced Manhattan motions are different.

Consequently, $M_L$ will be produced after a finite amount of time.

In the sequel of this section, three important propositions concerning the ARIADNE'S CLEW algorithm will be established. However, given the restricted length of this chapter, only sketches of proofs are proposed.

*Remark*: In fact Proposition 2 proves that any algorithm producing Manhattan motions without producing twice the same is complete. This is true either for an algorithm enumerating the Manhattan motions or for an algorithm drawing randomly the Manhattan motions (without drawing twice the same). Of course the ARIADNE'S CLEW algorithm is doing much, much, better than those two.

**Definition 4**: for a given $\varepsilon$, let us call the COMPLEXITY OF THE PROBLEM the minimum number C of identical tiles necessary to do a paving of the space, the biggest dimension of a tile being equal to $\varepsilon$.

**Definition 5**: let us call RESOLUTION R the number of landmarks generated by the ARIADNE'S CLEW algorithm to find a solution.

**Proposition 3**: resolution R is always inferior or equal to complexity C.

*Sketch of proof:*
- as long as $R < C$, two different landmarks may not be in the same tile given that the ARIADNE'S CLEW algorithm maximises the distance between the landmarks.

- for $R = C$, there is exactly one landmark in each tile.

- in that case, there exists a Manhattan motion starting at a distance of $\pi$ less than $\varepsilon$ (starting at the landmark in the same tile as $\pi$) and ending at a distance of $\tau$ less than $\varepsilon$ (ending at the landmark in the same tile as $\tau$).

*Remark*: In practice, experiences prove that $R \ll C$. There are two main reasons for this. First, most of the time, SEARCH stops EXPLORE after the generation of just a few landmarks. Second, the ARIADNE'S CLEW algorithm adapts locally its resolution to the surrounding free space, generating a lot of landmarks where narrow doors or corridors have to be found and generating just a few of them when in an open free space.

## 4.6 Parallel Implementation

It is possible to design a massively parallel implementation of the ARIADNE'S CLEW algorithm with 5 embedded levels of parallelism (see Figure 4.10):

1 - At top level of parallelism, SEARCH and EXPLORE may run in parallel. EXPLORE produces a landmark while SEARCH exploits the previous one. SEARCH stops EXPLORE as soon as it reaches the target.

2 - Both SEARCH and EXPLORE need to run a genetic algorithm which can be implemented in parallel as described in Section 2.

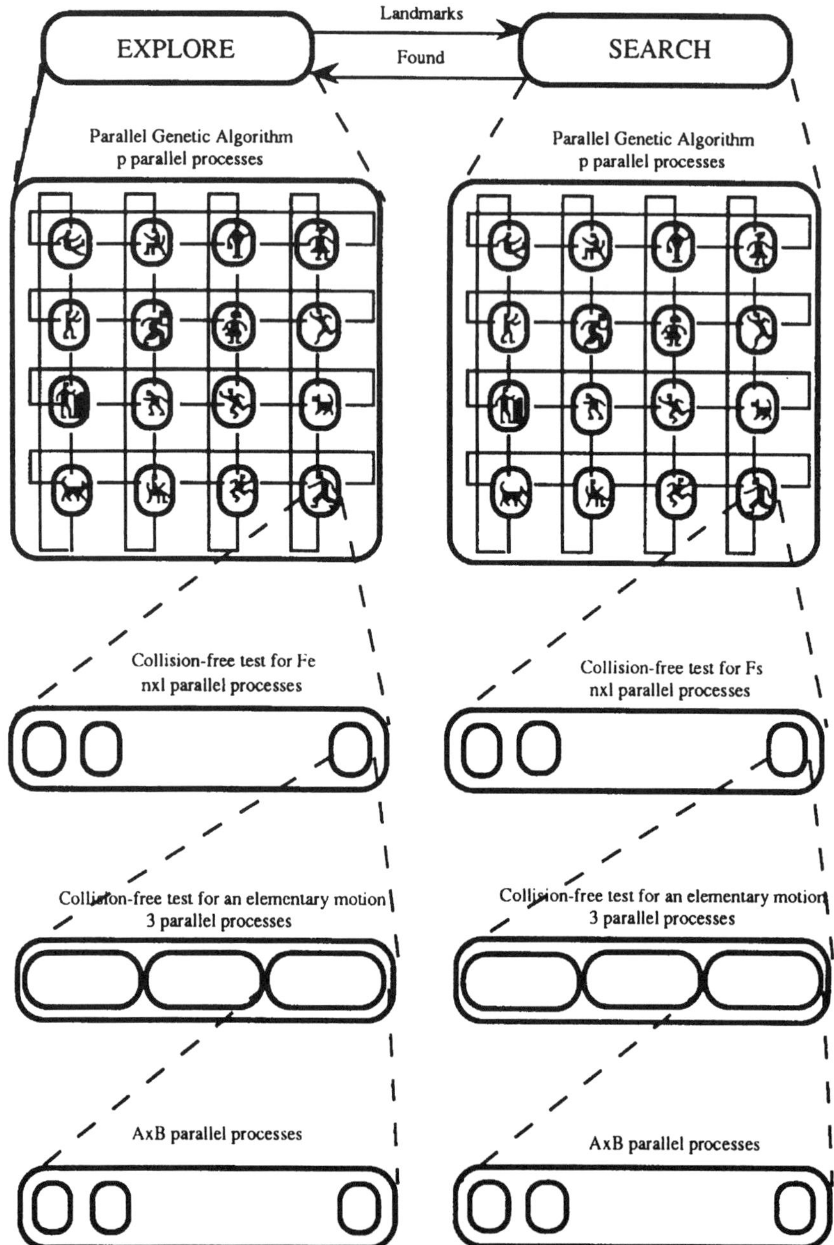


Figure 4.10

3 - $F_S$ and $F_e$, the evaluation functions of SEARCH and EXPLORE, consist mainly in testing collision on paths. This may be done by NxL parallel processes, where N is the number of degrees of freedom and L the length of the considered Manhattan motions.

4- Each of these NxL processes may be further decomposed as three parallel processes testing, respectively, vertex-to-plan, plan-to-vertex and edge-to-edge collisions.

5 - Finally, each of these tests needs AxB parallel processes where A is the number of elements in the first set of the test (A is the number of vertices, plans or edges) and where B is the number of elements in the second set (B is the number of plans, vertices or edges).

The methodology used to implement this on a parallel machine consists in writing the application fully in parallel as if there were as many processors available as the number of processes. Of course, in practice, this is not the case. However, we use languages and tools (particularly the PAROS parallel operating system and the PARX communication kernel developed in the SUPERNODE II project, see [8]) which allows us to conceive a parallel program independently of the architecture of the target machine.

We implement this on SUPERNODE machine made of 128 Transputers.

### 4.7 Conclusion, Results and Perspectives

We have presented a general method to search a continuous configuration space. This method is implemented using a minimisation technique based on parallel genetic algorithms. We have demonstrated the validity of the method on a set of complex path planning problems. Finally, we proposed a massively parallel implementation of the method which permits on-line re-planning.

Our experimental set-up used to test our algorithm for an actual six degrees of freedom arm is represented on Figure 4.11. Robot I is under the control of the MegaNode (128 T800 Transputers parallel machine) running a parallel implementation of the ARIADNE'S CLEW algorithm. Robot II is used as a dynamic obstacle: it is manually controlled via KALI (KALI is a robot control software initially developed at McGill University). First we use a CAD system called ACT which permits precise geometrical description of the arms and obstacles and which is able to present 3D simulations. We compile this representation into a special format which is downloaded to the MegaNode. A final position is then specified to Robot I, the MegaNode quickly (2 seconds) produces a plan which assume Robot II is standing still, should the position of Robot II change under manual control, Robot I stops and the MegaNode (re)computes another path.

Robor I
Robot II
KALI
68030
Bus VME
(VxWorks)
KALI
68030
Bus VME
(VxWorks)
Sun 4
(Unix)
server
VxWorks
Ethernet
Genetic Algorithm
(SEARCH)
Genetic Algoritm
(EXPLORE)
Sun 3
(Unix)
server
Mega-Node
Mega-Node
128 Transputers
(Parix)
ACT
Robot simulation package
Silicon Graphics
(Unix)

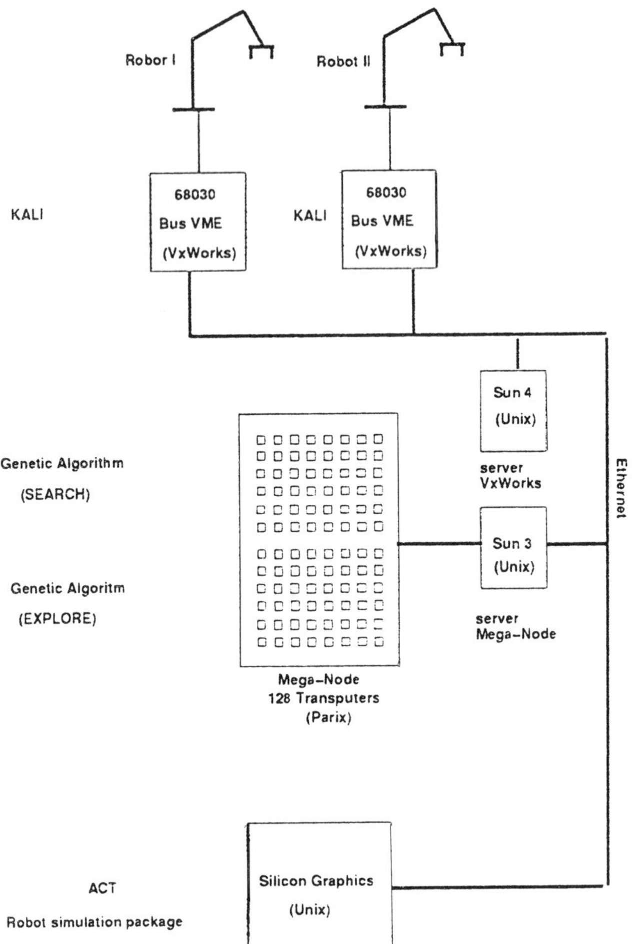

Figure 4.11

**Bibliography**

[1] J.-C. Latombe; "Robot motion planning"; Ed. Kluwer Academic Publisher, 1991.

[2] J.H. Holland; "Adaptation in natural and artificial systems"; Ann Arbor: Univ. of Michigan Press, 1975.

[3] D.E. Goldberg; "Genetic algorithms in search, optimization, and machine learning"; Addison-Wesley, 1989.

[4] E.-G. Talbi and P.Bessiere; "A parallel genetic algorithm for the graph partitioning problem"; ACM Int. Conf. on Supercomputing, Cologne, Germany, June 1991.

[5] E-G. Talbi and T. Muntean; "A parallel genetic algorithm for process-processors mapping", Int. Conf. on High Speed Computing II, Montpellier, M.Durand and F. El Dabaghi (Editors), Elsevier Science Pub., North Holland, pp.71-82, Oct 1991.

[6] T. Lozano-Percz, J.L. Jones, E. Mazer & P.A. O'Donnell; "HANDEY A robot task planner"; the MIT Press, 1992

[7] J.M. Ahuactzin, E-G. Talbi, P. Bessiere and E. Mazer; "Using genetic algorithms for robot motion planning"; ECAI92, Vienna, Austria, 1992

[8] T. Muntean, N. Gonzalez and Y. Langue; "PARX Kernel for the PAROS parallel operating system"; ESPRIT '91; Kluewer Academic Publishers, 1991.

## Chapter 5

**Michael de la Maza**
Numinous Noetics Group
Room NE43-815
Artificial Intelligence Laboratory
Massachusetts Institute of Technology
545 Technology Square
Cambridge, MA 02139

mdlm@ai.mit.edu

# The Boltzmann Selection Procedure[1]



**Abstract**
This chapter investigates Boltzmann selection, a tool for controlling the selective pressure in optimizations using genetic algorithms. An implementation of variable selective pressure, modeled after the use of temperature as a parameter in simulated annealing approaches, is described. The convergence behavior of

[1]This chapter is an elaboration of two previously published papers. de la Maza and Tidor (1992) describe Boltzmann selection and its application to a problem in molecular biology, while de la Maza and Tidor (1993) develop the theoretical analysis of Boltzmann selection.

0-8493-2529-3/95/$0.00 + $.50

optimization runs is illustrated as a function of selective pressure; the method is compared to a genetic algorithm lacking this control feature and is shown to exhibit superior convergence properties on a small set of test problems. An empirical analysis is presented that compares the selective pressure of this algorithm to a standard selection procedure.

Then, in order to understand these results in a broader context, an analytical discussion of selection procedures used in genetic algorithms is presented. A unified framework for discussing and comparing procedures is developed and used to compare proportional, Boltzmann, power law, and sigma truncation selection procedures. Two properties, translation and scale invariance, are defined and studied for each of these procedures. Selective pressure is investigated for proportional and Boltzmann selection. It is proven that, for a normal distribution of individuals in the optimization space, proportional scaling decreases selective pressure during the course of an optimization run.

## 5.1 Introduction

A number of problem solving methods in current use are based on paradigms derived from natural phenomena. Examples include simulated annealing, neural networks, and genetic algorithms. The first of these is modeled after physical systems that are remarkably successful at finding global optima by sampling a potential energy surface as the temperature is slowly reduced [16]. At higher temperatures, relatively larger excursions over the potential energy surface are permitted. During cooling, the system evacuates less favorable optima and becomes trapped in the neighborhood of more favorable ones; the amount of parameter space sampled effectively decreases with the temperature, and the system generally converges to very good local solutions. Simulated annealing has been implemented using both first-derivative methods, in which equations of motion on the potential energy (or general optimization) surface are integrated to produce the search path [22], and Monte Carlo methods, which do not require derivative information [19]. In both cases a temperature parameter is used to control the optimization. Artificial neural networks, inspired by the highly interconnected, relatively simple, non-linear processing units found in biological nervous systems, are proving useful in areas of machine learning and pattern recognition. Genetic algorithms are based on the same principles of natural selection that describe the evolution of sizable biological populations over time scales covering a large number of generations. A fitness function describes the success of each member of the population in terms of that member's parameters (genetic makeup or "genes" ); the fitness is a direct measure of an individual's reproductive potential, which follows in some measure the imperative, "survival of the fittest" [6]. Mechanisms for creating diversity are also incorporated, including, but not limited to, mutation and crossover.

Genetic algorithms are atypical in that many solutions are followed in parallel and these are recombined in search of improved ones. The evolutionary aspect provides for the elimination of trial solutions that are relatively unsuccessful, but a variety of selection criteria are possible. The quality of the overall result and the computational effort required depend critically on the selection criteria used. Here we compare the standard proportional scaling method with a new Boltzmann-based protocol. As an illustration, note that one extreme selection scheme would allow only copies of the fittest individual to survive. Variability would be introduced only by mutation (and, if so desired, by crossover of mutant siblings); this would correspond to a highly parallel Monte Carlo search, but at zero temperature (i.e., a simple "always improving" optimization). While this might be efficient to perfect the best optimum once it had been located, it would be extraordinarily inefficient for most problems at the start of an optimization. In fact, for small enough mutational steps in the parameter space, it would lead to the local optimum closest to the fittest individual in the starting population. This corresponds to a "zero tolerance" evolutionary system, in which the slightest advantage of one individual over another results in the loss of the less fit individual from the gene pool. The other extreme would be an "infinitely tolerant" environment; i.e., one that permits all individuals to survive to reproduction, regardless of fitness. If the total number of individuals in the population is fixed, this corresponds to a random walk in the space with no preference for optima. Good solutions that are found are likely to be lost to mutation and crossover. Between these two extremes lies a continuum of evolutionary tolerance. Early in an optimization, it would be useful to have a high tolerance, so that the search is carried out over a large portion of the space (like the initially high temperature used for simulated annealing) and a large variety of individuals are retained in the population so that, even if they, themselves, are not of high fitness, they might donate to a crossover that produces an exceptionally fit individual. Later in the procedure, when the major optima have been located and partially refined, it would be reasonable to eliminate the lesser optima and concentrate on refining the better ones, so a lower tolerance would be useful.

Genetic searches generally converge from a heterogeneous starting population of random individuals to a more homogeneous population in which the individuals are nearly identical. Once the population has converged to near homogeneity, a predominantly local search is performed. The selection scheme can be used to exert control over the rate of convergence of the population, as is shown in this work using a function with multiple optima.

There is a modest literature on methods of controlling the selective pressure in a genetic algorithm, and some systematic studies have been performed [3,24,1,17]. We feel that it is important [1] to develop methods in the genetic algorithm that allow specific control of the selective pressure, and [2] to study the best ways of varying the selective pressure during the course of a genetic algorithm run to obtain rapid convergence to an optimal solution.

We have implemented a genetic algorithm using such a scheme for varying the evolutionary tolerance of the environment with Boltzmann scaling. The plan of the rest of this chapter is as follows. Section 2 describes Boltzmann selection and compares it to proportional selection on two test problems. Section 3 explores the scale and translation invariance of several selection procedures and proves a theorem which might explain why Boltzmann selection outperforms proportional selection. Section 4 summarizes related work and Section 5 concludes the chapter. The purpose of Section 2 is to provide the reader with two examples of how Boltzmann selection can be used to optimize functions. Each step in the approach to both problems is described in detail so the reader can adapt these methods to other optimization problems. Section 3 then provides theoretical justification for the use of Boltzmann selection.

## 5.2 Empirical Analysis

### 5.2.1 Framework

This section describes a general framework for defining selection procedures. Here we use it to define the proportional selection procedure which will be empirically compared to Boltzmann selection on two test problems. Later, in Section 3 this definition will be used to prove properties about several selection procedures.

Definition 2.1 defines a selection procedure in terms of four functions. To determine the number of copies of each individual in one generation that will be propagated into the next, the four functions are composed as follows: W(P(F(U(X)))), where U is the problem dependent objective function, F is called the fitness function, P usually produces a vector, $\vec{R}$, of probabilities multiplied by the number of individuals in the population, and W is usually the roulette wheel procedure [10]. The values produced by U are often called raw fitnesses. In some sense, this choice of functions is arbitrary; however, we feel that it captures our intuitions about the salient components of selection procedures. Notice that the fitness function used by the genetic algorithm, F, is not always the same as the function to be optimized, U. The key transformation, F(U(X)) produces the fitnesses from the objective function. Frequently this involves simply adding a constant so the fitness function is positive. Defining proportional selection, which is one of the most widely used selection procedures [15,10], will give an intuitive understanding of the definition.

**Definition 2.1** *We define a selection procedure to be a quadruple (W, P, F, U) where*

$$W:\vec{R}\rightarrow\vec{I}$$

$$P:\vec{R}\rightarrow\vec{R}$$

$$F:\vec{R}\rightarrow\vec{R}$$

$$U:X\rightarrow\vec{R}$$

*and where X is a population, R is the set of reals, and I is the set of integers.*

Definition 2.2 defines proportional selection. Note that the transformation from objective function to fitness is the identity function. The P used here is what gives proportional selection its name. This P is used in many selection procedures, including proportional selection, power law selection [9], and sigma truncation [8]. These other procedures are all different, however, because they use different functions for F. While these selection procedures all proportionally scale the fitness function, traditional proportional selection is the only one that proportionally scales the objective function. In Section 3, we will briefly consider two other selection procedures, rank selection [3] and tournament selection [4], that are substantially different.

**Definition 2.2** *The proportional selection procedure is:*

$U(X)$ *is the problem dependent objective function*

$$F(\vec{R})=\vec{R}$$

$$P_i(\vec{R})=\frac{\vec{R}_i}{\langle\vec{R}\rangle}$$

$W(\vec{R})$ *is the roulette wheel procedure* [12]

*where the notation $P_i$ is used to indicate the ith element of P and $\langle \vec{R} \rangle$ is a scalar equal to the average of all of the elements in $\vec{R}$ in the current generation.*

This high-level description leaves out some implementation details which are specified in the Lisp code listing in Figure 5.1.

## 5.3 Introduction to Boltzmann Selection

Having described proportional selection in terms of our definition, we now turn to an explanation of Boltzmann selection.

In an equilibrated simulated annealing ensemble, the probability of visiting a point in optimization space, *Xj,* is,

$$P(X_j) = \frac{e^{-u(X_j)/T}}{\sum_i e^{-u(X_i)/T}}$$

where the minus sign in the exponent is necessary because a minimization is performed, T is the temperature, the numerator contains the Boltzmann weighting term, and the denominator is a normalization factor. The Boltzmann function has the property that at higher temperatures the system visits more of phase space, whereas at lower temperatures the probability of visiting points more unfavorable than the global minimum is lower. We have implemented an analogous equation in the selection step of our genetic algorithm. Definition 2.3 specifies Boltzmann selection (not to be confused with Boltzmann tournament selection, which is defined in [11]). The T parameter in the definition is a variable that corresponds to evolutionary tolerance (analogous to temperature in simulated annealing) and the plus sign is changed to minus when minimization, rather than maximization, is desired.

```
(defun Boltzmann-selection (individual-list temperature)
       (let,
              ((beta (/ temperature))
              (get_exp (mapcar #(lambda (anindiv)
                     (list
                     (exp (* beta (first anindiv)))
                     (second anindiv)))
                     individual-list))
              (total (reduce #'+ (mapcar #'first get_exp))
              (boltzmann_fitness
                     (mapcar #'(lambda (anindiv)
                            (list
                            (/ (first anindiv) total)
                            (second anindiv)))
                     get_exp)))
              (roulette-wheel boltzmann_fitness)))
```

Figure 5.1: Implementation of Boltzmann selection in Lisp. This procedure implements the fitness function, F, of Boltzmann selection. The two inputs to the Boltzmann-selection procedure are a list of individuals and the temperature. Each element of the list of individuals is itself a list with two components. The first component is $U_i(X)$ and the second is the genotype of the individual $(X_j)$. The procedure ends by calling the `roulette-wheel` procedure which selects the individuals that will be propagated into the next generation. This roulette wheel function should take as input a list each of whose elements is a list of two components. The first component is $P_j(X)$ and the second component is the genotype of the individual $(X_j)$. The function of this roulette wheel procedure is identical to the one used in proportional selection and the same code could be used for both. The Lisp code given here has been tested on a Sun SPARCstation running Lucid Common Lisp (version 4.1) and Austin Kyoto Common Lisp (version 1.530).

**Definition** 2.3 *The Boltzrnann selection procedure is:*

$$F_i(U(X)) = e^{U_i(X)/T}$$

*and U, P, and W are defined as in proportional selection.*

The Boltzmann formulation provides a number of attractive features. The result of the selection step is independent of overall translational shifts in the optimization surface. Multiplying the optimization surface by a constant, so that U'(X) = cU(X), which corresponds to changing the units in which U(X) is measured, is also invariant so long as the parameter, T, which has the same units as U(X), is similarly scaled. Moreover, there is no requirement that the optimization function be non-negative, as there is with proportional selection, since the exponential provides an appropriate transformation. The proof of the translational invariance of Boltzmann selection will be given in Section 3.

### 5.3.1 Experiments with Boltzmann Selection

In this section we first describe two model problems used to compare Boltzmann and proportional scaling, we then explain how a tolerance schedule for the Boltzmann GA was chosen and present comparative results showing faster convergence for the Boltzmann GA. Finally, we provide an empirical analysis that illustrates that a genetic algorithm with proportional scaling increases, rather than decreases, evolutionary tolerance as the point of completion nears (contrary to what one might wish).

#### 5.3.1.1 Description of Model Problems

**Molecular Biology Problem**. This section describes a problem, inspired by molecular biology, in which a pattern must be built that distinguishes between functional and non-functional protein sequences.

A database of instances, composed of the twenty letters used to represent the twenty amino acids, is divided into positive and negative classes. A random pattern is generated and the same pattern is embedded in a random location in all of the positive instances. The goal is to find this pattern or an acceptable substitute. In addition to containing any character that can appear in the instances, the substrings, also called individuals, can contain a "don't care" symbol, which matches any character.

A typical database is shown in Table 5.1. All of the positive instances contain the substring RIEY while none of the negative instances do. Each database has ten instances, each having a 0.5 probability of being in the positive class.

| Instance | Class |
|---|---|
| K D G R W E G G G H W V M A R M | negative |
| N H I T R H Y V Q P C C C Y D K | negative |
| T I C N V S D Q W L F F K L W S | negative |
| E E P S R I E Y T I M G I E V T | positive |
| V P Y K P C P K H S L S G A F K | negative |
| R I E Y R W P V K V R H Q N Y G | positive |
| V A H R C K N W Q M T W I T H Q | negative |
| T L Q F Y K E N D L T K C G L K | negative |
| R C W Y K N A Y I G Q Y V C P H | negative |
| G V M Q G T R I E Y Y F F C G S | positive |

Table 5.1: Typical database. Each instance has sixteen letters. The instances in the positive class all contain the sequence "RIEY".

The optimization function, *U(X),* is a measure of the difference between how well an individual matches the positive instances and how well it matches the negative instances. The score of individual *Xj* is calculated using,

$$U_j(X) = \frac{1}{|P|}\sum_{I_k \in P} \max(match(X_j, I_k)) - \frac{1}{|N|}\sum_{I_k \in N} \max(match(X_j, I_k))$$

where I is the set of all instances, P is the subset containing |P| positive instances and N is the subset containing |N| negative instances. The matching function returns a list of numbers that indicate how well an individual matches each substring of an instance. A point is given for each character that correctly matches and half a point is given for the "don't care" symbol. For example, the individual **INE,* when matched against the instance *THISISFINE,* returns 0.5 when matched against *THIS;* 0.5 when matched against *ISIS;* and 3.5 when matched against *FINE.*

Both the Boltzmann and proportional GAs shared the following properties. There were three recombination operators: crossover, mutation, and shift. The crossover operator was a traditional 1-point crossover. Mutation was accomplished by randomly switching exactly one character in an individual to another character. The shift operator performed a cyclic permutation. To create the next generation from the present one, first a selection step (either Boltzmann or proportional selection) was performed, creating generation i + 1/2 from generation i. Each of these individuals was examined in turn and one of the three operators was chosen (at random in the ratio crossover:mutation:shift of 2:1:1) and applied to create an individual for generation i + 1. In the case of crossover, an individual in generation i + 1/2 was crossed over with any of the individuals in that generation (including itself, producing the identity transformation) with equal probability.

The shift operator was introduced because many runs converged to a local optimum that was a cyclic permutation away from the global optimum (correct answer). With mutation and crossover alone, the rate of moving from the local optima to the global optimum is negligible because it requires crossing deep valleys. The cyclic permutation shift operator crosses these valleys in a single step.

The mutation rate (25%) seems deceptively high. For individuals with eight characters, each character was mutated with an average probability of 3.125%. If the twenty-one characters are represented as bits, then approximately 4.4 bits are needed to represent each character. Thus, the mutation rate per bit is approximately 0.7%, which is similar to that of other genetic algorithms.

**F2 Function.** Deb and Goldberg's F2 function [9] is:

$$F2(x) = \sin^6(5\pi x)\exp\left[2\ln 2(\frac{x-0.1}{0.8})^2\right]$$

On the interval [0.0, 1.0], F2 has five peaks, each one smaller than the previous one (see Figures 5.4, 5.5, and 5.6).

Individuals for both the Boltzmann and proportional GAs were composed of three decimal digits and represent a value between 0.000 and 0.999 (inclusive). The optimization function was simply the value of F2 for the x value encoded by the individual. The population consisted of 100 individuals. The 1-point crossover rate was 90% and the mutation rate was 10%. The mutation operator added a uniform random number between 0.1 and -0.1 to the individual. To create the next generation from the present one, first a selection step (either Boltzmann or proportional scaling) was performed, creating generation i + 1/2 from generation i. Each of these individuals was processed in turn and one of the two operators was chosen (at random in the ratio crossover:mutation of 9: 1) and applied to create an individual for generation i + 1. In the case of crossover, an individual in generation i + 1/2 was crossed over with any of the individuals in that generation (including himself, producing the identity transformation) with equal probability.

5.3.1.2 Finding the Initial Tolerance

The appropriate initial tolerance value was determined by performing a series of experiments. The tolerance schedule is shown in Figure 5.2. This tolerance schedule was chosen by adapting a successful simulated annealing cooling schedule to genetic algorithms. The tolerance is constant for the first ten generations and then ramps down over the next thirty generations to a final value. The final tolerance was set to be 0.5.

Experiments using the molecular biology problem with a four character pattern were used to determine the initial tolerance. Ten initial tolerances were tested: 0.5, 1.25, 2, 3.5, 5, 6.5, 8, 9.5, 11, and 12. The number of generations required for convergence (see Section 3.3.1) was recorded; the results are shown in Figure 5.3. Each experiment was repeated eight times; the numbers shown are averages.

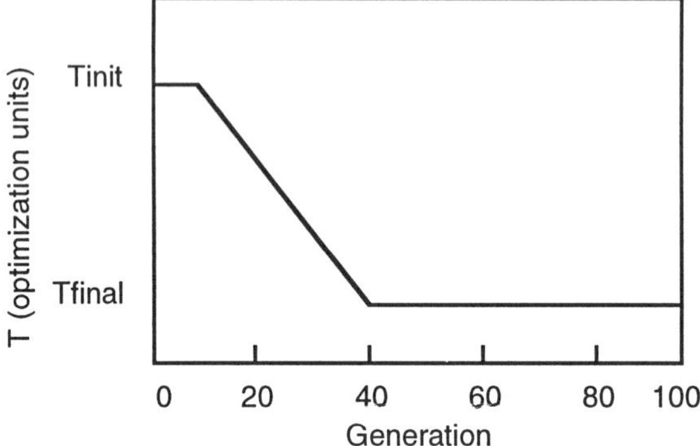


Figure 5.2: The tolerance schedule used in the Boltzman selection genetic algorithm. The tolerance was set to $T_{init}$ for the first ten generations, ramped down to $T_{final}$ over thirty generations, and maintained at $T_{final}$ until completion. In all runs, $T_{final}$ = 0.5 optimization units.

The U-shaped curve in Figure 5.3 is in accordance with our intuition about how the initial tolerance should affect search behavior. If the initial tolerance is too high, then the genetic algorithm spends too much time performing a random search and requires a long time to focus on the few good solutions. If the initial tolerance is too low, then the genetic algorithm performs a local search around the individual with the highest fitness in the initial population and, therefore, risks never finding the solution.

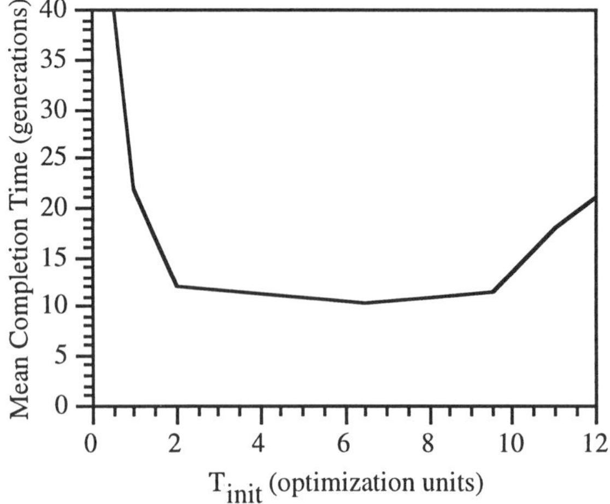

Figure 5.3: The average number of generations to completion of the Boltzmann scaling genetic algorithm for the problem of pattern length four. Experiments that required more than fifty generations to complete were stopped at fifty generations and combined to compute the average as if they had completed in fifty generations.

On the basis of these results, an initial tolerance of 4 was chosen for the next series of experiments. Unless otherwise noted, this tolerance schedule was used for all of the problems discussed in this section. The observation that other optimization surfaces were searched reasonably quickly with the same schedule suggests that the method is robust with respect to small changes in the tolerance schedule.

### 5.3.2 Comparison

This section compares Boltzmann scaling and proportional scaling on a small set of molecular biology problems and Deb and Goldberg's F2 function [7].

#### 5.3.2.1 Molecular Biology Problem

The results of comparing the Boltzmann and proportional GAs are shown in Table 5.2. The first and second columns give the number of characters in the instances and in the patterns, the third shows the size of the population, the fourth indicates how many times each experiment was performed and the fifth and sixth give the average number of generations for the Boltzmann and proportional GAs to converge. For this purpose convergence is defined as finding a pattern that is a perfect match in each of the positive instances ("don't care" matches everything) but in none of the negative instances. Note that the GA is not required to find the optimal or characteristic pattern. The last column is the result of applying a one-tailed statistical test: $z=(\mu_1-\mu_2)/\sqrt{\sigma_1^2/n_1+\sigma_2^2/n_2}$. If this number is greater than 2.326 then the Boltzmann GA is better than the proportional GA at the $p < 0.01$ level. If it is greater than 2.576 then it is significant at the $p < 0.005$ level. A one-tailed (rather than two-tailed) test was used to show that the performance of the Boltzmann GA was superior to (rather than different from) the performance of the proportional GA.

| Instance Length | Pattern Length | Population Size | Runs | Boltzmann | Proportional | Stat |
|---|---|---|---|---|---|---|
| 25 | 4 | 100 | 49 | 12.0 | 16.3 | 2.4 |
| 35 | 6 | 100 | 44 | 12.0 | 25.8 | 6.5 |
| 50 | 8 | 100 | 36 | 16.9 | 34.2 | 6.7 |

Table 5.2: Results of comparison between Boltzmann GA and proportional GA. The first two columns give the length of the instance and pattern. The third column shows the size of the population of individuals. The Runs column indicates how many times each experiment was repeated. The Boltzmann and Proportional columns show the average number of generations needed for each algorithm to converge. The final column gives the result of applying a significance test to the results.

The results are clear. On all three versions of this problem, the Boltzmann GA is far superior to the proportional GA.

Figure 5.4 shows the progress of the top individual in a Boltzmann scaling experiment. At generation 0, the score is very low and the individual does not match the target pattern, "RIEYGKSD", very well. But after a series of mutations, crossovers, and shifts, the instance is perfectly aligned with the target pattern at generation 19. After this point, the top individual is changed, one position at a time, until it matches the target pattern perfectly. Note that in collecting the data for Table 5.2 this run would have been considered to converge at generation 34, when the pattern matches all of the positive instances and none of the negative instances.

```
(.......%%%%%%%%............................)
(rvftsdtRIEYGKSDawvqekhmkwiqfyprfateshkyiiitgvscvp)Gen      Score
(...........cvSwiwwk ........................)0           1.60
(.............Svswiwwk ......................)1           1.60
(......swIwYGfg.............................)2-3         5.80
(.....gswIwYGf..............................)4           7.60
(.....gnwIwYGf..............................)5           8.80
(.........wYGKSswi..........................)6          13.80
(........IwYGKSsw...........................)7-13       24.20
(........IwYGKSs*...........................)14        25.60
(......s*IwYGKS.............................)15-18     27.20
(.......*IwYGKS*............................)19-22     30.00
(.......*IhYGKS*............................)23-33     31.40
(.......*IEYGKS*............................)34-42     43.60
(.......RIEYGKS*............................)43        52.00
(......*RIEYGKS.............................)44-48     52.00
(.......RIEYGKS*............................)49-53     52.00
(.......RIEYGKSD............................)54        63.40
```

Figure 5.4: Simple Boltzmann scaling experiment. On each line the top individual, the generation number, the number of perfect alignments with the positive instances, and the score of the top individual is shown. The top line shows a positive instance with the target region, "RIEYGKSD", in capitals and the rest of the string in lower case. Each line shows the top individual and where it matches the instance. When a letter matches with the target sequence it is capitalized. "*" is the "don't care" character. The complete data base had five positive and five negative instances, so the maximum number of perfect alignments is five.

5.3.2.2 F2 Function

Two experiments were performed using the F2 function [7] to explore the properties of tolerance. The experiments differed only in the distribution of the initial population. The first experiment, performed with a population randomly distributed around the middle peak, demonstrates that the proportional GA does not allow individuals to jump from the middle peak to the second highest peak and then onto the highest peak, while the Boltzmann GA does. It also illustrates how the Boltzmann GA searches the F2 space. The second experiment, performed with a random initial population, shows how tolerance affects the search of the

Boltzmann GA and compares it to how the proportional GA searches the F2 space.

In the first experiment, the 100 individuals were randomly distributed between 0.400 and 0.600. The middle peak is at approximately 0.5. Figure 5.5 shows a snapshot of the proportional GA and Boltzmann GA populations after 50 generations have passed. Notice that the proportional GA was not able to move any individuals from the middle peak, while the Boltzmann GA fully explored the second highest peak and had an individual on the highest peak. Figure 5.6 shows a time series of the progress of the Boltzmann GA. The population of individuals began, at generation 0, with the 100 individuals on the middle peak. By generation 23, some of the individuals began to explore the second highest peak. At generation 60, there were few individuals left on the middle peak, many individuals on the second highest peak, and a few individuals on the highest peak. By generation 90, almost all of the individuals were on the highest peak.

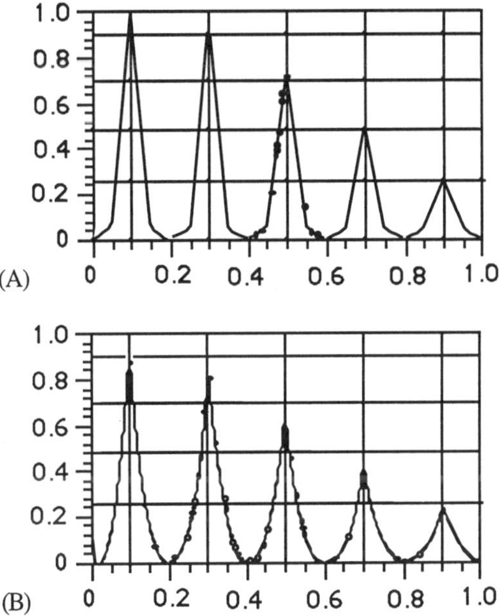


Figure 5.5: The proportional and Boltzmann GA populations at generation 50. (A) proportional GA population, (B) Boltzmann GA population. The initial population was randomly distributed between 0.400 and 0.600. The individuals are represented by small circles and the F2 function is the dark, continuous line. These graphs show the population immediately after the recombination operators have been applied and before the scaling operation has been done. Notice that none of the individuals in the proportional GA have been able to escape the local optimum of the middle peak.

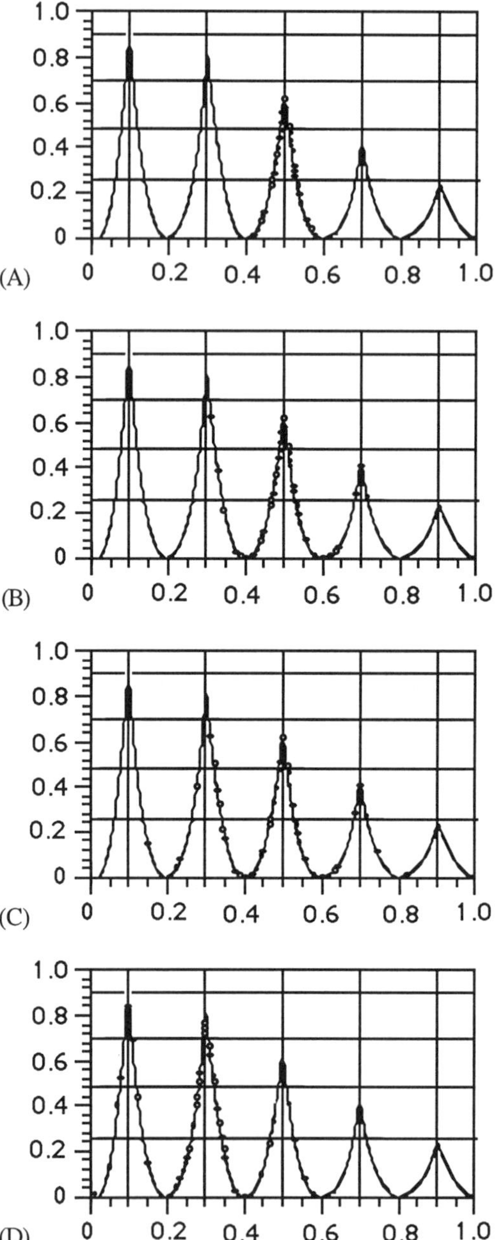
1.0
0.8
0.6
0.4
0.2
0
(A)
0 0.2 0.4 0.6 0.8 1.0
(B)
(C)
(D)

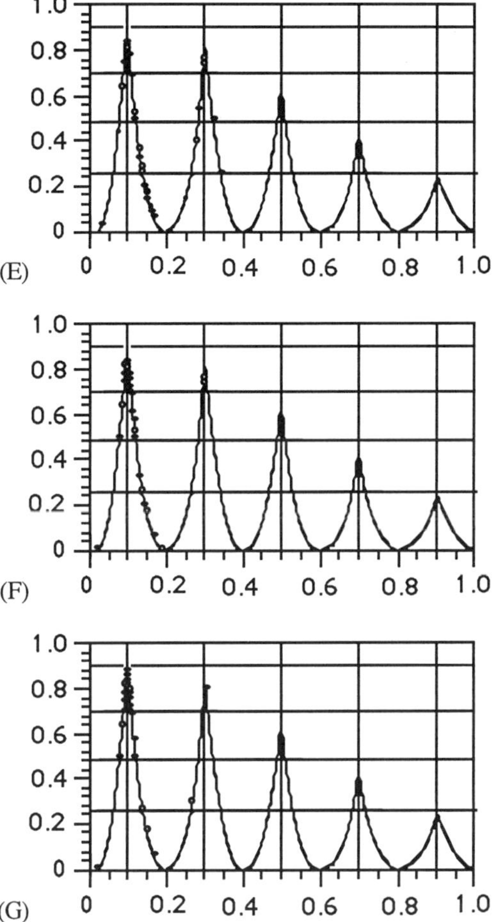


Figure 5.6: Boltzmann GA population time series. The initial population was randomly distributed between 0.400 and 0.600. The individuals are represented by small circles and the F2 function is the dark, continuous line. These graphs show the population immediately after the recombination operators have been applied and before the scaling operation has been done. Each graph shows the population at a different generation: (A) generation 0, (B) generation 23, (C) generation 49, (D) generation 60, (E) generation 70, (F) generation 80, (G) generation 90.

The second experiment, with a random initial population, demonstrates that the behavior of the Boltzmann GA can be altered by changing tolerance. The first graph in Figure 5.7 shows the distribution of individuals in the Boltzmann GA subject to a constant tolerance of 10. The second graph repeats the same experiment but with a tolerance of 1. As expected, in the experiment with the higher tolerance, the individuals were comparatively more distributed throughout the space than in the experiment with the lower tolerance. The lower tolerance caused more copies of the highest fitness individuals to be made and therefore there was much more pressure to explore the highest peak than the other peaks. For purposes of comparison, the same experiment done with the proportional GA is also shown.

### 5.3.3 Tolerance in the Proportional GA

Given the formalism that has been presented to modify evolutionary tolerance, it is possible to study how the proportional GA sets an effective tolerance value at a given generation by choosing the tolerance that minimizes,

$$\sum_j \left[ \frac{U_j(X)}{\sum_i U_i(X)} - \frac{e^{U_j(X)/T}}{\sum_i e^{U_i(X)/T}} \right]^2$$

where $U_j(X)$ is the score of individual j and T is the tolerance.

Minimizing this function gives the tolerance which best characterizes the behavior of the proportional GA in the framework of the Boltzmann GA. For runs of the molecular biology problem, the function was minimized using the golden section search described by Press et al.[21].

The results are shown in Figure 5.8. They indicate that in the proportional GA the effective tolerance increases, rather than decreases, as a function of the number of generations. This result, which runs contrary to both intuition and theory, strongly suggests that the traditional proportional scaling technique may need reconsideration.

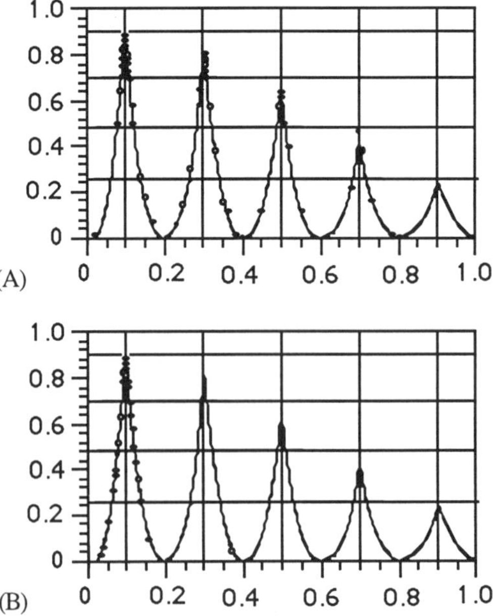

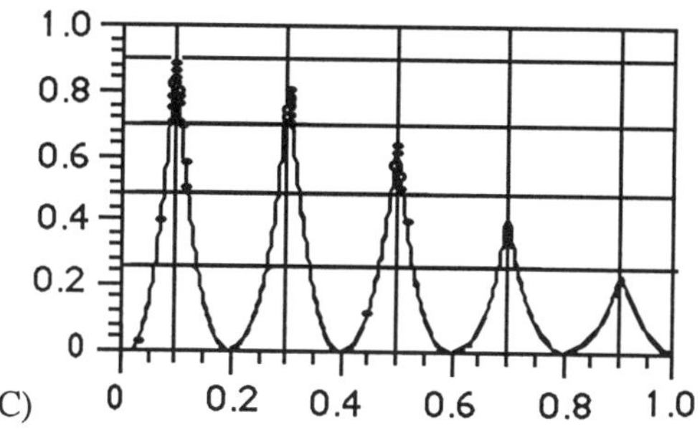


Figure 5.7: Random initial population. The initial population was randomly distributed between 0.000 and 0.999. The individuals are represented by small circles and the F2 function is the dark, continuous line. These graphs show the population immediately after the recombination operators have been applied and before the scaling operation has been done. (A) Boltzmann GA population at generation 20 with a tolerance of 10, (B) Boltzmann GA population at generation 20 with a tolerance of 1, (C) proportional GA population at generation 20. As expected, the individuals in (A) are comparatively more distributed than the individuals in (B).

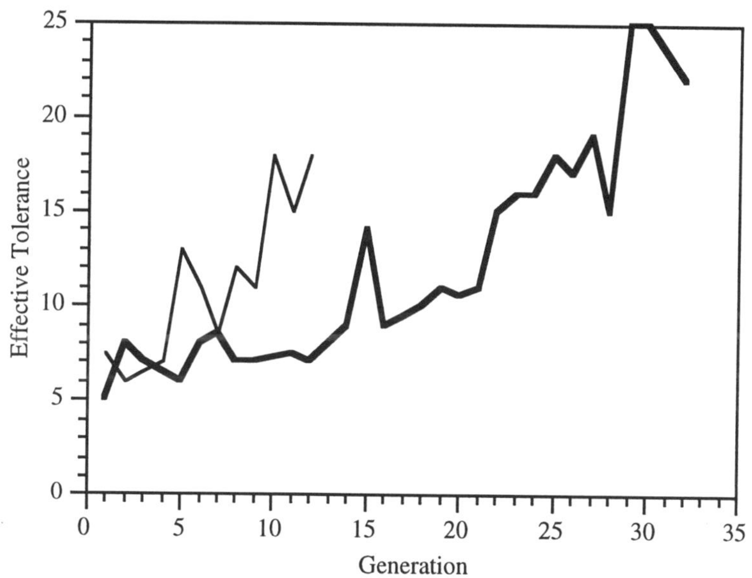


Figure 5.8: Effective tolerance in proportional GA. The dark line is for an experiment in which the Boltzmann GA outperformed the proportional GA; the light line is for an experiment with the opposite outcome. Both experiments are for patterns of length eight.

## 5.4 Theoretical Analysis

### 5.4.1 Definitions of Scale and Translation Invariance

Section 5.2 gives empirical evidence showing that a genetic algorithm with Boltzmann selection converges faster than an algorithm with proportional selection. Moreover, the parameter T in the above transformation is a variable parameter that can be used to control selective pressure during the course of a genetic algorithm run. At the end of this theoretical analysis section, we will give an analytical result that may account for this difference in convergence.

The primary properties of selection procedures that we will explore are scale and translation invariance. As we discuss below, these invariances are important properties not only because we feel intuitively that an optimal algorithm should follow the same search path on a simple transformation of a problem as on the original problem, but also because selective pressure should be carefully chosen by the user and the algorithm and not imposed by the objective function.

**Definition 3.1** *A selection procedure is scale invariant exactly when:*

$$P(F(U(X))) = P(F(kU(X)))$$

*where $k > 0$ and X is an arbitrary population.*

Intuitively, a selection procedure is scale invariant if multiplying the objective function by a constant does not change the values produced by P.

**Definition 3.2** *A selection procedure is translation invariant exactly when:*

$$P(F(U(X))) = P(F(U(X) \oplus \vec{C}))$$

*where C is a vector of identical constant elements Ci, $\oplus$ is vector addition, and X is an arbitrary population.*

Intuitively, a selection procedure is translation invariant if adding a constant to the objective function does not change the values produced by P.

In the next section we show that the Boltzmann selection procedure is translation invariant, but not scale invariant.

### 5.4.2 Scale and Translation Invariance of some Selection Procedures

In this section we explore the scale and translation invariance of four selection procedures. Because proportional selection is the best known and most widely used selection procedure, we carefully examine its lack of translational invariance.

#### 5.4.2.1 Proportional Selection

Proportional selection is the most widely used selection procedure. We show that the proportional selection function is scale invariant, but not translation invariant, and we examine the nature of its lack of translational invariance.

**Observation 3.3** *The proportional selection procedure is scale invariant.*

*Proof:*

$$P_i(F(kU(X))) = \frac{F_i(kU(X))}{\langle F(kU(X))\rangle}$$
$$= \frac{kU_i(X)}{\langle kU(X)\rangle} = \frac{U_i(X)}{\langle U(X)\rangle}$$
$$= \frac{F_i(U(X))}{\langle F(U(X))\rangle} = P_i(F(U(X)))$$

**Observation 3.4** *The proportional selection procedure is not translation invariant*

*Proof:*

$$P_i(F(U(X))) = P_i(F(U(X)\oplus\vec{C})) \Leftrightarrow$$

$$\frac{F_i(U(X))}{\langle F(U(X))\rangle} = \frac{F_i(U(X)+\vec{C}}{\left\langle F(U(X))\oplus\vec{C}\right\rangle} \Leftrightarrow$$

$$\frac{\left\langle U(X)\oplus\vec{C}\right\rangle}{\langle U(X)\rangle} = \frac{U_i(X)+\vec{C}_i}{U_i(X)} \Leftrightarrow$$

$$1+\frac{\left\langle\vec{C}\right\rangle}{\langle U(X)\rangle} = 1+\frac{\vec{C}_i}{U_i(X)} \Leftrightarrow$$

$$\frac{\left\langle\vec{C}\right\rangle}{\langle U(X)\rangle} = \frac{\vec{C}_i}{U_i(X)} \Leftrightarrow$$

$$\langle U(X)\rangle = U_i(X)$$

In general, the last equation is false.

The same result has been shown by Grefenstette and Baker [2]. In some sense, the proportional selection function is trivially not translation invariant because a negative constant Ui can be chosen such that $(U_i(X) + \vec{C}_i) < 0$. But this proof shows that proportional selection is not translation invariant even when $\vec{C}_i > 0$.

We can further investigate the role of this constant by asking how it affects selective pressure. We prove that as the constant increases, selective pressure decreases. For this purpose we define selective pressure on an individual to be the fitness of the individual divided by the average fitness of the population [24].

**Theorem 3.5** *If $F(\vec{R})$ is bounded, then*

$$\lim_{C\to\infty} \frac{F_i(\vec{R})+c}{\left\langle F(\vec{R})+c \right\rangle} = 1$$

*Proof.*

$$\lim_{C\to\infty} \frac{F_i(\vec{R})+c}{\left\langle F(\vec{R})+c \right\rangle} = \lim_{C\to\infty} \frac{\frac{F_i(\vec{R})}{c}+1}{\frac{\left\langle F(\vec{R}) \right\rangle}{c}+1} = 1$$

Intuitively, the theorem says that as the constant c increases the differences among individuals are blurred and therefore there is less and less selective pressure. This result is of particular importance because adding a constant to an objective function is recommended in the genetic algorithm literature as a way to make negative fitness functions positive (see, e.g., [10,20]). This theorem says that the choice of this constant can greatly affect the selective pressure, and therefore the performance, of the genetic algorithm. Michalewicz makes a similar, albeit more informal, argument [20]. Thus, this seemingly cosmetic change in the fitness function, F, can have wide-ranging consequences.

Moreover, as a genetic algorithm progresses, individuals will become increasingly fit. Intuitively, this might act like adding a constant to the fitness function which decreases selective pressure [24].

The following theorem makes the effect of the constant on selective pressure more precise.

**Theorem 3.6** *If F is bounded and greater than zero and c > 0, then*

$$\frac{F^{\min}+c}{F^{\max}+c} > 1 - \frac{F^{\max}}{c}$$

*where $F^{min}$ is the fitness assigned to the least fit individual and $F^{max}$ is the fitness assigned to the most fit individual in a particular generation.*

*Proof.*

$$\begin{aligned}
\frac{F^{\min}+c}{F^{\max}+c} &> \frac{c-F^{\max}}{c} \Leftrightarrow \\
cF^{\min}+c^2 &> c^2-(F^{\max})^2 \Leftrightarrow \\
cF^{\min} &> -(F^{\max})^2
\end{aligned}$$

This last equation is clearly true.

If $F^{min}$ and $F^{max}$ are interpreted to be the lower and upper bound on F, then the theorem can be used to make general statements about an entire run. For example, assume that the constant c is an order of magnitude greater than the upper bound on F, then the least fit individual in the population will never be less than 1 - 1/10 = 0.9 as fit as the most fit individual in the population.

We now consider Boltzmann selection [17], power law selection [9], and sigma truncation selection [8].

5.4.2.2 Boltzmann Selection

**Observation 3.7** *The Boltzmann selection procedure is not scale invariant.*

*Proof:*

$$P(F(U(X))) = P(F(kU(X))) \Leftrightarrow$$

$$\frac{e^{kU_i(X)/T}}{\left\langle e^{kU(X)/T}\right\rangle} = \frac{e^{U_i(X)/T}}{\left\langle e^{U(X)/T}\right\rangle} \Leftrightarrow$$

$$e^{(k-1)U_i(X)/T} = \frac{\left\langle e^{kU(X)/T}\right\rangle}{\left\langle e^{U(X)/T}\right\rangle}$$

This last equation is clearly false.

Although Boltzmann selection is not scale invariant, any changes in scale can be offset by multiplying the temperature parameter by the scaling constant, k.

Observation 3.8 *The Boltzmann selection procedure is translation invariant.*

*Proof.*

$$P_i(F(U(X)\oplus\vec{C})) = \frac{e^{(U_i(X)+\vec{C}_i)/T}}{\left\langle e^{(U(X)\oplus\vec{C})/T}\right\rangle}$$

$$= \frac{e^{\vec{C}_i/T}e^{U_i(X)/T}}{\left\langle e^{\vec{C}/T}e^{U(X)/T}\right\rangle} = \frac{e^{U_i(X)/T}}{\left\langle e^{U(X)/T}\right\rangle}$$

$$= P_i(F(U(X)))$$

Thus, the Boltzmann selection procedure is translation invariant but not scale invariant.

5.4.2.3 Power Law Selection

**Definition 3.9** *The power law selection procedure is*

$$F_i(U(X)) = U_i(X)^b$$

*where b is a constant and U, P, and W are defined as in proportional selection.*

**Observation 3.10** *The power law selection procedure is scale invariant.*

*Proof:*

$$P_i(F(kU(X))) = \frac{F_i(kU(X))}{\langle F(kU(X))\rangle}$$

$$= \frac{(kU_i(X))^b}{\langle (kU(X))^b\rangle} = \frac{k^b (U_i(X))^b}{k^b \langle (U(X))^b\rangle}$$

$$= \frac{F_i(U(X))}{\langle F(U(X))\rangle} = P_i(F(U(X)))$$

**Observation 3.11** *The power law selection procedure is not translation invariant.*

*Proof:*

$$P_i(F(kU(X))) = P_i(F(U(X) \oplus \vec{C})) \Leftrightarrow$$

$$\frac{F_i(U(X) \oplus \vec{C})}{\left\langle F(U(X) \oplus \vec{C}\right\rangle} = \frac{F_i(U(X))}{\langle F(U(X))\rangle} \Leftrightarrow$$

$$\frac{(U_i(X) + \vec{C}_i)^b}{\left\langle (U(X) \oplus \vec{C})^b\right\rangle} = \frac{(U_i(X))^b}{\langle (U(X))^b\rangle} \Leftrightarrow$$

$$\frac{\langle (U(X))^b\rangle}{\left\langle (U(X) \oplus \vec{C})^b\right\rangle} = \frac{(U_i(X))^b}{(U_i(X) + \vec{C})^b}$$

The last equation is false.

5.4.2.4 Sigma Truncation Selection

**Definition 3.12** *The sigma truncation selection procedure is:*

$$F_i(X) = g(U_i(X) - (< U(X) > -c\sigma))$$

*where*

$$g(x) = \begin{cases} x \text{ if } x > 0 \\ 0 \text{ otherwise} \end{cases}$$

*$\sigma$ is the standard deviation of U(X) in a particular generation, c is a small constant, and U, P, and W are defined as in proportional selection.*

**Observation 3.13** *The sigma truncation selection procedure is translation invariant.*

*Proof:* Note that the standard deviation of $U(X)$ is equal to that of $(U(X) \oplus \vec{C})$.

$$
\begin{aligned}
&P_i(F(U(X)\oplus\vec{C}))\\
&=\frac{g((U_i(X)+\vec{C}_i)-(<U(X)\oplus\vec{C}>-c\sigma))}{\left\langle g((U(X)+\vec{C})-(<U(X)\oplus\vec{C}>-c\sigma))\right\rangle}\\
&=\frac{g(U_i(X)-(<U(X)>-c\sigma))}{\langle g(U(X)-(<U(X)>-c\sigma))\rangle}\\
&=P_i(F(U(X)))
\end{aligned}
$$

The proof of the translation invariance of sigma truncation is very different from the proof of the translation invariance of the Boltzmann fitness function. In the Boltzmann proof, the structure of P is exploited to cancel the constant. In the sigma truncation proof, the structure of P is irrelevant because $F_i(U(X)) = F_i(U(X)\oplus\vec{C})$. We formalize this notion in the following definition.

**Definition 3.14** *A selection procedure is strongly translation invariant exactly when:*

$$F_i(U(X)) = F_i(U(X)\oplus\vec{C})$$

Intuitively, a selection procedure is strongly translation invariant when the invariance is independent of the particular choice of P. It is easy to show that strong translation invariance implies translation invariance with respect to P. Strong scale invariance can be analogously defined.

### 5.4.3 Rank Selection and Tournament Selection

Rank selection [3] is both scale and translation invariant because the relative position of an individual in a sorted list of raw fitness is not affected by a translation in the raw fitness or a scaling change in the raw fitness. Similarly, the outcome of a head to head competition between the raw fitnesses of two individuals is not changed by a scaling change or a translational change, so tournament selection [4] is both scale and translation invariant.

Since tournament selection and rank selection are interested only in qualitative comparisons of fitness, rather than quantitative numerical values, they are insensitive to many other transformations in the fitness function. For example, cubing the objective function will not change tournament and rank selection.

### 5.4.4 Understanding the Relationship between Proportional and Boltzmann Selection

The relationships among selection procedures have been studied empirically by many researchers (e.g., [13,1,17] and others). Typically, two identical genetic algorithms, differing only in the type of selection procedure they employ, are tested on a variety of test functions. Here, we are interested in analytically exploring the relationship between proportional and Boltzmann selection.

We propose a general technique for analytically comparing two selection procedures. All of the selection procedures that we have studied, with the exception of proportional fitness, have problem dependent parameters. The relationship between two selection procedures can be studied by explaining how to set these parameters so that one selection procedure acts like the other. In this particular case, the Boltzmann selection procedure has an extra parameter, T. We are interested in understanding how to set the parameter T so that Boltzmann selection is most like proportional selection. By "most like" we mean the setting of T such that the difference between W(P(F(U(X)))) and W'(P'(F'(U'(X)))) is minimized, where (W, P, F, U) and (W', P', F', U') define the two selection procedures. The following theorem explains how to set the parameter T in order to achieve this goal:

**Theorem 3.15** *If U(X) has a normal distribution at a particular generation, then Boltzmann selection is most like proportional selection when*

$$\beta e^{\sigma^2\beta^2} = \frac{1}{\mu}$$

where $\beta = 1/T$, p is < U(X) >, and $\sigma^2$ is the variance of U(X).

The proof of this theorem can be obtained from the author whose address is supplied at the beginning of this chapter. Notice that since $\beta e^{\sigma^2\beta^2}$ is a strictly increasing function of $\beta$, the exact value of $\beta$ can be found by using a simple binary search strategy (Cormen, Leiserson et al. 1990). If $\mu$ increases as a function of time (there is considerable empirical evidence that it does; see, e.g., (Goldberg 1989)) and $\sigma^2$ does not decrease rapidly as a function of time, then $\beta$ will decrease as a function of time. This corresponds to an increase in the temperature parameter, T. If T is interpreted as controlling selective pressure (when T is high, selective pressure is low and vice versa), then the proportional selection procedure decreases selective pressure with time. In the field of simulated annealing, T decreases with time and therefore selective pressure increases with time. This may partially explain why Boltzmann selection out performs proportional selection on some problems (de la Maza and Tidor 1992).

### 5.5 Discussion and Related Work

We have implemented Boltzmann scaling on the optimization function to select the number of offspring each individual in the current population contributes to the next generation; the procedure outperforms a standard proportional scaling method on the small set of problems we have investigated. A broader range of problems should be used to test the generality of this result. The tolerance schedule is robust enough that the same schedule was used successfully for problems of different sizes and correspondingly different scales in optimization space. These results show that, for the molecular biology problem, many Boltzmann experiments completed with a correct solution before the decrease in tolerance that occurred after generation ten and nearly all completed before the schedule leveled off again after generation forty.

One possibility that we have not investigated, but which is used in biological systems, is to vary population size. In high tolerance periods the size of the population could be allowed to increase, and in low tolerance periods it could be forced to decrease. The advantage of such an approach is that more low fitness individuals could be retained for use in crossover during critical stages of the optimization, though it is not clear whether the benefits of this outweigh the computational overhead.

A refinement of our method that we have considered is to eliminate all duplicates in the population before applying Boltzmann selection and adjusting the selection to restore the fixed population size, as would be required by a strict interpretation of the Boltzmann equation. The current distribution of fitness after selection is biased somewhat more toward fit individuals than the refined method would be, but we expect that any benefit would be small relative to the cost of finding and eliminating duplicates. Moreover, biological systems, particularly those with larger genomes, have no such mechanism. Rather, they use a suite of genetic operators that tend to keep exact duplicates as a low probability event.

Whitley [23] reports using an exponential selection protocol for a genetic algorithm and found that this increased problems of premature convergence. This contradicts our results and suggests that the use of a reasonable evolutionary tolerance schedule is important. It should be noted that the evolutionary tolerance corresponds roughly to the acceptable range of scores, in optimization units, between the best and worst individuals kept after selection; thus, it is expected to vary with the scale of optimization space and the use of trial runs to choose useful parameters is valuable.

Goldberg [11] describes a Boltzmann tournament scheme in which the population of individuals converges to a Boltzmann distribution. The method was developed so that genetic algorithms could benefit from the asymptotic convergence properties enjoyed by simulated annealing and so that simulated annealing procedures might be efficiently implemented on parallel machine architectures. The algorithm includes a non-genetic "anti-acceptance" step that effectively converts between Boltzmann and uniform distributions. Our goal here is to achieve faster convergence to the global optimum rather than to a specific distribution. We use Boltzmann scaling to control the approach to this optimum by varying selective pressure through the tolerance (or its physical analogue, temperature). Indeed, this is found to improve convergence over proportional scaling on at least this set of problems. Moreover, proportional scaling appears to increase, rather than decrease, effective tolerance during the course of an optimization.

Back and Hoffmeister [1] study the performance of a genetic algorithm as a function of selective pressure on the less fit individuals in the population (referred to as "extinctiveness"). For a unimodal objective function, they find optimum performance with strong selective pressure, which produces relatively little genetic diversity and a gradient-directed search. In contrast, for a multimodal objective function, they find optimum performance with weaker selective pressure, which produces more genetic diversity and exploration of the search space.

The parameters of their genetic algorithm were not modified during the course of a run and they comment that, without knowing the character of the objective function, it is difficult to choose the proper search strategy. One approach to solving this problem is to use a hybrid strategy that is initially more explorative and then becomes more directed as the run proceeds, as is generally done in simulated annealing [17].

Whitley [24] gives an informal argument that explains why proportional selection decreases selective pressure with time and proposes the use of rank-based selection, because it does not rely on the relative arbitrariness of the objective function to define the selective pressure. We have illustrated the use of a transformation from objective function to fitness and identified invariance properties of this key transformation for a number of selection procedures. We have pointed out that Boltzmann selection is invariant to translations and that a simple parameter, $T$, can be used to control scaling. Moreover, this parameter can be used to vary selective pressure during a run to switch from a more explorative to a more directed search strategy. Theorem 3.15 proves that proportional selection decreases selective pressure with time. In simulated annealing and in Boltzmann selection, selective pressure increases with time.

Baker [2] studies selection algorithms with the goal of overcoming the premature convergence problem. Grefenstette and Baker [14] examine how selection procedures interact with implicit parallelism and argue that the usual application of the $k$-armed bandit problem to genetic algorithms may be flawed, a view which is disputed by Goldberg and Deb [12]. Grefenstette [13] extends the results of Baker and Grefenstette and strongly argues that because the genetic algorithm has access to a biased sample, instead of an unbiased sample, of points, the standard understanding of implicit parallelism needs to be reexamined. Goldberg and Deb [12] hint that proportional selection may not maintain selective pressure as the point of convergence nears, a suggestion that is supported by Theorem 3.15.

## 5.6 Conclusion

This chapter has illustrated the implementation of a procedure for genetic selection based on Boltzmann scaling of the optimization function and empirically demonstrated that it leads to convergence to the correct solution in fewer generations than traditional proportional scaling on a small set of problems. Furthermore, it was proved that proportional scaling, contrary to intuition and annealing methods, actually increases evolutionary tolerance during the experiment.

Translation and scale invariance are powerful properties to examine for selection procedures. Intuitively, it may be desirable for an optimization procedure to solve a problem equally well whether it is expressed in feet or meters and in the Gregorian or the Chinese calendar. If the procedure itself is not translation and scale invariant, parameters could be available that can be adjusted for each problem. Presumably, setting these parameters properly will result in similar solutions in similar times for different translations and scalings of the same problem.

**References**

[1] Back, T. and F. Hoffmeister (1991). Extended selection mechanisms in genetic algorithms. *Fourth International Conference on Genetic Algorithms.* 92-99.

[2] Baker, J.E. (1989). *An Analysis of the Effects of Selection in Genetic Algorithms*. Vanderbilt University, Nashville. Ph.D., Thesis.

[3] Baker, J. E. (1985). Adaptive selection methods for genetic algorithms. *International Conference on Genetic Algorithms and Their Applications.* 101-111.

[4] Brindle, A. (1981). Genetic algorithms for function optimization. University of Alberta, Canada. Technical, 81-2.

[5] Cormen, T. H., C. E. Leiserson, et al. (1990). *Introduction to Algorithms.* Cambridge, MA, MIT Press.

[6] Darwin, C., Ed. (1951). *The Origin of Species by Means of Natural Selection; or, The Preservation of Favoured Races in the Struggle for Life.* London, Oxford University Press.

[7] Deb, K. and D. E. Goldberg (1989). An investigation of niche and species formation in genetic function optimization. *Third International Conference on Genetic Algorithms.* 42-50.

[8] Forrest, S. (1985). Documentation for PRISONERS DILEMMA and NORMS programs that use the genetic algorithm. Unpublished manuscript.

[9] Gillies, A. M. (1985). *Machine learning procedures for generating image domain feature detectors.* University of Michigan, Ann Arbor.

[10] Goldberg, D. (1989). Genetic Algorithms in Search, Optimization and Machine Learning. Reading, MA, Addison-Wesley.

[11] Goldberg, D. (1990). "A note on Boltzmann tournament selection for genetic algorithms and population-oriented simulated annealing." *Complex Systems* **4**(4): 445-460.

[12] Goldberg, D. and K. Deb (1991). "A comparative analysis of selection schemes used in genetic algorithms." *Foundations of Genetic Algorithms* : 69-93.

[13] Grefenstette, J. (1991). "Conditions for implicit parallelism." *Foundations of Genetic Algorithms*. 252-261.

[14] Grefenstette, J. and J. Baker (1989). How genetic algorithms work: A critical look at implicit parallelism. *Third International Conference on Genetic Algorithms*. 20-27.

[15] Holland, J. D. (1975). *Adaptation in Natural and Artificial Systems*. Ann Arbor, MI, University of Michigan Press.

[16] Kirkpatrick, S., J. C.D. Gelatt, et al. (1983). "Optimization by simulated annealing." *Science* **220**: 671-680.

[17] de la Maza, M. and B. Tidor (1992). Increased flexibility in genetic algorithms: The use of variable Boltzmann selective pressure to control propagation. *Research: New ORCA CSTS Conference: Computer Science and Operations Developments in Their Interfaces*. 425-440.

[18] de la Maza, M. and B. Tidor (1993). An analysis of selection procedures with particular attention paid to proportional and Boltzmann selection. *Fifth International Conference on Genetic Algorithms*. San Mateo, Morgan Kaufmann.

[19] Metropolis, N., A. W. Rosenbluth, et al. (1953). "Equation of state calculations by fast computing machines." *Journal of Chemical Physics*. **21**: 1087-1092.

[20] Michalewicz, Z. (1992). *Genetic Algorithms + Data Structures = Evolution Programs*. Berlin, Springer-Verlag.

[21] Press, W. H., S. A. Teukolsky, et al. (1992). *Numerical Recipes in C: The Art of Scientific Computation*. Cambridge University Press.

[22] Verlet, L. (1987). "Computer "experiments" on classical fluids. I. Thermodynamical properties of Lennard-Jones molecules." *Physical Review* **159**: 98-103.

[23] Whitley, D. (1987). Using reproductive evaluation to improve genetic search and heuristic discovery. *Second International Conference on Genetic Algorithms*. 108-115.

[24] Whitley, D. (1989). The GENITOR algorithm and selection pressure: Why rank-based allocation of reproductive trials is best. *Third International Conference on Genetic Algorithms*. 116-121.

## Chapter 6

**Shumeet Baluja**
School of Computer Science
Carnegie Mellon University
Pittsburgh
Pennsylvania 15213-3890

baluja@cs.cmu.edu

# Structure and Performance of Fine-Grain Parallelism in Genetic Search



**Abstract**

Within the parallel genetic algorithm framework, there currently exists a growing dichotomy between coarse-grain and fine-grain parallel architectures. This chapter attempts to characterize the need for fine-grain parallelism, and to introduce and compare three models of fine-grain parallel genetic algorithms (GAs). The performance of the three models is examined on seventeen test problems and is compared to the performance of a coarse-grain parallel GA. Preliminary results indicate that the massive distribution of the fine-grain parallel GA and the modified population topology yield improvements in speed and in the number of evaluations required to find global optima.

0-8493-2529-3/95/$0.00 + $.50

## 6.1 Introduction

Since Holland's pioneering work [Holland, 1975], there have been many variations of the simple genetic algorithm. The development of genetic algorithms has been driven by the goal of maintaining the balance of diverse sampling and efficient focusing. With regards to parallelism, there have been two stages of development beyond the genetic algorithms (GAs) proposed by Holland. The first is the coarse-grain parallel genetic algorithm, in which several large populations are evolved in parallel with very little interaction. The second stage is the fine-grain parallel genetic algorithm, in which numerous small, constantly interacting populations are evolved in parallel.

### 6.1.1 The Motivation Behind Parallelism

Explained simply, a parallel genetic algorithm (pGA) divides a single large population into smaller subpopulations. Each of the subpopulations runs a separate genetic algorithm either independently or with limited interactions with other subpopulations. One motivation for this division is the potential increase in speed through the assignment of each processor, of a multi-processor system, to evolve a single population. However, a more interesting motivation stems from the observation that, after some period of evolution, the majority of the chromosomes in a single population will become very similar. Genetic diversity will be lost, and recombination thereafter may not be productive. One method of addressing this problem is to evolve subpopulations independently. As GAs are randomized algorithms, independent evolutions are likely to explore different portions of the search space. If the functions to be optimized for each subpopulation are the same, each subpopulation should reveal closely competitive, yet unique results.

The amount of interaction between subpopulations can be a critical factor in determining a pGA's effectiveness. Eliminating interaction between subpopulations effectively makes dividing a larger population similar to performing several GA runs with smaller populations. With too much interaction, the benefits of subpopulations are lost. Good chromosomes from one subpopulation quickly spread to other subpopulations, and the evolutions no longer remain independent. Cohoon et al. suggest that members of subpopulations be swapped after the subpopulations begin to reach equilibrium [Cohoon, 1988]. The results of Whitley and Starkweather, Tanese, and Grosso, also support limited interactions [Whitley and Starkweather, 1990] [Tanese, 1989] [Grosso, 1985].

Although some of the pGAs differ in many parameter settings, [Whitley and Starkweather, 1990] [Pettey, 1989] [Cohoon, 1988], several important factors remain consistent in the majority of them: they evolve a relatively small number of subpopulations, and each subpopulation contains a large number of chromosomes. The pGAs described to this point are referred to as coarse-grain parallel genetic algorithms (cgpGAs). One of the drawbacks of cgpGAs is that after a subpopulation converges to an equilibrium state, the introduction of new material may not be effective. The new material may not be incorporated because of its incompatibility with the existing information. A reason for incompatibility may be as simple as two subpopulations may evolve answers to opposite sides of a large hamming cliff, or in more general terms, that two subpopulations may find good solutions which, when combined, reveal a worse solution.

### 6.1.2 Massive Parallelism

Massive Parallelism in genetic algorithms has been used in at least two different contexts. In the first context, parallelism refers to the machine architecture on which the GA is run. Parallelism is employed to achieve a gain in speed, and to allow much larger population sizes to be evolved in reasonable amounts of time [Forrest and Perelson, 1990]. The second context, and the one explored throughout the remainder of this paper, is one in which numerous small, constantly interacting subpopulations are evolved in parallel, with localized mating rules. Another related application of massive parallelism can be found in [Hillis, 1990]. Hillis used a massively parallel architecture to co-evolve parasites with chromosomes.

Fine-grain parallel genetic algorithms (fgpGAs) addressed some of the problems found in cgpGAs. One way to conceptualize the modified form of parallelism is to view the populations as overlapping, with a portion of the constituents of one population also being constituents of one or more other subpopulations. In an analogous manner to biological natural selection, in which a population is typically composed of relatively independent subpopulations which interact, recombination occurs between two chromosomes from within localized neighborhoods [Davidor, 1991] [Spiessens and Manderick, 1991] [Muhlenbein, 1989] [Schleuter, 1990]. The constant interaction between subpopulations helps to alleviate the problems of recombining incompatible solutions. It is difficult for a subpopulation to exist in a state of equilibrium until all of its neighboring subpopulations reach equilibrium.

A potential drawback of the fgpGA architecture is that local optima can quickly spread through the entire population. Since there is constant swapping between subpopulations, the possibility of independent evolutions may be hindered. Further, because the size of subpopulations is small, the schemata represented in strong local optima can quickly dominate all of the genetic information in individual subpopulations. In practice, this problem is partially overcome by limiting the amount of swapping between subpopulations. Another factor which can reduce the detrimental effects of constant swapping is the large number of subpopulations. Subpopulations which are a large distance apart may evolve unique chromosomes in a manner similar to cgpGAs. This has been termed isolation by distance [Collins and Jefferson, 1991]. In the next section, three fgpGA structures are examined which vary with respect to how swapping between subpopulations is implemented. The effects of the speed of information flow, which is dependent upon the amount of interaction between subpopulations, will be discussed throughout the remainder of this paper.

## 6.2 Three Fine-Grain Parallel GA Topologies

Using the fine-grain parallel subpopulation structure, three topologies were examined. The first implementation uses a circularly linked linear ordering of subpopulations. Each subpopulation evolves only 2 chromosomes per generation. These 2 chromosomes are chosen from a group of 10 chromosomes. The group of 10 is comprised of 1 chromosome from each of the four immediate left and 1 chromosome from each of the four immediate right subpopulations, and the 2 chromosomes which were evolved in the subpopulation during the previous generation. See Figure 6.1. Each chromosome selected from the neighbors is chosen *randomly* from the two evolved at each neighbor. The fitness of every chromosome is calculated relative to the other chromosomes in the group of 10. Two chromosomes from the set of 10 are probabilistically chosen for recombination, based upon their relative fitness. The other 8 chromosomes are discarded. In the subsequent generation, the two "children" chromosomes produced (through crossover and mutation of the selected parents) are available for recombination, either by the subpopulation on which they are located, or by its neighbors.

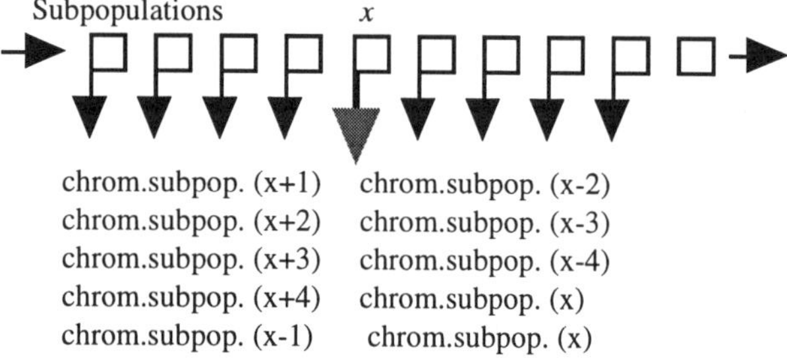


chrom.subpop. (x+1) chrom.subpop. (x-2)
chrom.subpop. (x+2) chrom.subpop. (x-3)
chrom.subpop. (x+3) chrom.subpop. (x-4)
chrom.subpop. (x+4) chrom.subpop. (x)
chrom.subpop. (x-1) chrom.subpop. (x)

Figure 6.1: The architecture of subpopulations, arrangement #1. Subpopulation *x* shown enlarged. The subpopulations form a circular list.

The second implementation uses a two dimensional toroidal array of subpopulations. As in the previous implementation, each of the subpopulations evolve 2 chromosomes which are chosen from a group of 10. The 8 immediate neighboring subpopulations donate to the group of 10. See Figure 6.2. The remainder of the procedure is completed in the same manner as described above.

In the third implementation, a linear ordering of subpopulations is used once again. One chromosome from each of the 3 immediate left and one chromosome from the 4, 5, 6 subpopulations from the right contribute to the group of 10 chromosomes. See Figures 6.3 and 6.4. Figure 6.4 gives a pictorial example of how the genetic information flows through the series of GAs. Unlike the first implementation, the immediate right subpopulations are not used. Since there are 4 remaining positions in the group of 10, they are filled with 2 copies of each of the chromosomes evolved in the previous generation. This structure includes only 6 neighbours to examine the effects of reducing the spread rate of chromosomes and introducing a bias to the chromosomes evolved within each subpopulation. After the selection of 10 chromosomes, the fgpGA proceeds in the same manner as described earlier. This model was chosen because it allows a faster spread of chromosomes than the first implementation, and a slower rate than the second implementation.

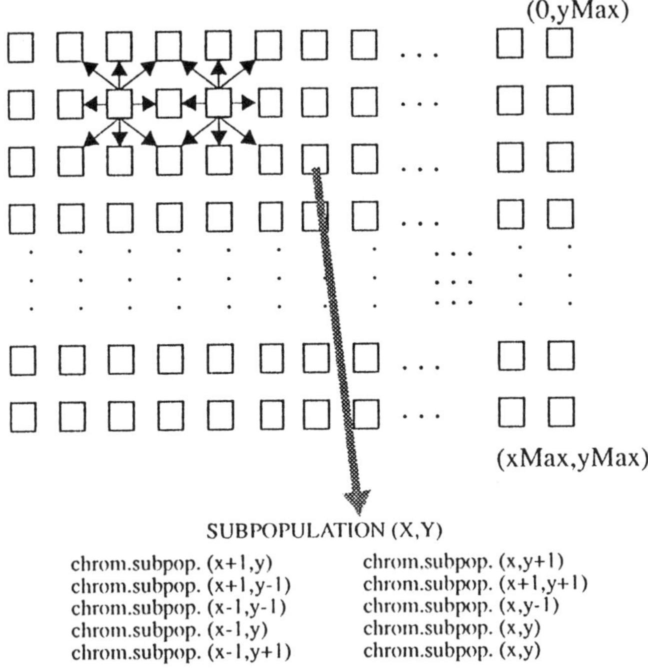


Figure 6.2: Each subpopulation contributes one of its two chromosomes to each of its 8 nearest neighbors. The composition of subpopulation(x,y) is shown. Both of the chromosomes evolved at subpopulation(x,y) are included. The subpopulations form a toroid.

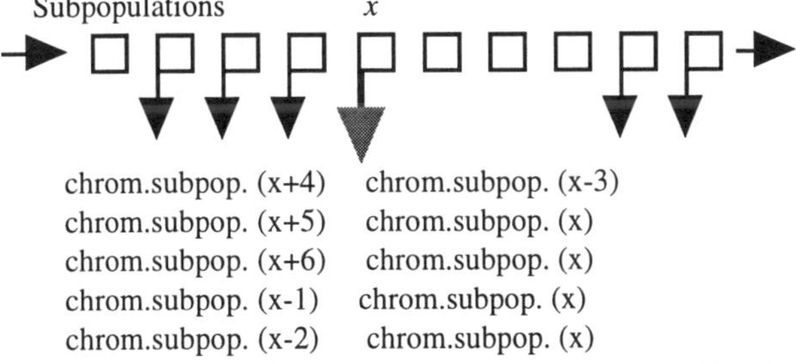


Figure 6.3: The architecture of subpopulations arrangement #3, subpopulation *x* shown enlarged. The subpopulations form a circular list.

Making the large assumption that the best chromosomes are not lost during crossover or mutation, the maximum spread rate of a good chromosome varies per implementation. In arrangement #1, the linear ordering, assuming 4096 subpopulations, the lower bound on the number of generations for all of the subpopulations to see the best chromosome is 512-1, or 511 generations. In arrangement #2, the toroid, the minimum number of generations to get from any subpopulation to the furthest away, is half the diagonal of the square. Assuming a 64 * 64 toroid, within approximately 31 (32-1) generations, the chromosome could be in all of the subpopulations. Using arrangement #3, a compromise between the first two with regard to speed, the number of generations is approximately 455. All of the -1 factors arise since in the first generation a good chromosome is found, it can be included in its neighbor's selection of 10 chromosomes. These estimates are only the lower bound of the spread of the best chromosome. The chromosomes will not generally spread this fast. For these speeds to be achieved, the chromosome must be reselected for recombination at each generation, and none of the valuable schemata can be destroyed by crossover and mutation. Further, this also assumes that no better chromosomes are found before full spreading. In the experiments performed, the chromosome in the neighboring subpopulations was selected randomly from the two which were evolved in the neighbor. However, always selecting the best, or using a probabilistic scheme of selection may also work well. Another topology, termed the "ladder" population structure, has been explored by Muhlenbein and Schleuter [Muhlenbein, 1989] [Schleuter, 1990].

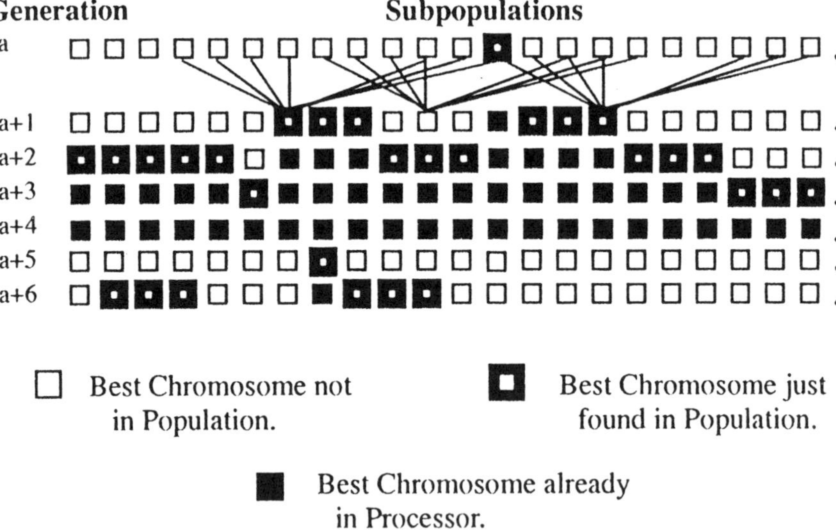


Figure 6.4: Making the large assumption that the best chromosome is not lost during crossover or mutation, the above diagram depicts how the best chromosome could spread through the populations, using fgpGA configuration #3. In generation a+5, a new best chromosome is found. The connections shown are the subpopulations from which a chromosome is included. These connections do not change through the run of the GA. Each subpopulation has similar connections to those shown.

## 6.3 Performance of fgpGAs and cgpGAs

### 6.3.1 Description of Algorithms Compared

Four algorithms were tested, the three implementations of fgpGAs described in the previous section, each with 4096 subpopulations, and a 40 subpopulation cgpGA described in this section [Baluja, 1993]. All used a constant 1% mutation rate and two point crossover. The evolution was generational, and crossover took place with each set of parents. An alternative to the generational model, the steady state model, is explored in [Whitley and Starkweather, 1990]. All algorithms also used a modest form of elitist selection, in which the single best chromosome in generation *a* replaced the worst chromosome in generation $a+1$. Elitist selection was performed within each subpopulation. In the fgpGAs, the best chromosome was selected from each group of 10 chromosomes, and replaced the worst chromosome from the group of 10 in the next generation. Elitist selection does not ensure that a particular chromosome will be selected for recombination, only that it will be a candidate for selection.

The cgpGA was very loosely based upon the cgpGA described in [Whitley and Starkweather, 1990]. Forty subpopulations were evolved. Each subpopulation contained 100 chromosomes, for a total of 4000 chromosomes evaluated simultaneously. Assuming a circular ordering of subpopulations, after every 100 generations, the best chromosome from each subpopulation migrated to a subpopulation *e* subpopulations away, where *e* is defined to be the number of generations divided by 100 that have passed. Because the population was a set size, the migrating chromosome replaced the worst chromosome in the target subpopulation.

In an attempt to efficiently map these algorithms onto the hardware architecture on which these tests were attempted, the MasPar MP-I, the fgpGA evaluated 8192 chromosomes per generation, while the cgpGA evaluated only 4000.

### 6.3.2 The Problems Attempted

*DeJong's Test Suite:* This test suite is comprised of five minimization problems commonly used to test the effectiveness of GAs [DeJong, 1975]. The functions were encoded using standard binary code.

*Subset Sum:* The problem can be stated as follows: given S elements, each of a possibly unique weight, is there a subset of S that adds up exactly to an arbitrary number, T? This problem was implemented as a 120-bit chromosome. Each bit represented a unique object, assigned a random integer weight between 1 and 200. The weight T was selected to be either 1/4, 1/20, or 1/40 of the sum of the weights of the objects. The object was to find the group of sets whose weights add exactly to T. The sum was guaranteed to be divisible by 4, 20 and 40, respectively; note that this does not guarantee a set of weight T.

*All Ones:* Three versions of the all-ones problem were attempted [Syswerda, 1989]. The first was the straight all-ones problem. The objective is to find the chromosome which contains a 1 in each bit position. The second version contains bits which are meaningless. This problem was encoded as a 180 bit problem, but only the first 120 bits were counted toward the evaluation. The optimal solution to this problem is 120. The third is the contiguous bits problem. Points are only given for 1's which have at least one other neighbor which also has a value of 1.

*Fully and Partially Deceptive Order 4:* The fully deceptive problem is a 40 bit maximization problem. The problem was defined in Whitley and Starkweather's paper GENITOR II [Whitley and Starkweather, 1990]. The problem is comprised of 10 subproblems, each 4 bits long. The subproblems use the lookup table shown in Table 6.1. The partially deceptive problem uses the same evaluations, with the exceptions of the reversed evaluations for 1111 and 0101.

| Chrom | Evl | Chrom | Eval |
|---|---|---|---|
| **1111** | 30 | **0110** | 14 |
| **0000** | 28 | **1001** | 12 |
| **0001** | 26 | **1010** | 10 |
| **0010** | 24 | **1100** | 08 |
| **0100** | 22 | **1110** | 06 |
| **1000** | 20 | **1101** | 04 |
| **0011** | 18 | **1011** | 02 |
| **0101** | 16 | **0111** | 00 |

Table 6.1: Order 4 fully deceptive problem.

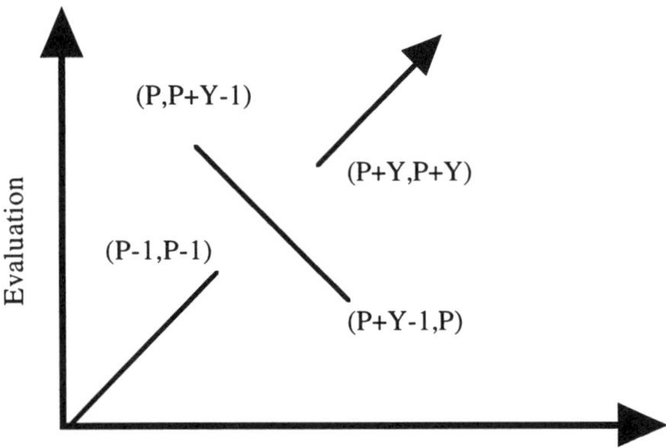


Figure 6.5: The gap problem. o(x) is the number of ones per chromosome. The gap size is Y. The starting point of the gap is P.

Both the fully deceptive and partially deceptive problems were attempted using two orderings of bits. The first encoding is block encoding: the placement of the 4 bits which comprise a subproblem are located next to each other. For example, the first subproblem has bits in position 1,2,3,4. The second encoding is interleaved: the bits in each subproblem are uniformly spread throughout the chromosome. For example, the first subproblem has bits in positions 1, 11, 21, 31. Using two point crossover, the first encoding is easier for the GA to solve than the second.

*The Gap Problem:* This is a maximization problem [Liepins and Baluja, 1991] shown in Figure 6.5. The gap function f(x), with gap of size Y, starting at point P, with o(x) being the number of ones in the bit string, is defined by:

$$f(x) = \begin{cases} 2P + Y - o(x) - 1, if(P \le o(x) \le P + Y - 1) \\ o(x), if((o(x) < P) \vee (o(x) > P + Y - 1)) \end{cases}$$

This problem was tested on a 120 bit chromosome string. Gap sizes of 20 and 25 were tried with the starting gap point, P=60.

### 6.3.3 Results and Discussion

Table 6.2 shows the results of the 17 test problems; they are the average of 10 runs per problem for each algorithm. One of the difficulties inherent in comparing parallel genetic algorithms with each other, and with traditional GAs, is choosing the best criteria [Baluja, 1993]. Criteria which measure performance of the GA by the fitness of the best individual through the run of the algorithm are biased in favor of larger parallel GAs. If the number of evaluations performed is chosen as the criterion, parallel GAs often do not perform well, as parallel GAs may perform a lot of repetitive search. However, the quality of solutions evolved by pGAs have been shown empirically to be better than single population GAs in a variety of problems [Petty, 1989] [Tanese, 1989]. The measure used in this study is the number of generations to find the optimal solution and the number of evaluations per generation. However, using the optimal solution as a stopping criterion raises another issue: GAs find regions of good performance very quickly; the majority of the time is spent locating relatively small improvements in search of the optimal solution. For example, when the evaluation curves of DeJong f4 are examined, it is clear that the vast majority of the time between generations 200-700 is spent making very small improvements, see Figure 6.6. As stated by Forrest and Mitchell "it could be argued that the GA is more suited to finding good solutions quickly rather than finding the absolute best" [Forrest and Mitchell, 1993]. The results in this study certainly agree with this.

The ability of good chromosomes to spread rapidly through the population contributed to the success of the fgpGAs. A sample run, shown in Figure 6.7, displays the number of subpopulations that contain chromosomes which have evaluations equal to the best chromosome in the entire population. These chromosomes are candidates for selection in their respective groups of 10. This does not imply that all the chromosomes are exactly the same, nor does it imply that they will be chosen for recombination. The sudden drops of the number of populations, in Figure 6.7, represent generations in which a better chromosome was found. The actual spread rate does not match the fastest possible spread rates mentioned in Section 6.2. Although a good chromosome can be immediately accessed by its neighbors as soon as it is found, for more than the immediate neighbors to incorporate the chromosome, it must again be selected for recombination. Further, if it is selected, valuable schemata must not be destroyed by crossover or mutation operators. Although the populations which surround the immediate neighbors will incorporate the children chromosomes into their population, for them to spread the chromosome further, they must also select the children chromosomes for recombination. However, the evaluation of the children chromosomes may not be as good as the original chromosome. Further, if the crossover and mutation operations have destroyed valuable schemata, the children produced may not be preserved by elitist selection.

| Test Function | ***fgpGA*** Linear Order | ***fgpGA*** 64*64 Array | ***fgpGA*** Linear Skip | ***cgpGA*** 40 sb pop. |
|---|---|---|---|---|
| DeJong Function #1 | 32.0 | 29.8 | 30.6 | 79.0 |
| DeJong Function #2 | 40.0 | 38.6 | 43.9 | 111.8 |
| DeJong Function #3* | 22.0 | 19.5 | 20.4 | 64.5 |
| DeJong Function #4 | See Figure 6.6 | | | |
| DeJong Function #5** | 17.9 | 18.0 | 17.8 | 18.0 |
| Subset Sum (1/4) | 21.0 | 14.0 | 12.0 | 35.6 |
| Subset Sum (1/20) | 68.0 | 55.0 | 65.0 | 344.5 |
| Subset Sum (1/40) | 95.4 | 76.8 | 87.8 | 629.0 |
| All-Ones | 114.0 | 90.5 | 107.7 | 648.2 |
| Sparse All-Ones*** | 134.0 | 94.8 | 113.3 | 342.0 |
| Contiguous All-Ones | 131.0 | 90.8 | 111.0 | 609.1 |
| Fully Deceptive (A) | 90.0 | 57.5 | 78.4 | 305.9 |
| Fully Deceptive (B) | 1220 (4) | 742.5 | 942.2 (5) | 1634.7 |
| Partially Deceptive (A) | 39.0 | 32.0 | 38.0 | 95.1 |
| Partially Deceptive (B) | 70.0 | 53.0 | 75.0 | 252.5 |
| Gap Problem (Size 20) | 161.2 | 126.3 | 164.3 | 675.9 |
| Gap Problem (Size 25) | 816.4 (5) | 441.2 | 699.1 (9) | 776.0 |

Table 6.2: Results for the 17 test problems. Each entry represents the average number of generations to find the optimal solution. The fgpGA evaluated 8192 chromosomes per generation, while the cgpGA evaluated 4000. A number in parentheses indicates that the optimal solution was only found the specified number of times, out of 10. The fgpGA was allowed 1400 generations, the cgpGA was allowed 3000.

* The stopping criterion for DeJong's F3 was an evaluation of -30.
** The stopping criterion for DeJong's F5 was an evaluation of 0.998004.
*** Due to memory restrictions, this problem was attempted with 90 significant bits, and 30 extra bits (cgpGA only). The fgpGA runs were full size (120 significant bits, 60 extra bits).
Portions of this table appear in [Baluja, 1993].

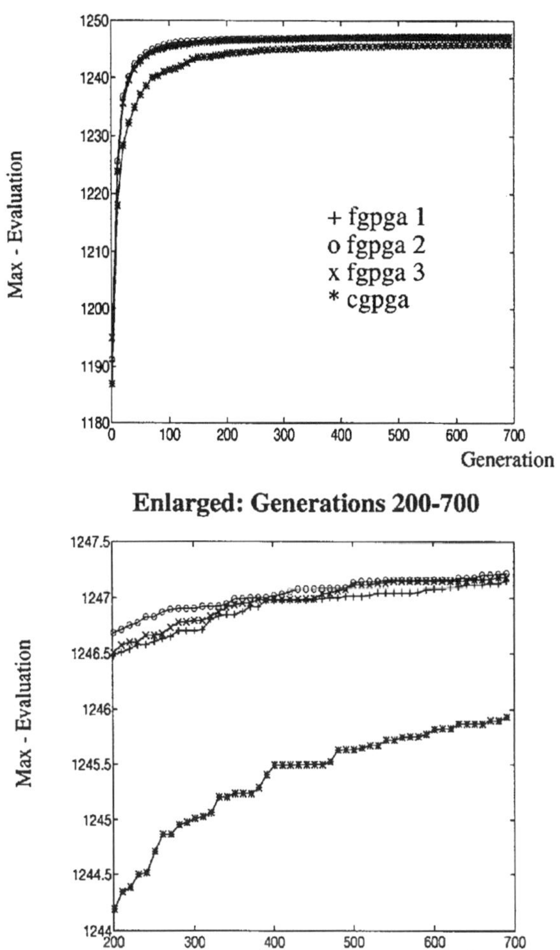


Figure 6.6: Average evaluations for 10 runs of the cgpGA and the fgpGAs on DeJong's F4, including the random Gaussian factor. The cgpGA was run with 50 chromosomes per subpopulation and 80 subpopulations. In the last 500 generations, very little improvement was made. [Baluja, 1993].

The different success rates of the fgpGAs on the Deceptive - Order 4 problem and the Gap(25) problem for the three fgpGAs illustrate the significant role subpopulation interaction has in performing successful search. It is interesting to note that in both of these problems, the cgpGA and the fgpGA (implementation 2) did the best; the other two implementations of the fgpGA did poorly. A possible explanation is that the structure of these problems benefits from larger population sizes. Since the fgpGA-2 has the fastest spread rate, it simulates a larger population more closely than the other implementations. One of the immediate plans for future research is to examine the performance on these two problems in greater detail.

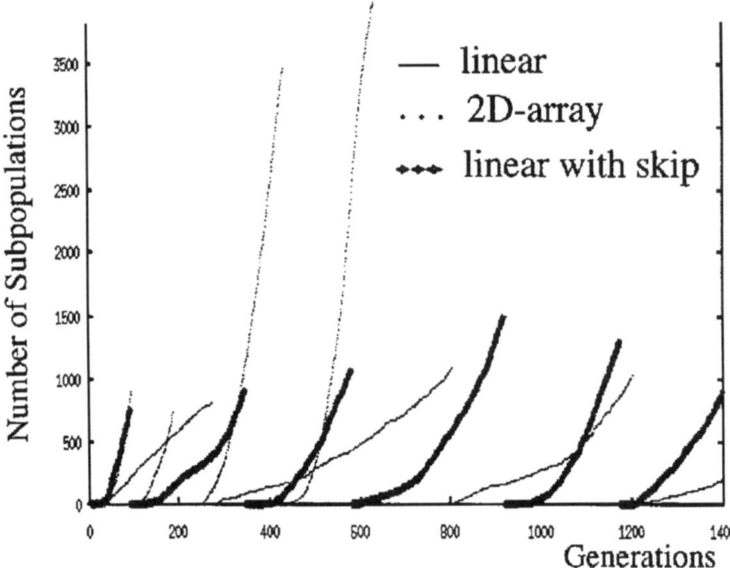


Figure 6.7: The number of populations which contain the best chromosome using the fgpGAs to optimize the order 4 fully deceptive problem, interleaved. The sudden drops in the number of subpopulations represent a new best solution found in one of the subpopulations. There are a total of 4096 subpopulations. The 2D array architecture found the optimal at approximately generation 600. The others did not find the optimal in 1400 generations.

The parameters in the cgpGA and fgpGA were not tuned per problem. It is suspected that with a little tuning, both types of GAs could significantly improve performance. However, to measure the ability of the algorithms to perform on a variety of problems without parameter tuning, the parameters were held constant throughout all of the test runs.

## 6.4 Future Directions

To evaluate fine-grain parallelism in more detail, both harder problems and different population topologies should be explored.

### 6.4.1 Test Problems

The test problems attempted in this study comprise a fairly standard test suite of problems which aid in quantifying the effectiveness of GA models. However, many of these problems were designed to test the abilities of single population GAs, and do not reveal the potential of parallel GAs. For future testing of the fgpGA topologies, both harder problems ahd multi-objective problems should be attempted.

One of the harder problems tested should be the Traveling Salesperson Problem. This would help quantify the differences in performance of this system and the systems developed by Muhlenbein and Schleuter [Muhlenbein, 1989] [Schleuter, 1990]. They have extensively explored the TSP problem with parallel GA systems and have achieved very promising results. The functions termed *Tanese Functions* by S. Forrest and M. Mitchell [Tanese, 1989][Forrest and Mitchell, 1993] should also be attempted. These functions have proven to be very hard to optimize genetically, but are susceptible to hill climbing techniques.

Parallel GAs lend themselves to multi-objective optimization problems. The evaluation criterion of each population can reflect different objectives. When members of separate subpopulations are mixed, the children produced may be strong with respect to more than a single objective. Multi-objective problems have been explored in variants of cgpGAs by [Husbands, 1991] and [Cohoon, 1988]. Fine-grain parallel GAs also offer the ability to perform multi-objective optimization. It will be very interesting to see how the placement of objectives in subpopulations affects the abilities of the GAs. For example, all of the subpopulations with one objective could be placed close to each other, so that 'inner' subpopulations are surrounded only by others which have the same objective. Alternatively, the objectives could be assigned to the subpopulations in an interleaved manner. The formation and assimilation of niches will certainly play an integral role in the abilities of the GA to successfully optimize each of the objectives. Niche formation has been studied in massively parallel architectures by [Davidor, 1991].

### 6.4.2 Subpopulation Interaction

The massive distribution of the fgpGAs allows flexibility in the design of the interactions between populations. Three important issues which need to be resolved are: with which other subpopulations each subpopulation should interact, what the interaction should be, and how often the interactions should occur. For the problems which were tested, the 2D array topology worked well. This topology allowed for a rapid flow of genetic information, which is desirable in easy problems as good solutions can rapidly propagate. However, for harder problems, fast flow may not be a desirable property. A slower flow may prove its worth in the cases in which independent evolutions are needed to successfully optimize the function. Experimenting with time-varying and adaptive flows might also achieve impressive results; however, this may add another level of complexity to fgpGA design.

The three fgpGAs presented vary with respect to with which subpopulations may interact. The frequency and type of interactions (simply selecting one at random from the two evolved at the neighbor) have remained constant. However, the configuration used may be far from optimal. Fine-grain parallel genetic algorithms, and parallel genetic algorithms in general, still encompass a level of complexity which is not fully understood. The empirical results here are presented with the hope that they may help form insights into more rigorous models of the interactions in parallel GAs.

**Acknowledgments**

I would like to thank Dean Pomerleau, Stephen Smith, Todd Jochem, and Chuck Thorpe for their many helpful comments and suggestions throughout the development of this paper. This paper is dedicated to the memory of Dr. Gunar Liepins.

This research was partly sponsored by Defense Advanced Research Projects Agency, under contracts "Perception for Outdoor Navigation" (contract number DACA76-89-C0014, monitored by the U.S. Army Topographic Engineering Center) and "Unmanned Ground Vehicle System" (contract number DAAEO7-90-C-R059, monitored by TACOM). It was also partially sponsored by the National Science Foundation, under NSF Contract BCS-9120655, titled "Annotated Maps for Autonomous Underwater Vehicles", and the NSF grant titled "Massively Parallel RealTime Computer Vision". The views and conclusions contained in this document are those of the author and should not be interpreted as representing the official policies, either expressed or implied, of the Defense Advanced Research Projects Agency, the National Science Foundation, or the U.S. Government.

**References**

Baluja, S. (1993) The Evolution of Genetic Algorithms: Towards Massive Parallelism. To Appear in P.E. Utgoff, ed., *Machine Learning: Proceedings of the Tenth International Conference.* Morgan Kaufmann Publishers, San Mateo, CA.

Baluja, S. (1992) A Massively Distributed Parallel Genetic Algorithm. CMU-CS-92-196R. School of Computer Science, Carnegie Mellon University.

Caruana, R. and J. Schaffer (1988) Representation and Hidden Bias: Gray Vs. Binary Coding for Genetic Algorithms. *Proceedings of the 5th International Conference on Machine Learning.* Morgan Kaufmann. Los Altos. CA. June 1988 152-161.

Cobb, H. (1990) An Investigation Into the Use of Hypermutation as an Adaptive Operator in Genetic Algorithms Having continuous, Time Dependent Nonstationary Environments. NCARAI Library. AlC-90-00 l.

Cohoon, J.P., S.U. Hedge, W.N. Martin and D. Richards (1988), Distributed Genetic Algorithms for the Floor Plan Design Problem. Technical Report TR-88-12. School of Engineering and Applied Science, Computer Science Department, University of Virginia.

Collins, R. and D. Jefferson (1991) Selection in Massively Parallel Genetic Algorithms. *Proceedings of the Fourth International Conference on Genetic Algorithms.* Morgan Kaufmann, San Mateo, CA.

Davidor, Y (1991) A Naturally Occurring Niche and Species Phenomenon: The Model and First Results. *Proceedings of the Fourth International Conference on Genetic Algorithms.* Morgan Kaufmann, San Mateo, CA.

DeJong, K.A. (1975) *An Analysis of the Behavior of a Class of Genetic Adaptive Systems.* (Doctoral dissertation, University of Michigan). Dissertation Abstracts International 36-10, 5140B.

DeJong, K.A. and W. Spears (1990) An Analysis of MultiPoint Crossover. NCARAI Library. AIC-90-014.

Eshelman, L. (1990). The CHC Adaptive Search Algorithm: How to have safe search when engaging in nontraditional genetic recombination. *Foundations of Genetic Algorithms,* Bloomington, IN.

Forrest, S. and A. Perelson (1990) Genetic Algorithms and the Immune System. *Parallel Problem Solving from Nature,* H.P. Schwefel and R. Manner, Eds. Springer-Verlag, Berlin.

Forrest, S. and M. Mitchell (1993) What Makes a Problem Hard for a Genetic Algorithm? Some Anomalous Results and Their Explanation. To Appear in *Machine Learning.*

Goldberg, D.E. (1989) *Genetic Algorithms in Search, Optimization, and Machine Learning.* Addison-Wesley. Grosso, P. (1985) *Parallel Subcomponent Interaction in a Multilocus Model.* Ph.D. Dissertation. Computer and Communication Sciences, University of Michigan.

Hillis, D. (1990) Co-evolving Parasites Improve Simulated Evolution as an Optimization Procedure. *Physica D.* 42. 228-234. North-Holland, Amsterdam.

Holland (1975) *Adaptation in Natural and Artificial Systems.* Ann Arbor: The University of Michigan Press.

Husbands, E, E Mill and S.Warrington (1991) Genetic Algorithms, Production Plan Optimisation and Scheduling. *Parallel Problem Solving from Nature,* H.P. Schwefel and R. Manner, Eds. Springer-Verlag, Berlin.

Ingber, L. and B. Rosen (1992) Genetic Algorithms and Very Fast Simulated Reannealing: A comparison. To be published in *Mathematical and Computer Modelling.*

Liepins, G.E. and S. Baluja (1991) apGA: an Adaptive Parallel Genetic Algorithm. *Computer Science and Operations Research, New Developments in Their Interfaces,* Balci, Sharda and Zenios, Eds. Pergamon Press, 1992.

Liepins, G.E. and M.D. Vose (1990) Representational Issues in Genetic Optimization, *Journal Expt. Theor. Artificial Intelligence,* 2, 101 - 115

Muhlenbein, H. (1989) Parallel Genetic Algorithms, Population Genetics and Combinatorial Optimization. *Proceedings of the Third International Conference on Genetic Algorithms.* Morgan Kaufmann, San Mateo, CA.

Schaffer, J.D., R.A. Caruana, L.J. Eschelman, and R. Das (1989). A Study of Control Parameters Affecting Online Performance of Genetic Algorithms for Function Optimization, In J.D. Schaffer (Ed.) *Proceedings of the Third International Conference on Genetic Algorithms.* Morgan Kaufmann, San Mateo, CA.

Schleuter, M.G. (1990), Explicit Parallelism of Genetic Algorithms through Population Structures. *Parallel Problem Solving from Nature,* H.P. Schwefel and R. Manner, Eds. Springer-Verlag, Berlin.

Spiessens, P. and B. Manderick (1991) A Massively Parallel Genetic Algorithm: Implementation and First Results. *Proceedings of the Fourth International Conference on Genetic Algorithms.* Morgan Kaufman, San Mateo, CA.

Syswerda, G. (1989) Uniform Crossover in Genetic Algorithms, In J.D. Schaffer (Ed.) *Proceedings of the Third International Conference on Genetic Algorithms.* Morgan Kaufmann, San Mateo, CA.

Tanese, R. (1989). Distributed Genetic Algorithms. In J.D. Schaffer (Ed.) *Proceedings of the Third International Conference on Genetic Algorithms.* Morgan Kaufmann, San Mateo, CA.

Whitley, D. and T. Starkweather (1990). GENITOR II: a Distributed Genetic Algorithm, *Journal Expt. Theor. Artificial Intelligence,* 2, 189-214.

## Chapter 7

**Kelvin K. Yue**
Department of Computer Science

**David J. Lilja**
Department of Electrical Engineering
University of Minnesota
200 Union Street S.E.
Minneapolis, MN 55455

yue@cs.umn.edu
lilja@ee.umn.edu

# Parameter Estimation for a Generalized Parallel Loop Scheduling Algorithm



## Abstract

Algorithms that dynamically schedule parallel loop iterations in a shared-memory multiprocessor have been proposed to balance the processors' workload while maintaining low scheduling overhead. However, none of the existing strategies perform well for all types of loops on all types of system architectures. We present a generalized loop scheduling algorithm that can be adjusted to match the loop characteristics to the system environment. A new method of simulation using the Genetic Algorithm is developed to determine appropriate scheduling parameters. This approach allows us to quickly choose sets of scheduling parameters for different loops executing on different systems. Stochastic simulations show that our parameterized strategies perform at least as well as the best existing algorithms for different combinations of loop iteration characteristics and system assumptions. Our generalized strategy is thus more robust than existing strategies.

0-8493-2529-3/95/$0.00 + $.50

## 7.1 Introduction

Since the body of a loop may be executed multiple times, exploiting loop-level parallelism is an effective means of increasing performance in a shared-memory multiprocessor system [9]. Parallel loop scheduling algorithms, such as chunk scheduling [8], self-scheduling [3], guided self-scheduling [12], factoring [7], and trapezoid self-scheduling [13], have been proposed to evenly distribute the workload among the processors while maintaining low scheduling overhead. However, the performance of these scheduling algorithms is sensitive to the loop characteristics and the system architecture so that no single algorithm performs well for all types of loops on all types of system architectures [14].

In this chapter, we propose a generalization of the current parallel loop scheduling algorithms in which the scheduling characteristics are parameterized. By using this generalized algorithm, we can quickly adjust the scheduling strategy to match the loop characteristics to the system environment. As the combinations of scheduling strategies, loop characteristics, and system environments are enormous, a new simulation method involving the Genetic Algorithm is developed to estimate the scheduling parameters needed to achieve good performance.

The use of the Genetic Algorithm for multiprocessor scheduling has been previously proposed [6, 11], but these methods depend on knowing *a priori* precise task information, such as the order of the tasks' execution, the task arrival times, the *exact* execution times, and the dependences between tasks. These methods then generate a *schedule* specific to this set of tasks. These methods are not feasible for loop-level parallelism since the time needed for finding a schedule may be longer than the loop execution time and, in many cases, the loop characteristics are unknown until run-time. Instead of finding a specific schedule, our proposed method uses the Genetic Algorithm to find appropriate values for the parameters of the generalized scheduling algorithm to produce a specific scheduling strategy or algorithm. This algorithm, then, is used at run-time to dynamically generate the actual schedule for executing the loop iterations.

Two new scheduling strategies are found using this method, one of which is suitable for scheduling loops with small iteration execution time variances, while the other is suitable for loops with large variances. They perform as well as, or better than, existing algorithms. Since the scheduling parameters of our algorithms can be adjusted based on the changes in the loop characteristics or system environments, our generalized method is more robust.

This chapter is organized as follows: Section 7.2 provides background information on existing parallel loop scheduling strategies. Section 7.3 presents our methodology for finding scheduling parameters using the Genetic Algorithm, while Section 7.4 discusses the simulated results of applying this strategy to loop-level parallelism. Section 7.5 concludes the chapter.

## 7.2 Current Scheduling Algorithms

In this section, the current techniques for scheduling Doall loop iterations on a shared-memory multiprocessor system, such as that shown in Figure 7.1, are reviewed. A performance comparison of these algorithms is also presented.

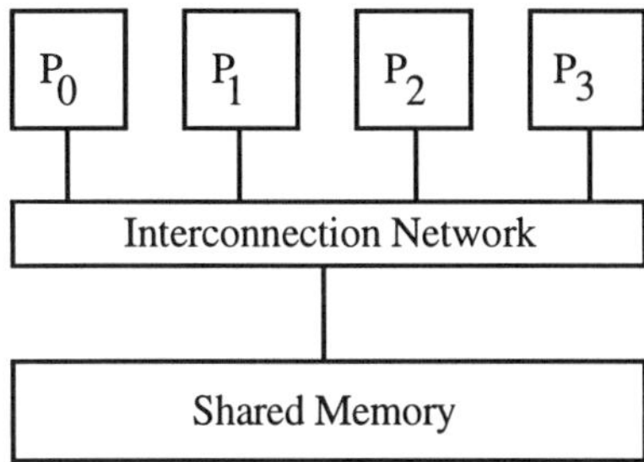


Figure 7.1: Shared memory multiprocessor architecture.

A Doall loop is the simplest form of parallelizable loop. In this type of loop each iteration is independent of the other iterations so that the iterations can be executed concurrently as independent tasks. An example of a Doall loop is:

```
DO i=I,N
    a(i) = b(i) + c(i)
END DO
```

The iterations of a Doall loop are assigned to the processors to execute based on some loop scheduling strategy. There are two main categories of scheduling algorithms: static and dynamic [9]. *Static scheduling,* or *prescheduling,* assigns iterations to the processors at compile time. Each processor *knows* exactly which iterations it should execute before the program is invoked and, therefore, there is no scheduling overhead. For example, the compiler could assign iterations to the processors based on the processor number so that processor 0 executes iterations 1, $P$ + 1, 2$P$ + 1,..., processor 1 executes iterations 2, $P$ + 2, 2$P$ + 2,..., and so on, where $P$ is the number of processors. The main disadvantage of static scheduling is *load imbalance* [2]. This unequal distribution of work to the processors can be caused by differences in the iteration execution times, or by differences in the number of iterations each processor executes. Since the *schedule* of iteration execution is fixed at compile-time, it cannot be adjusted based on the dynamically varying workload of the processors.

*Dynamic scheduling* assigns iterations to processors at run-time and can therefore adjust the schedule to the processors' workload. *Self-scheduling* is the simplest form of dynamic scheduling. With self-scheduling, each idle processor obtains the index of the next iteration it should execute by accessing a shared work queue. By taking one iteration at a time, this algorithm balances the workload very well, but the scheduling overhead is large since the shared work queue must be accessed once for each iteration.

To reduce the scheduling overhead, *chunk scheduling* assigns groups of iterations as a single unit to the processors. Kruskal and Weiss [8] analyzed load imbalances with this strategy and proposed the optimal chunk size to be $\left[\left(\sqrt{2}Nh\right)/\left(\sigma P\sqrt{\log P}\right)\right]^{2/3}$, where $N$ is the number of iterations, $P$ is the number of processors, $\sigma$ is the standard deviation of the distribution of iteration execution times, and $h$ is the scheduling overhead. They assume that the central-limit theorem holds for the iteration execution times, which is valid only when $N$ is large.

Another approach to reduce load imbalance while maintaining low scheduling overhead is to decrease the chunk size as the program executes. There are two strategies for decreasing the chunk size: *linear* decreases and *nonlinear* decreases. *Guided self-scheduling* (GSS) [12] decreases the chunk size nonlinearly by allocating iterations with a chunk size equal to $\lceil R/P \rceil$, where $R$ is the number of iterations remaining to be executed. This algorithm allocates large chunk sizes at the beginning of a loop's execution to reduce the scheduling overhead. As the number of iterations remaining to be executed decreases, smaller chunks are allocated to balance the load.

The *factoring* scheduling algorithm (FS) [7] is similar to GSS except that it allocates iterations in batches of P equal-sized chunks. After a batch is scheduled, the new chunk size is calculated to be $\lceil R/(\mathrm{x}P) \rceil$, where R is the number of iterations remaining, and x typically is chosen to be 2. The initial chunk size for FS is smaller than GSS. As a result, it has more iterations remaining at the end of the loop's execution to balance the load. However, FS requires many more scheduling steps than GSS. To reduce the number of scheduling steps, *safe self-scheduling* [10] proposes to use an x factor smaller than 2 so that more iterations will be allocated per chunk. However, the calculation of the $x$ factor for safe self-scheduling requires knowing not only the maximum and minimum iteration execution times, but also the probability of branching for the conditional statements in the loop. Safe self-scheduling may be less robust than factoring or guided self-scheduling since these characteristics typically are not known until runtime.

*Trapezoid self-scheduling* (TSS) [13] decreases the chunk size linearly to achieve a better tradeoff between the scheduling overhead and the distribution of the processors' workload compared to the nonlinear strategies. The number of chunks, $C$, is equal to $[2N/(f + l)]$ and the chunk size is decreased by a factor of $(f - l)/(C - 1)$ at each scheduling step, where typically $f = N/(2P)$ and $l = 1$. TSS does not allocate chunks as large as GSS in the beginning, and it does not require as many scheduling steps as FS. However, the linearly decrementing chunk size may create large load imbalances if the execution time differences between the last few chunks are large.

To summarize, one-iteration-at-a-time self-scheduling can perfectly balance the workload but it generates a large scheduling overhead that adds directly to the overall execution time. Chunk scheduling, on the other hand, requires minimum overhead, but it produces greater load imbalance. Guided self-scheduling, factoring, and trapezoid self-scheduling use a variable chunk size to tradeoff load imbalances with the scheduling overhead. However, the performance of these algorithms is sensitive to the characteristics of the loop and the system environment so that no single algorithm performs best in all cases [14, 15]. For instance, if the variance in iteration execution times is large, GSS may not balance the workload well since it does not save enough single-iteration chunks until the end [7, 13]. Factoring saves enough single-iteration chunks to balance the load, but with small variances in iteration execution times, these chunks cause extra scheduling overhead [9]. Trapezoid self-scheduling assigns small initial chunks, as does factoring, and it requires fewer scheduling steps than GSS [15], but the difference in execution time between the last few chunks might be large due to the linear decrement in the chunk size. This large difference may create correspondingly large load imbalances [7].

### 7.3 A New Scheduling Methodology

In the previous section, we reviewed five dynamic scheduling algorithms and concluded that no single algorithm produces the best performance in all cases. To match the scheduling algorithms to the loop characteristics and system environments, one can exhaustively try all of the strategies for all types of loops on all types of systems. However, this is obviously infeasible, if not impossible.

We propose a generalization of all of these scheduling algorithms in which the scheduling characteristics are parameterized and, therefore, can be easily adjusted to match the scheduling algorithm to both the individual loop and the system architecture. We also develop a new simulation methodology that uses the Genetic Algorithm as a heuristic search engine to choose appropriate parameters for the generalized scheduling strategy.

This section details the generalization of the scheduling algorithms and presents the simulation methodology. The implementation of the Genetic Algorithm is also described.

#### 7.3.1 A Generalized Loop Scheduling Algorithm

As discussed in Section 7.2, there are two primary types of dynamic scheduling algorithms: those that use a fixed chunk size based on the total number of iterations, and those that use a variable chunk size based on the remaining number of iterations. The first step of the generalization is to define a pargreeter $X$ which is equal to $N$, the total number of iterations, if the scheduling strategy uses a fixed chunk size. Otherwise, $X$ is equal to $R$, the remaining number of iterations, if the strategy uses a variable chunk size.

Notice that the chunk size for the current scheduling algorithms is related to the total number of processors, P. For instance, the chunk size for chunk scheduling is $\lceil N/P \rceil$, for GSS it is $\lceil R/P \rceil$, for FS it is $\lceil R/2P \rceil$, and the initial chunk size for TSS is $\lceil N/2P \rceil$. Therefore, the chunk size for our generalization is in terms of $aX/fP$. The parameter f is used to represent the factoring size giving $f = 1$ for CS and GSS, and $f = 2$ for FS. We also introduce another adjustment factor, $a$, to make our generalization more versatile by not limiting the scheduling algorithm to only integer factors.

To include all of the possible chunk sizes while allowing the chunk size to be decremented either linearly, as in TSS, or nonlinearly, as in GSS and FS, the parameter $\ell$ is introduced and the generalization is refined to $aX/fP - \ell$ For fixed-sized scheduling algorithms, or for variable-sized scheduling algorithms with a nonlinearly decreasing chunk size, $\ell$ is used as a *refining* factor. For instance, if chunk scheduling is used where the chunk size is determined to be some integer value that cannot be calculated with only $aX/fP$, then $\ell$ is set to a constant value to adjust the chunk size to the desired value. On the other hand, if a linearly decrementing chunk size strategy is used, $\ell$ is a function of the scheduling step. In TSS, for example, $\ell = i \times \left(\frac{N}{2P} - 1\right) \Big/ \left\lceil \frac{2N}{2P+1} - 1 \right\rceil$ where $i$ is the current scheduling step. As the execution proceeds, the number of the scheduling step is increased, which causes the chunk size to decrease linearly.

In our generalization, we also include a parameter, $m$, for a minimum chunk size feature as suggested in [12]. If the calculated chunk size is smaller than $m$, a chunk size of $m$ is used instead. Also, the parameter $C$ is the number of chunks with the same size that are scheduled before the chunk size is recalculated. In FS, $C$ is equal to $P$, while for the other scheduling algorithms, $C$ is always 1. Note that $C$ can take on any value in our generalization.

The generalization of loop scheduling algorithms is summarized as follows:

The number of iterations per chunk, $K$, is determined by:

$$\begin{cases} K = \left\lceil \frac{a}{f}\frac{X}{P} - \ell \right\rceil & \textit{if } K > m, \\ K = m & \textit{otherwise}, \end{cases}$$

and $C$ batches of the same chunk size, $K$, are scheduled before $K$ is recalculated.

In the above expression, $a$ and $f$ are the adjusting factors, $X$ is equal to $N$, the total number of iterations, if the strategy uses a fixed chunk size, or $X$ is equal to $R$, the number of iterations remaining to be executed, if the strategy uses a variable chunk size, $P$ is the number of processors, $\ell$ is the linear decrement factor, and $m$ is the minimum chunk size allowed. The following table shows the parameter values that will produce specific scheduling algorithms:

| Algorithm | $C$ | $a$ | $f$ | $X$ | $\ell$ | $m$ |
|---|---|---|---|---|---|---|
| Self Scheduling | 1 | $P$ | $N$ | $N$ | 0 | 1 |
| Chunk Scheduling | 1 | # | $a$ | $N$ | 0 | 1 |
| Guided Self-Scheduling | 1 | # | $a$ | $R$ | 0 | 1 |
| Factoring | $P$ | # | $2a$ | $R$ | 0 | 1 |
| Trapezoid Self-Scheduling | 1 | # | $a$ | $N$ | $\delta$ | 1 |

The symbol # represents any positive integer and $\delta = i\left(\frac{N}{2P} - 1\right) \Big/ \left\lceil \frac{2N}{2P+1} - 1 \right\rceil$ where $i$ is the current scheduling step. For CS, GSS, FS, and TSS, we can choose any positive integer for parameter $a$ by properly choosing the corresponding parameter $f$. In addition to the values shown in the table, self-scheduling can also be represented with the parameters $a = f = 1$, $X = N$, $\ell = (N/P) - 1$, $C = 1$, $m = 1$.

Based on this generalization, we have shown that we can parameterize the different existing loop scheduling strategies. This generalization allows us to select the desired scheduling algorithm, and it allows us to produce completely new scheduling algorithms, by choosing the appropriate parameters. As a result, we can easily adjust the scheduling algorithm to match the loop characteristics to the system environment.

### 7.3.2 Parameter Estimation

To utilize the generalized scheduling algorithm, a quick and simple method for matching the scheduling parameters to the loop characteristics and the system environment is needed. As previously mentioned, it is infeasible, if not impossible, to exhaustively test all the parameter combinations to determine which one generates the best performance. Therefore, we develop a new simulation methodology that uses the Genetic Algorithm (GA) as the means to determine appropriate parameters.

Our simulation consists of two modules: the GA engine and the multiprocessor simulator (Figure 7.2). The GA engine generates possible scheduling strategies and then sends them to the multiprocessor simulator for evaluation. The multiprocessor simulator simulates a shared memory multiprocessor environment executing a Doall loop based on the given scheduling strategies. It returns a measure of the performance of each strategy to the GA engine, which then creates new strategies based on the simulated performance of the previous strategies. In the following subsections, the implementations of the GA engine and the multiprocessor simulator are presented in detail.

#### 7.3.2.1 GA Engine

The Genetic Algorithm (GA) [5] has been applied to a wide variety of areas, ranging from artificially intelligent machine learning to gas pipeline control systems, since it was first introduced

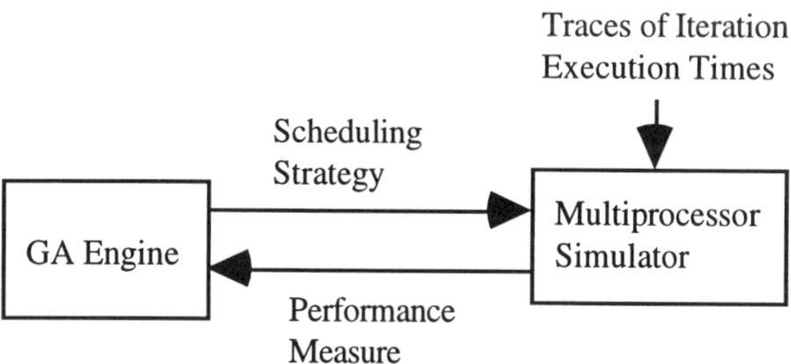


Figure 7.2: Simulation environment for estimating the parameters of the generalized scheduling algorithm.

approximately twenty years ago. It has been proven to be a robust and efficient algorithm for searching and optimization problems [4]. GA is based on the concept of natural selection and adaptation and the idea of *survival of the fittest.* GA is different from other optimization and search algorithms in four important characteristics [4]:

1. GA works with a coding of the parameter set, not the parameters themselves.

2. GA searches from a population of points, not just a single point.

3. GA uses payoff (objective function) information, not derivatives or other auxiliary knowledge.

4. GA uses probabilistic transition rules, not deterministic rules.

These characteristics of the Genetic Algorithm combine both exploration and exploitation in the searching process [1]. Unlike the hillclimbing search, which is simply exploitation, GA explores new domains in the search space and will not be limited to local maxima. Unlike random search, GA uses the known results to guide it to a better solution, thereby making the search process more efficient. Moreover, GA is more feasible than a brute-force trial-and-error method since it does not try every possible parameter combination in the search space. Therefore, we think that the Genetic Algorithm can be used in our simulation to find estimates of the scheduling parameters based on the system environment.

In the following subsections, the genetic operations of GA are reviewed and the representation and implementation of our generalized scheduling algorithm in the Genetic Algorithm framework is presented.

**Implementation** The parameters for the scheduling strategy are represented in the chromosome format shown in Figure 7.3. The parameters $\ell, f, a, C$, and $m$ are described in Section 7.3.1. Their binary representations are decoded into integer values in the simulator. The bits $X$ and $\ell$ are condition bits. If bit $X$ is one, $N$, the total number of iterations, is used. Otherwise, $R$, the remaining number of iterations, is used. If bit $\ell$ is one, the linear decrement strategy is used. GA does not impose any specific rules in designing or coding of a chromosome, and the quality of the resultant solutions does not depend on the arrangement of the parameters within the chromosome due to the robustness of the Genetic Algorithm [4].

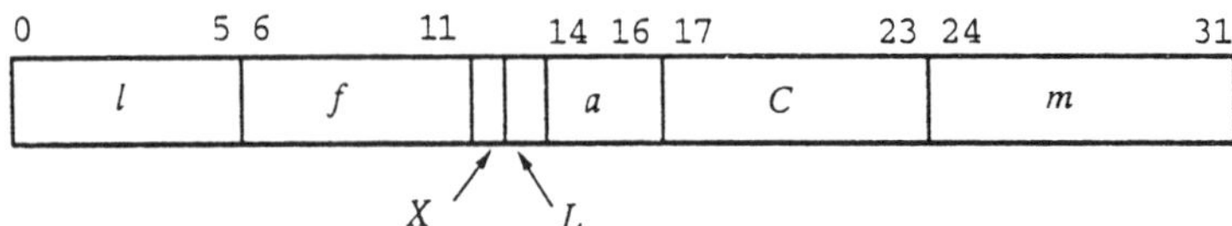


Figure 7.3: Chromosome representation for the generalized loop scheduling strategy.

The fitness function for a chromosome is the parallel execution time efficiency of a Doall loop executed using the scheduling strategy encoded in the chromosome. It is calculated as:

$$E = \frac{Speedup}{P} = \frac{Sequential\ Run-time}{P \times Parallel\ Run-time}$$

The sequential run-time is the sum of all iteration execution times, excluding the scheduling overhead, which is equivalent to the execution time for the Doall loop when it is executed on a sequential machine. The parallel run-time is the total execution time of the last processor to finish executing.

#### 7.3.2.2 Multiprocessor Simulator

After a set of scheduling parameters is generated by the GA engine, its performance then needs to be evaluated. In this experiment, we use a simple stochastic simulation model. It is possible, however, to use a more complicated simulation, or even a real multiprocessor system, for the performance evaluation.

This module simulates a shared memory multiprocessor system with $P$ processors, all of which execute at the same speed. When a processor is idle, it locks the loop index variable to determine the next chunk of iterations it will execute. It then unlocks the loop index variable and begins executing the iterations. The number of iterations a processor assigns itself at each scheduling step, i.e., the size of a chunk, is determined by the scheduling strategy. When two or more processors attempt to simultaneously access the loop index, the one with the smallest processor identification number is allowed to go first. The delay introduced by this contention adds directly to the execution time of the stalled processors. The specific scheduling strategy used in the simulator is dynamically configured according to the information sent from the GA Engine.

The execution times of the iterations are generated by a random number generator with a normal (Gaussian) distribution. The mean and variance of the iteration execution times are specified based on the types of the loops [14]. Again, it is possible to use more complicated methods to generate traces of the iteration execution times, but we use the simplest method to demonstrate our scheduling strategy. At the end of the simulation, the efficiency of the scheduling strategy is calculated and returned to the GA engine where it is used as the fitness value of the chromosome that defines the given scheduling strategy. The following algorithm summarizes the simulation environment:

```
/* Initialization */
randomly generate the initial population

FOR each chromosome in the population
   /* begin multiprocessor simulation */
    simulate the scheduling strategy
    measure the efficiency
   /* end multiprocessor simulation */
   use the efficiency value as the fitness of the
   chromosome
```

```
ENDFOR

DO until (population converges) or (no. of generations >
predefined value)
   /* Selection Phase */
   select the chromosomes with the highest fitness values

   /* Reproduction Phase */
   generate the new chromosomes using crossover operator
   and mutation operator.

   /* Evaluation Phase */
   FOR each chromosome in the new population
      /* begin multiprocessor simulation */
       simulate the scheduling strategy
       measure the efficiency
      /* end multiprocessor simulation */
      use the efficiency value as the fitness of the
      chromosome
   ENDFOR
ENDDO
```

## 7.4 Results

We use the simulation methodology discussed in the previous section to match scheduling algorithms to the loop characteristics while varying the number of processors, the number of iterations, the scheduling overhead, and the variance in iteration execution times. The GA engine found that scheduling algorithms that use a fixed chunk size, and algorithms that use a variable chunk size with linear decrement, do not perform as well as the other algorithms. The GA engine discovered two new scheduling algorithms that perform as well as, or better than, existing scheduling algorithms. We call these two new algorithms *CS-2* and *FS-alt* due to their similarity to chunk scheduling and factoring, respectively. CS-2 is similar to chunk scheduling in that it uses $N/P$ iterations per chunk, except it saves $2P$ single-iteration chunks to balance the load at the end. FS-alt is similar to factoring except that it uses a larger *factor* and it allocates fewer chunks per batch. In this section, these two new scheduling algorithms are presented and their performance is compared with the other scheduling algorithms.

### 7.4.1 New Scheduling Algorithms

CS-2 allocates iterations with two different chunk sizes it has $2P$ chunks with 1 iteration per chunk and $P$ chunks with $\lceil N/P - 2 \rceil$ iterations per chunk. When the loop execution begins, each processor acquires a chunk with $\lceil N/P - 2 \rceil$ iterations and starts executing. Near the end of the execution, the single-iteration chunks are used to dynamically balance the workload among the processors. The total number of scheduling steps, that is, the number of accesses to the shared work queue, for CS-2 is $3P$. CS-2 is suitable for loops with small variances in iteration execution times. Standard chunk scheduling cannot balance the variation well for this type of loop while other dynamic scheduling algorithms require too many scheduling steps and, thus, are too costly for effectively balancing this small variation. For loops with large variances in execution times, CS-2, similar to standard chunk scheduling, does not perform well compared to other dynamic scheduling algorithms.

FS-alt, on the other hand, performs well for loops with large variances. It is similar to factoring, except that it uses a factor of 5/6 instead of 1/2, and it uses *P*/2 chunks per batch instead of *P* chunks per batch. FS-alt improves on the performance of factoring by allocating larger chunks at the beginning of the execution and, therefore, reducing the number of scheduling steps. To compensate for the larger chunk sizes, FS-alt allocates *P*/2 chunks per batch allowing it to save enough small chunks to balance the processors' workload at the end of the loop's execution. The following table shows the parameter values for CS-2 and FS-alt from our generalization of the loop scheduling algorithm.

| Algorithm | $C$ | $a$ | $f$ | $X$ | $\ell$ | $m$ |
|---|---|---|---|---|---|---|
| CS-2 | $P$ | # | $a$ | $R$ | 2 | 1 |
| FS-alt | $P/2$ | 5 | 6 | $R$ | 0 | 1 |

### 7.4.2 Performance Comparisons

Figure 7.4 shows the speedup values of the two GA-generated scheduling strategies compared to the current algorithms. The speedup is measured on a $P$ = 16 processor system executing a Doall loop with an average iteration execution time of 100 cycles. The overhead for scheduling a chunk of iterations is 10% of the mean iteration execution time, or 10 cycles. The total number of iterations, N, is set to 500 iterations and 5000 iterations while the variance in iteration execution times is changed from 5 cycles to 70 cycles.

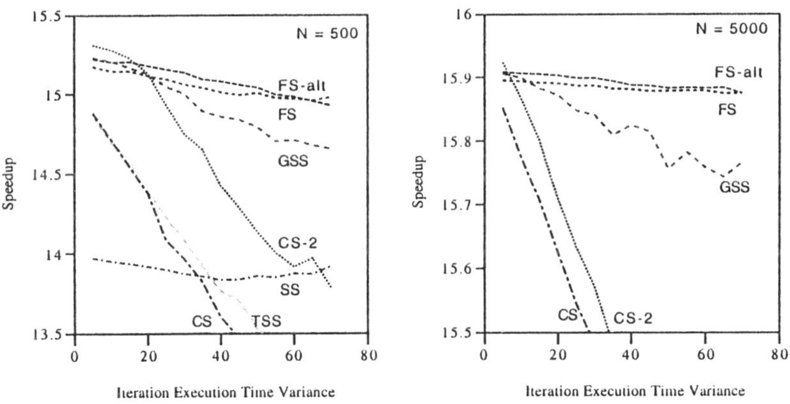


Figure 7.4: Speedup comparisons of the different scheduling algorithms.

As shown in Figure 7.4, CS-2 performs better than the other scheduling algorithms when the iteration execution time variances are small. As the variance increases, CS-2's performance decreases with the same rate as standard chunk scheduling. FS-alt performs slightly better than FS in both cases, but the difference between the two in the $N = 5000$ case is quite small. The speedup of SS and TSS are less than 15.5 when $N = 5000$ and are not shown in the figure. The large scheduling overhead and poor load balancing are the causes of the poor performance for SS and TSS, respectively.

The comparisons shown in Figure 7.4 are based on the assumption that all the scheduling algorithms have the same scheduling overhead. It suggests that the two GA-generated scheduling algorithms slightly improve the overall performance. The scheduling overhead for some algorithms is much lower than for the others, however. For instance, SS requires only a `Fetch&Add` operation to obtain the next iteration while FS requires a more complicated calculation. To eliminate this factor, Figure 7.5 compares the total number of scheduling steps for all of the algorithms with different values of $N$, the total number of iterations, on a 16-processor system. The total number of scheduling steps for SS is not shown in the figure since it is simply the total number of iterations. This figure shows that FS requires the most scheduling steps, and the number of scheduling steps increases at a faster rate than the others as the total number of iterations increases. FS-alt schedules iterations in a fashion similar to FS, but it requires fewer total scheduling steps. CS-2 requires at least twice as many scheduling steps as CS, but, as shown in Figure 7.4, it balances the workload more evenly than CS.

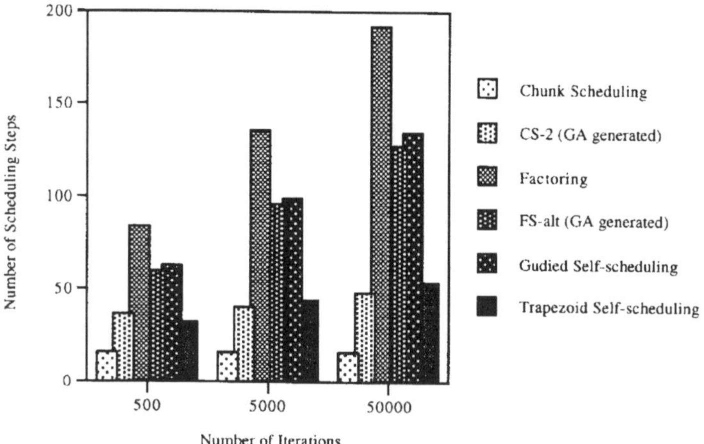


Figure 7.5: Comparison of the number of scheduling steps.

The total number of scheduling steps not only directly contributes to the scheduling overhead, but it also relates to the network and memory contention since, as the number of scheduling steps increases, the chance of two or more processors trying to access the shared loop index at the same time increases as well. When one processor is accessing the loop index, all the other processors which need to obtain additional work at the same time must wait. Figure 7.6 compares the processor execution times divided into three different categories: the *execution time,* which is the time the processor spends executing the iterations, the *overhead,* which is the time the processor spends calculating the chunk size and accessing the shared loop variables, and the *contention time,* which is the time the processor is idle waiting to access the shared variables or waiting for synchronization. The sum of these three times is equal to the parallel execution time of the Doall loop using the specific scheduling algorithm. We set the average iteration execution times to 100 cycles and vary the number of processors (*P*), the total number of iterations (*N*), the scheduling overhead (*O*), and the iteration execution time variance (*V*).

In Figure 7.6(a), we have a 500-iteration Doall loop with a variance of 10 cycles executing on a 16-processor system with a scheduling overhead of 10 cycles. SS produces the largest scheduling overhead while, as expected, CS has the smallest. The GA-generated algorithms, CS-2 and FS-alt, both have a small scheduling overhead compared to the others. CS-2, FS, FS-alt, and GSS have similar contentions and, therefore, the algorithms with lower scheduling overhead, i.e., CS-2 and FS-alt, have the lower total parallel runtime. We use Figure 7.6(a) as the comparison baseline as we alter the system parameters. In Figure 7.6(b), the total number of processors (*P*) is doubled. The average execution time for all of the algorithms is halved as more processors share the same amount of work. The scheduling overhead per processor is decreased since the processors do not need to obtain work from the shared work queue as many times as in the baseline case. The contention, or the processor idle time, is increased, however, since more processors are competing to access the shared work queue. An opposite effect occurs when the number of iterations (*N*) is doubled, as shown in Figure 7.6(c). In this case, the processors spend more time executing the iterations since the workload per processor is increased. The scheduling overhead is also increased since more iterations need to be assigned, but the contention time is decreased since there is more work for the processors and the processors are less likely to wait idle.

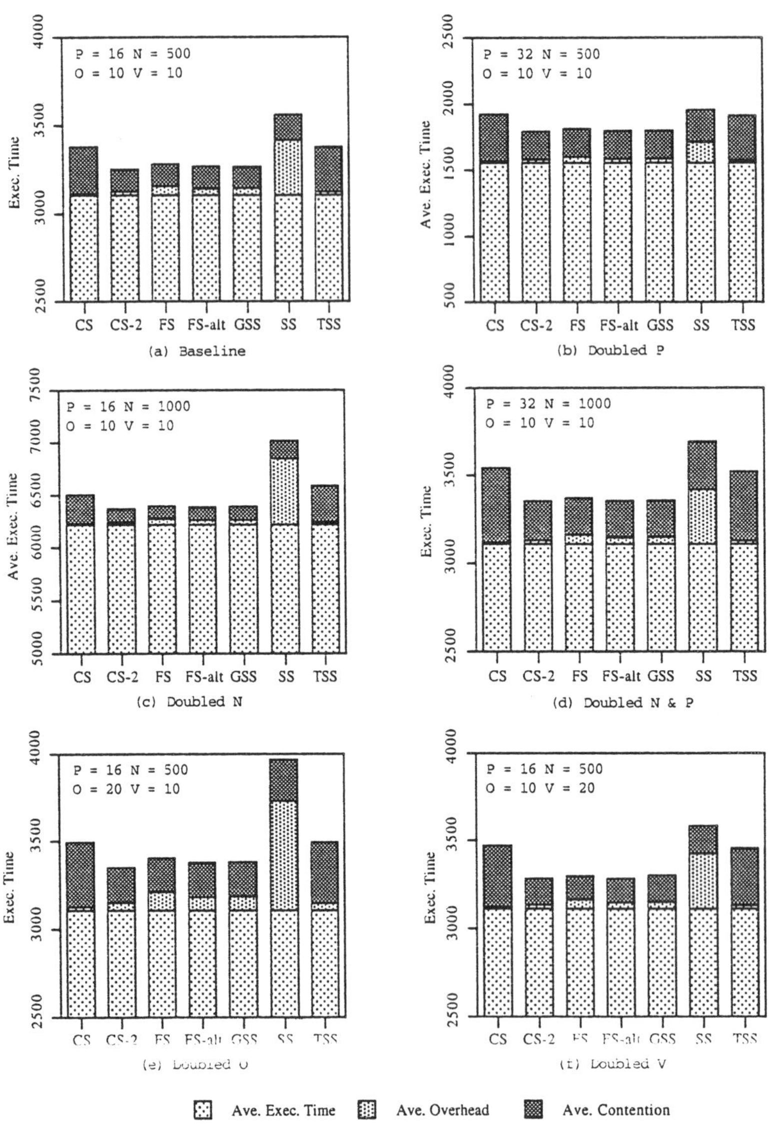


Figure 7.6: Breakdown of processor execution times.

To compare the scalability of the scheduling algorithms, both the number of processors ($P$) and the number of iterations ($N$) are doubled. The contention of all of the algorithms is increased from the baseline since more processors are sharing the single shared work queue. Self-scheduling (SS) suffers the most as it has to access the shared work queue for each iteration and there are more iterations to be executed, and more processors to compete for the work queue. CS-2, FS-alt, FS, and GSS also have increased contention time, but they still outperform the other algorithms. CS-2 and FS-alt have relatively little scheduling overhead comparing to FS and GSS. As a result, these two scheduling algorithms have the shortest overall parallel execution times.

From Figure 7.5, it is seen that FS requires more scheduling steps than any of the other algorithms except SS. The effect of this factor on the overall performance becomes more obvious when the overhead per scheduling step (*O*) is doubled, as shown in Figure 7.6(e). Both the scheduling overhead and the contention time for all of the scheduling algorithms are increased compared to the baseline in Figure 7.6(a), since it takes longer to access the shared work queue, and since the competing processors must wait idle longer. The performance of FS degrades more than FS-alt and CS-2 because of its much greater number of scheduling steps. CS-2 requires fewer scheduling steps than FS-alt, and, therefore, the overall parallel execution of CS-2 is less than that of FS-alt.

To compare the algorithms' sensitivity to variances in the iteration execution times, we measure the execution time after doubling the variance (*V*). Both CS and TSS produce longer parallel execution times due to the larger load imbalance induced by the larger variance. The performance of SS remains the same since it always balances the workload perfectly. The increases in contention times for CS-2 and GSS are larger than that of FS and FS-alt because they do not have enough single-iteration chunks to balance the workload at the end of the loop's execution. However, the overall execution times for CS-2 and FS-alt are still less than the others, as shown in Figure 7.6(f).

In this section, we have compared the performance of the GA-generated algorithms, CS-2 and FS-alt, with the current scheduling algorithms. The results show that CS-2 has smaller scheduling overhead than the other algorithms and that it outperforms the others when the iteration execution time variance is small. FS-alt, on the other hand, has a larger scheduling overhead than CS-2, but it produces better load balance when the iteration execution time variance is large.

## 7.5 Conclusion

In this chapter, a generalized scheduling algorithm is proposed in which the scheduling strategy is parameterized and so can be adjusted to match the loop characteristics and the system environment. A new simulation methodology using the Genetic Algorithm is developed to find appropriate parameters for this generalized scheduling. Two new scheduling algorithms, CS-2 and FS-alt, were discovered using this simulation methodology. CS-2 is similar to CS, but it improves the load balancing capability while maintaining a low scheduling overhead. It is suitable for loops with small iteration execution time variances. FS-alt, on the other hand, performs well for loops will large variances. It reduces the scheduling overhead of FS by using a larger factor and a smaller batch size.

Based on simulated performance comparisons, the newly discovered algorithms perform as well as, or better than, the existing algorithms. Since we can further *fine tune* the parameters of our generalized scheduling algorithm by interfacing the GA engine to a real multiprocessor system, or to a system of some other architectural design, our scheduling algorithm is more robust than current algorithms. Another possible use of the generalized scheduling algorithm is to have a dedicated processor executing the GA engine to adjust the scheduling parameters dynamically based on the status of the system and the loop execution. A variety of other techniques can be used to determine appropriate values for the parameters of this generalized loop scheduling algorithm.

**References**

[1] David Beasley, David R. Bull, and Ralph R. Martin. *An Overview of Genetic Algorithms: Part 1, Fundamentals,* volume 15 of *University Computing,* pages 58-69. Inter-University Committee on Computing, University of Cardiff, Cardiff, *CF2* 4YN, UK, 1993.

[2] Carl J. Beckmann and Constantine D. Polychronopoulos. The effect of scheduling and synchronization overhead on parallel loop performance. In *International Conference on Parallel Processing,* volume II: Software, pages 200-204, 1989.

[3] Zhixi Fang, Pen-Chung Yew, Peiyi Tang, and Chuan-Qi Zhu. Dynamic processor selfscheduling for general parallel nested loops. In *Proc. 1987 International Conference in Parallel Processing,* August 1987.

[4] David E. Goldberg. *Genetic Algorithms in Search, Optimization, and Machine Learning.* Addison-Wesley, Reading, Massachusetts, 1989.

[5] John Holland. *Adaptation in Natural and Artificial Systems.* University of Michigan Press, Ann Arbor, Michigan, 1975.

[6] Edwin S.W. Hou, Nirwan Ansari, and Hong Ren. A Genetic Algorithm for Multiprocessor Scheduling. *IEEE Transactions on Parallel and Distributed Systems,* 5:113-120, February 1994.

[7] Susan Flynn Hummel, Edith Schonberg, and Lawrence E. Flynn. Factoring - a method for scheduling parallel loops. *Communciations of the ACM,* 35:90 101, August 1992.

[8] Clyde P. Kruskal and Alan Weiss. Allocating independent subtasks on parallel processors (extended abstract). In *International Conference on Parallel Processing,* pages 236 240, 1984.

[9] David J. Lilja. Exploiting the parallelism available in loops. *Computer,* pages 13-26, February 1994.

[10] Jie Liu, Vikram A. Saletore, and Ted G. Lewis. Scheduling parallel loops with variable length iteration execution times on parallel computers. In *ISMM 5th International Conference on Parallel and Distributed Computing Systems,* pages 83-89, October 1992.

[11] Hirak Mitra and Parameswaran Ramanathan. A genetic approach for scheduling nonpreemptive tasks with precedence and deadline constraints. In *26th Hawaii International Conference on System Sciences,* volume 2, pages 556-564, 1993.

[12] C. Polychronopoulos and D. Kuck. Guided self-scheduling: A practical scheduling scheme for parallel supercomputers. *IEEE Transactions on Computers,* C-36:1425-1439, December 1987.

[13] Ten H. Tzen and Lionel M. Ni. Trapezoid self-scheduling: A practical scheduling scheme for parallel compilers. *IEEE Transactions on Parallel and Distributed Systems,* 4:87-97, January 1993.

[14] Kelvin K. Yue and David J. Lilja. Categorizing parallel loops based on iteration execution time variances (submitted for publication). 1994.

[15] Kelvin K. Yue and David J. Lilja. Scalability analysis for parallel loop scheduling algorithms (submitted for publication). 1994.

## Chapter 8

**M. O. Odetayo**
**D. Dasgupta**
Department of Computer Science
D Montfort University
Leicester LE1 9BH
U.K.

odetayo@cs.unm.edu

# Controlling a Dynamic Physical System Using Genetic-Based Learning Methods



## Abstract

This chapter presents two different approaches of designing genetic-based controllers for an unstable physical system (a simulated pole-cart system). One approach induces *rule-base controller* using a simple genetic algorithm (GA) and the other evolves *neuro-controller* applying a recently developed Structured Genetic Algorithm (sGA) which appears to offer improvements over a simple GA approach. The control task here is a typical unstable, multi-output, dynamic system in which a pole is supposed on a controllable cart, and the controller must keep the pole upright (within a specified vertical angle) and the cart within the limits of the given track. In this chapter, we first describe a simple GA based learning method for inducing control rules, and then demonstrate the evolvability of neuro-controller using a Structured GA.

0-8493-2529-3/95/$0.00 + $.50

## Introduction

When building a controller for a dynamic system, traditional control theory requires a mathematical model to predict the behaviour of the system. In many cases this cannot be done, either because the system is too complicated or because insufficient information about its environment is available. The pole balancing problem is one such inherently unstable classical control problem. The complexity of the task is significant enough to make the problem interesting while still being simple enough to make it computationally tractable.

We have chosen this task for several reasons, they include:

Many learning algorithms [19][1][24][5][23] have solved the problem in the form considered here and therefore we have a good test bed for evaluating the effectiveness of the genetic algorithm-based learning method.

The task is regarded as an example of the inherently unstable, mutiple-output, dynamic systems present in many balancing conditions such as the aiming of a rocket thruster [1][4][13]. The system is non-linear and has multi-output interacting parameters that are to be controlled. It is inherently unstable — the system can only be controlled by an appropriate thrust at the base of the cart.

• The way the task is set up creates a genuinely difficult credit-assignment problem and therefore poses a great challenge to a learning method.

• The complexity of the problem could be increased to a desired level by balancing 1, 2, ..., etc. poles each on top of the next [19].

We shall discuss how the time to learn and the amount of computation required by GAPOLE compares with those taken by learning methods that have previously solved the problem [23]. We show that it copes well with changing conditions and that it is an effective alternative technique.

Genetic Algorithms (GAs) are a class of adaptive general purpose methods, for machine learning and optimisation, based on the principles of population genetics and natural evolution. However, not much has been done to determine their suitability as a machine learning and adaptive control tool for other more general applications. Knowledge in this field is being advanced rapidly, however.

This chapter is divided into two main sections. The first section discusses our experiments using a simple genetic algorithm and compares the performance with other classical AI methods. We developed and implemented a simple GA-based program (GAPOLE) for inducing control rules for a dynamic physical system: a simulated pole-cart system. The second section describes the use of a structured genetic algorithm for automatic designing neuro-controller for the same task.

In the following, we briefly describe the principle of genetic algorithms and develop a Simple GA-based learning system, called GAPOLE, and assign it the task of inducing control rules for a dynamic system — a simulated pole-cart system. The dynamics of the pole-cart system are not made available to the algorithm. The only evaluative feedback indicating how well it is performing is a failure signal which occurs when the system is out of control. That is, either when the cart has gone beyond the track limit or the pole has fallen past a predefined vertical angle. This presents a big challenge to a learning method as the effect of a wrong action may not be known until several steps later. Thus training information may be delayed making it difficult to correctly credit individual actions.

The GAPOLE program was used to derive (or 'breed') a species of controllers that give a specified level of performance. Comparison of its performance (time to learn and the amount of computation) with the best available alternatives showed that it compares well, but it is noteworthy that it performs well in a "noisy" and changing control environments.

## 8.1 The Control Task

The task is to move a wheeled cart, with a rigid pole hinged on top of it, along a bounded straight track without the pole falling beyond a predefined vertical angle and without the cart going off the ends of the track limits. This is achieved by applying a force of fixed magnitude (a 'bang-bang' force) to the left or right of the cart (Figure 8.1).

The state of the pole-cart system at any time t is specified by four variables:
$x$ = position of the cart on the track, where:
$\dot{x}$ = velocity of the cart.
$\theta$ = angle of the pole with the vertical.
$\dot{\theta}$ = angular velocity of the pole.

The pole-cart system was simulated using the following equations of motion derived by Anderson [1] with the given parameter values:

$$\ddot{\theta}_t = \frac{mg\sin\theta_t - \cos\theta_t[F_t + m_p L\dot{\theta}_t^2 \sin\theta_t]}{L[(4/3)m - m_p\cos^2\theta_t]}$$

$$\ddot{x}_t = \frac{F_t + m_p L[\dot{\theta}_t^2 \sin\theta_t - \ddot{\theta}_t \cos\theta_t]}{m}$$

where
$m_c$ = 1.0 kg = mass of the cart.
$m_p$ = 0.1 kg = mass of the pole.
$m = m_c + m_p$ = 1.1 kg = total mass of the system.
$L$ = 0.5 m = distance of centre of mass of pole to the pivot.
$g$ = 9.8 ms$^2$ = acceleration due to gravity.
$F_t$ = force applied to cart (of specified magnitude).

A time step of r = 0.02 seconds and the following discrete-time state equations were also used in the simulation.

$$x_{t+1} = x_t + \tau\dot{x}_t \quad \dot{x}_{t+1} = \dot{x}_t + \tau\ddot{x}_t$$
$$\theta_{t+1} = \theta_t + \tau\dot{\theta}_t \quad \dot{\theta}_{t+1} = \dot{\theta}_t + \tau\ddot{\theta}_t$$

The state space can be regarded as a four dimensional space and a state variable defines each dimension. The dynamics of the physical system being unknown to the controlling system; the only information for evaluating performance is a failure signal indicating that the pole-cart system is out of control. The pole-balancing problem consists of: 1) how to divide the quantity space of each variable into a small set of intervals; 2) what action to select for each combination of intervals describing the state of the pole-cart system in a way such that

- the poles are balanced i.e do not fall.
- the cart does not leave a predetermined limited track.

It is a good test bed for evaluating the effectiveness of investigating the learning performance of the genetic-based systems because:

- there is randomness in the task;
- the dynamics of the system are not known to the learning program; and
- it is a difficult and challenging task.

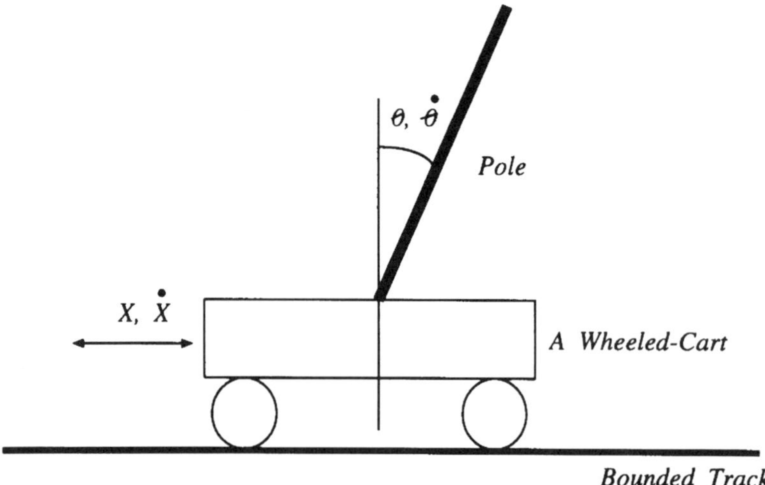


Figure 8.1: A pole-cart system on a bounded track.

The dynamics of the pole-cart system are unknown to the learning controller. The only information available to it at discrete time steps is either a vector indicating the current state of the system or a failure signal telling it that the system is out of control. In this experiment, the system is out of control when the cart has gone beyond ±2.4 meters from the centre or the pole has fallen beyond 12° from the vertical. These limits were employed for the three alternative methods for the same task [23].

The learning ability of the genetic-based system which can carry out a complex task was demonstrated by our work on the pole-balancing system[22]. Comparison of its performance with the best available alternatives showed that it compares well, but it is noteworthy that it is robust, and performs well in *noisy* and changing control environments.

In the following subsections, we briefly state the previous learning methods for the pole-cart problem and discuss the working principle behind GAs. We then give a detail description of the experiments for inducing simple GA-based control rules and the comparative results with other AI methods.

## 8.2 Previous Learning Algorithms for the Pole-Cart Problem

We review the three best learning algorithms — BOXES, AHC and CART — that have been applied to the problem of learning to control a simulated pole-cart system.

### 8.2.1 BOXES

Miche and Chambers [19] developed a program known as BOXES for learning to control the pole-cart system. They reduced the problem space into manageable proportion by partitioning it into disjoint regions called boxes. This was achieved by quantizing the four state variables. The quantization thresholds were predefined before the experiments started. They used a total of 225 partitions or boxes.

Each box is imagined as having a local demon that decides where the pole-cart system should move next (left or right) whenever it enters its box. In order to do this, a demon gathers data about its box through the following sets of variables [19]:

LL, the 'left life' of its box, which is a weighted sum of the 'lives' of left decisions taken on entry to its box during previous runs. (The 'life' of a decision is the number of further decisions taken before the run fails.)

RL, the 'right life' of its box.

LU, the 'left usage' of its box, which is a weighted sum of the number of left decisions taken on entry to its box during previous runs.

RU, the 'right usage' of its box.

TARGET is a figure supplied to every box by the supervising demon, to indicate a desired level of attainment, for example, a constant multiple of the current mean life of the system.

$T_1, T_2, ..., T_n$ times at which its box has been entered during the current run. Time is measured by the number of decisions taken in the interval being measured, in this case between the start of the run and a given entry to the box.

It uses a function of these variables to rate how good a decision to go right or left was. If the right value is greater than the left value then the demon would move the system to the right and vice versa.

Weighted averages of the lifetimes of the pole-cart after a decision to go left or right are used because estimates of the worth of a decision are inaccurate in the early stages of a trial (a trial is the period the system is kept under control from a starting position before failure occurs). For example, the direction in which the system should go whenever it enters a box may be correctly set, say to go right, but if the directions for the boxes around it are not properly set, then its decision to go right may appear bad [23].

Another interesting feature of the algorithm is its solution to the problem of getting stuck at a local peak. In order to prevent a box from settling down to a seemingly good direction (left or right), it compares an optimistic projection of how well each decision may perform [19][23]. For example the value of going left is determined as follows:

$value_L = LL + K \bullet TARGET^{LU+K}$, where K was set to 20.

### 8.2.2 AHC

AHC (Adaptive Heuristic Control Algorithm) partitioned the problem state space into predefined regions like the BOXES. However, unlike BOXES, learning takes place during a trial as well as at the end of it.

It uses four parameters to evaluate the performance of a box and to decide which action to take whenever the system enters the box. The parameters for a box are [23]:

- ACTION, a real number that is used to determine whether to go left or right.

- MERIT, a measure of its ability to predict the correct action.

- ELIGIBILITY determines if it is eligible to change from going one way to the other.

- FREQUENCY, a weighted count of the number of entries into the box.

A negative value of ACTION represents a tendency to go left, while a positive value represents a tendency to go right. The bigger the magnitude the greater the tendency. The actual direction is determined by mapping the action into a range from zero to one, then generating a random number between zero and one, and comparing the two. If the random number is less than the mapped number then go left, if not, go right.

### 8.2.3 CART

CART [5] does not partition the state space into regions. It has as its main goal the ability to accurately estimate the desirability of a state and therefore tries to avoid bad ones. Thus its main thrust is the search for a more desirable state.

CART chooses an action that is estimated to lead to a desirable state. At every step, it decides whether the same action as the last should be repeated or the action should be changed. If it is estimated that the pole-cart system will move to a more desirable state by continuing with an action, then the action is chosen; if not the other action is selected.

As learning progresses it gathers information that enables it to improve its ability to estimate the degree of desirability of the pole-cart states. It does this by labelling certain states in the trial as desirable or undesirable. This is achieved in three ways:

(1) It starts the learning process from the state when the pole is upright, the cart is centred, and the velocites (angular and cart) are zero. This initial state is labelled as a desirable state.

(2) An undersirable state is reached when the pole-cart is out of control (when the pole falls or cart has gone past the defined limits). The state immediately preceding the failure is labelled as undesirable, unless its degree of desirability is already less than -0.98.

(3) At the end of a trial that lasted more than 100 time steps, it backs up 50 states from the failure point; and from then on, searches for a state in which at least three of the state variables are approaching zero in magnitude, i.e., are approaching the starting state. This point is labelled as desirable.

The algorithm uses a function to interpolate from a chosen set of states (known as the training set) that have been labelled (as desirable or undesirable) in order to improve its accuracy in estimating the desirability of a pole-cart state. In order to do this, it views the desirability of a state as the height of a surface in five dimensional space. The first four dimensions represent the state, and the fifth represents its degree of desirability. The surface is changed after each trial as new points are used to evaluate the interpolating function.

## 8.3 Genetic Algorithms (GA)

Genetic Algorithms (GAs) are iterative adaptive general-purpose search strategies, based on the principles of population genetics and natural selection [14][12]. They simulate the mechanics of population genetics by maintaining a population of knowledge structures, analogous to the gene pool of a species, which is made to evolve.

An outline of the generic Genetic Algorithm is given below:

```
Initialise P(t=O); /, P(O) = initial population ,/
  Evaluate members of P(t);
   While (not termination condition)
     {
       Generate P(t+1) from P(t) as follows:
        {select individuals from P(t) on basis of fitness;
          recombine those selected;
        }
         t = t+1;
        evaluate members of P(t);
     }
```

A GA therefore learns by evaluating its knowledge structures using the fitness function, and forming new ones to replace the previous generation by breeding from more successful individuals in the population using the crossover and the mutation operators.

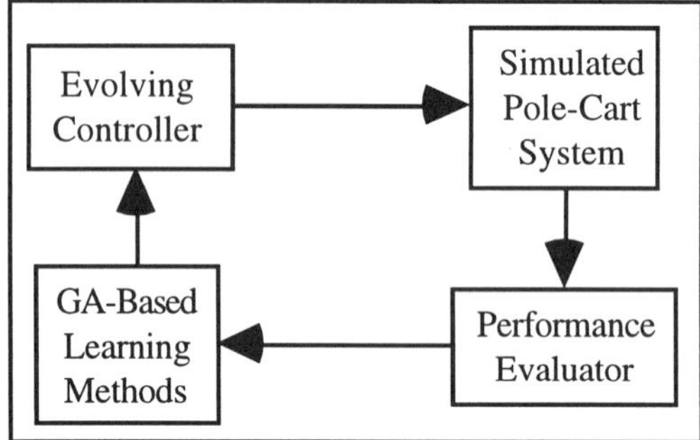


Figure 8.2: A genetic-based control system (GAPOLE).

## 8.4 Generating Control Rules Using a Simple GA

A simple GA-based learning program (GAPOLE) we developed for the pole-cart balancing problem described above. It consists of 4 components: A fixed population size of learning (rule-base) controllers; the learning algorithm; the performance evaluator; and the simulated pole-cart system. Their interaction is shown in Figure 8.2.

The following pseudo code shows how the chromosomal information (a set of directions) is used to control the pole-cart system:

```
while (state_of_pole != FALLEN and time_pole_held< MAX_HOLD)
 {Increment time_pole_held; move_system;}
```

### 8.4.1 Population of Learning Controllers

A learning controller is a set of production rules for controlling the pole-cart system. A production rule has a format as follows:

**condition** then **action**

The specified action will be performed when the condition is satisfied. A controller is regarded as a chromosome by the learning algorithm. We use the two names interchangeably without any loss of meaning.

### 8.4.2 Representation

As we stated in defining the control task, the state of the pole-cart system is specified by four real-valued variables. The state space can therefore be regarded as a four dimensional space. A state variable defines each dimension. At each point in the state space, the learning controller is required to decide whether the pole-cart system should go left or right so as to keep it under control. This implies that it has an infinite number of points and so the state space is reduced to manageable proportions by partitioning it into predefined regions as in [19] so that points within a region are mapped into the same decision.

We experimented with a number of partitions, taking full advantage of those used in [19]. The set of partitions we found that gave the best results and which we employed in our experiments are:

$x$ (cart position): - 2.4 to 2.4 metres [1 region]
$\dot{x}$ (cart velocity): $\infty$ $ms^{-1}$ to - 0.5, - 0.5 to 0.5, 0.5 to $\infty$ $ms^{-1}$ [3 partitions]
$\theta$ (pole angle): 12° to - 6, - 6 to - 1, - 1 to 0, 0 to 1, 1 to 6, 6 to 12° [6 partitions]
$\dot{\theta}$ (angular velocity): $-\infty^{\circ} s^{-1}$ to - 50, - 50 to 50, 50 to $\infty^{\circ} s^{-1}$ [3 partitions]

This creates a total of 54 regions (1*3*6*3). Conceptually a region can be regarded as a production rule with its condition part specified by the values of the state variables it covers and its action part specified by the direction of the pole-cart system whenever it is in that region. Thus the learning algorithm is required to evolve a set of 54 rules that will be able to keep the system under control.

We represent a sequence of these 54 regions (rules) — a controller — as a *chromosome* with a region regarded as a gene. A gene takes on a '1' indicating a left move or a '0' indicating a right move. At a time step, t, the pole-cart system will be at a gene (region) and the direction it moves next depends on whether the gene has a '1' (left move) or a '0' (right move). An individual chromosome, therefore, is made up of a string of '1's and '0's.

### 8.4.3 Performance Evaluator

The performance evaluator rates a chromosome (controller) by assigning it a fitness value. The value indicates how good the chromosome is in balancing the pole-cart system. The evaluator uses the length of the time (number of discrete time steps) that a chromosome holds the pole-cart system (from an initial position or state) without a failure as its fitness value.

## 8.5 Implementation Details

The population size affects the performance and efficiency of Simple Genetic Algorithm-based systems. A small size provides an insufficient sample, which makes them perform poorly. A large population size will undoubtedly raise the probability of the algorithm performing an informed and directed search.

However, a large population requires more evaluations per iteration or generation, possibly resulting in the method developing redundant controllers and becoming very slow especially when implemented serially. We experimented with different population sizes, they include 100, 150, 300 and 400. We found a population of 300 to be optimal for our task, but in general we assert that the optimum population size depends on the complexity of the domain, and in particular, on the shape of the fitness function.

We implemented a modified overlapping population in our simulations. First we do not replace a fixed percentage of the population from generation to generation, instead the percentage is allowed to vary dynamically within a fixed interval. Our aim is to strike a good balance between exploration and exploitation.

Secondly, we do not select those to be replaced randomly; we use the fitness value of an individual chromosome and the average fitness value of the population to determine whether or not a particular individual is to be replaced.

Since we do not generate a completely new population at each generation, some population members will continue unchanged into the next generation. This is at variance with the natural evolving process in which no member passes unchanged to the next generation. However, advantage can be taken of this 'immortality' when implementing a GA method on a computer system so as to minimise computation of evaluations of new entrants to the population. Also, in our application, we are interested in maintaining high performance levels as the algorithm learns to control the system; we therefore need to preserve the best rules so far discovered while we continue to search for better ones. This will ensure that the best information gained about the environment is not lost between generations. Our earlier simulations showed that it is difficult to preserve the best information with nonoverlapping populations.

Since we use a population of fixed size, some members have to be removed to make room for the newly generated ones. We decide on population members to be retained, if the termination condition has not been reached is as follows:

At the end of a generation, the average fitness of the current population is calculated. Individuals whose fitness values fall below the population average are replaced except when

(a) less than 20% of the population will survive (by 'survive' we mean continue into next generation unchanged) to the next generation; the best 20% of the population are retained;

(b) more than 60% of the population will survive; the best 60% are retained provided this has not been the case for more than 3 consecutive generations.

These measures are designed to discourage very few individuals from dominating the population, to ensure that adequate points are sampled in a generation and to increase diversity as soon as the algorithm detects that it has become low. The values of the parameters presented above were arrived at through experimentation.

Each offspring produced by crossover has a small probability (0.01) of being mutated. The number of offspring a survivor is allowed to produce by crossover is proportional to its fitness value. An individual is regarded as reproducing through crossover if it is the first of a couple to be chosen.

A mate is chosen randomly for a reproducing chromosome among the remaining survivors until all its children have been produced. When the number of children produced this way is less than the total needed, the remaining ones are produced by randomly choosing pairs from survivors for the crossover operation.

One of the main problems with implementing simple GA-based applications using finite population sizes is the possibility of premature convergence; that is, the possibility of the system converging onto suboptimal peak. Premature convergence takes place when population members are identical in their gene composition before the true optimum solution has been found [17]. That is, it occurs when there is a loss of diversity in the gene pool.

In order to minimise the loss of diversity, we introduced some innovative measures in our implementation that enable the program to dynamically alternate between exploiting the accumulated knowledge and exploring the solution space as the need arises. In that section, we specified that the percentage of population members retained to go unchanged ('survive') into the next generation varied between a minimum value (20%) and a maximum value (60%). In addition to the above measures, we introduced a new individual into a population only if it is different from every other member of the population by at least one bit. An offspring that is identical to a member present in a population is regarded as stillborn.

The GAPOLE is required to learn to control the simulated pole-cart system for 10,000 time steps continuously without a failure signal. A learning session is completed when at least one population member achieves this level of performance or when the total number of points sampled exceeds 100,000 points. No population sampled points close to this limit before at least one of its members achieved the required level of performance.

## 8.6 Experimental Results

The simulation program was written in C and to two sets of experiments were carried out on a Sequent Balance B8000 computer. Each set consists of running our GA program 50 times; each time initialising the Unix's random number generator with a new seed. Directions were randomly fixed for the chromosomes at the start of a run.

In the first set of experiments with a simple GA, a force of 10 Newtons was applied to the right or left (-10 for a left direction) of the base of the cart while a force of 5 Newtons was applied to right and 10 Newtons (-10) to left of the base of the cart in the second set of experiments.

| Pop. | Generations | | | Failures | | | Time taken(hr:min:sec) | | |
|---|---|---|---|---|---|---|---|---|---|
| Size | Min | Max | Mean | Min | Max | Mean | Min | Max | Mean |
| 100 | 6 | 544 | 28 | 472 | 28853 | 1643 | 2:46 | 6:37:21 | 23:24 |
| 150 | 2 | 42 | 14 | 259 | 3571 | 1301 | 0:41 | 1:56:39 | 16:13 |
| 300 | 3 | 21 | 9 | 729 | 3808 | 1842 | 1:48 | 2:14:31 | 17:04 |
| 400 | 4 | 21 | 9 | 1228 | 5297 | 2394 | 2:58 | 2:55:25 | 23:06 |

Table 8.1: Performance summary of GAPOLE for population sizes of 100, 150, 300 and 400 when pushing left and right with a force of 10 Newtons.

| Pop. | Generations | | | Failures | | | Time taken(hr:min:sec) | | |
|---|---|---|---|---|---|---|---|---|---|
| Size | Min | Max | Mean | Min | Max | Mean | Min | Max | Mean |
| 100 | 4 | 321 | 43 | 290 | 17260 | 2382 | 0:48 | 13:50:21 | 41:12 |
| 150 | 6 | 97 | 30 | 702 | 7612 | 2477 | 3:16 | 2:09:10 | 23:50 |
| 300 | 4 | 51 | 16 | 876 | 8324 | 2876 | 1:41 | 1:53:02 | 19:29 |

Table 8.2: Performance summary of GAPOLE for population sizes of 100, 150 and 300 when using a force of 5 Newtons to push right and a force of 10 Newtons to push left.

The results of the experiments on simulated pole-cart system are shown in Table 8.1 and Table 8.2.

Table 8.1 and Table 8.2 show that population size of 300 produced the best average computational times (17:04 and 19:29 minutes, respectively) for two experiment sets.

The simple GA-based learning program was regarded to have succeeded in balancing the pole-cart as soon as it was able to hold the system continuously without a failure signal for 10,000 discrete time steps. This is in line with the performance level set by Sammut [23].

Sammut [23] evaluated three best alternative learning algorithms (reviewed in the previous subsection) — BOXES [19], AHC [24] and CART [5] — that solved the pole-cart balancing task in the form considered in this chapter. He used 162 regions for BOXES and AHC in his experiments while we employed 54 for ours. CART does not divide the solution space into regions. These three algorithms are point-based (i.e., they generate, test and modify single solutions) while GAPOLE is population-based. At a generation or an iteration, therefore, they sample only one point in the solution space while the number of points sampled by our simple genetic algorithm-based program is equal to the number of new individuals introduced into the population at that generation. For our comparisons, a generation (or an iteration) when used for the point-based algorithms is equivalent to a trial or a sampled point.

The average number of generations and the average points sampled by the genetic-based method (GAPOLE), BOXES, AHC and CART to learn to balance the pole-cart system for 10,000 time steps using a force of 10 Newtons are in Table 8.3.

| | GAPOLE | BOXES | AHC | CART |
|---|---|---|---|---|
| Iterations | 9 | 225 | 90 | 13 |
| Points | 1842 | 225 | 90 | 13 |

Table 8.3: Average iterations and points sampled using a force of 10 Newtons. Averages for BOXES, AHC and CART were over 5 runs (Sammut [23]).

When the force applied to the pole-cart system was changed, i.e., a force of 5 Newtons was applied when going right and a force of 10 Newtons applied when going left, the average number of generations and the average points sampled by the four algorithms are given in Table 8.4.

| | *GAPOLE* | *BOXES* | *AHC** | *CART+* |
|---|---|---|---|---|
| Iterations | 16 | 837 | 2562 | terminated |
| Points | 2876 | 837 | 2562 | terminated |

Table 8.4: Average iterations and points sampled using 5 Newtons to push right, and 10 Newtons to push left. * The averages for AHC were taken over 4 runs. The fifth was stopped after it failed to achieve the performance level within 50,000 iterations (Sammut [23]) - the maximum number allowed. + CART terminated with a floating point exception (Sammut [23]).

The percentage increases in the number of generations and the number of points sampled by these algorithms when the force applied changed from being even to uneven is given in Table 8.5.

| | *GAPOLE* | *BOXES* | *AHC* | *CART* |
|---|---|---|---|---|
| Generations | 77.78% | 272.00% | 2746.67% | - |
| Points | 56.13% | 272.00% | 2746.67% | - |

Table 8.5: Percentage (%) increases in generations k points sampled when the force changed from even to uneven.

BOXES and AHC collect some statistical data for each box or partition in order to determine what their actions should be whenever the system is in that region. CART assumes that the control surface is described by a smooth function and thus cannot be used for surfaces that are discontinuous. Our technique neither keeps statistical information for each box nor assumes a particular type of surface. Also, it acts on the knowledge structures (controllers) syntactically. That is, it manipulates them without taking into consideration any interpretations given to these structures. Any knowledge structures can therefore be substituted for the population of controllers.

Our program uses a table you look up to make a control decision (the same holds true for BOXES) and thus makes the decision quickly. AHC takes a longer time to choose a control action since it revises the settings of its boxes after each step. CART computes two vectors and their inner product before it decides on a control action. As these calculations take quite sometime to perform, CART takes a considerably longer time to choose a control action than our technique.

AHC and CART are, however, able to learn during and after trials compared to GAPOLE that only learns after trials.

## 8.7 Difficulties with GAPOLE Approach

The GAPOLE simulation also showed that the evolution of viable rule-set (candidate solution) requires that parameters are restricted to a particular range, which alone is searched. Moreover, altering the direction of several boxes (regions) simultaneously in a generation could arrive at the solution point faster. But increasing mutation rates could be harmful to simple genetic algorithm-based systems as the search could degenerate to a random search with many non-viable offspring generated.

Another difficulty with the present (GAPOLE) approach is the partitioning of the search space (as in BOXES), the number of partitions to be used needs to be decided. If the number of partitions are too small than the control rules will be coarse and inaccurate, however, use of too many partitions may results in fine control action, but will be very difficult to induce control rules, using the GAPOLE approach, within a reasonable time and with reasonable effort. Moreover, static partitioning of state space appears to be inefficient for precise control in a dynamic system. An alternative genetic approach for generating a more robust viable controller will be investigated next for solving the problem.

## 8.8 A Different Genetic Approach for the Problem

In previous sections, we have seen that the pole balancing problem has often been used as an exercise in the control of dynamic systems and has been studied extensively by researchers in different fields. Other than classical control theory approaches, it has been solved mostly using different techniques of Artificial Intelligence (AI). They include machine learning [19], fuzzy logic [3, 20], qualitative modelling [16], neural networks [1, 2, 13], genetic algorithms [18, 25], etc.

In genetic approaches [21, 25], the problem state-space was divided (discretised) into a number of predefined partitions (as in the BOXES method). The genetic encoding was a binary string where each gene represented each partition and its value determined the appropriate action (push left or right). However, with these approaches the performance of the controller (or control rule) depends on thc number of partitions used and has difficulty in generalising the control rules [26].

For this pole-balancing problem, neural-based methods are widely used. The advantage of using neural networks is twofold: versatile mapping capabilities from input to output and its learning ability without explicit knowledge of mathematical basis of the system. It is widely recognised that the architecture of a neural network can have a significant impact on the network's function and processing capability. In most cases, predefined architectures are used for performing tasks with neural nets. Though genetic algorithms can replace the effort of human designers in determining network structures and also can be used for training predefined neural nets, but until recently, GAs were used for one or the other purposes (designing network structure or optimising neural net weights).

A combination of neural networks and genetic algorithms has also been used where a fixed network has been trained with genetic reinforcement learning [26]. The Genetic Cascade Learning algorithm was employed [15] to sequentially build the net to perform the task.

The method described below is an alternative neurogenetic approach, in which both the network architecture and its weights evolve together in an implicitly parallel fashion. In the remainder of this subsection we will give a brief description of Structured Genetic Algorithms (sGA) and then describe the application of sGA for the automatic design of neurocontrollers using genetic reinforcement learning.

## 8.9 The Structured Genetic Algorithm

The Structured Genetic Algorithm (sGA) [8][11] uses genetic redundancy and hierarchical genomic structures in its chromosome. Genes at different levels may be active *(on)* or passive *(off)* phenotypically. The primary mechanism for eliminating the conflict of redundancy is through higher level genes which act as switching operators for expressing genes at lower levels. The model also uses conventional genetic operators and the *survival of the fittest* criterion to evolve increasingly fit individuals. These characteristics allow the model to solve complex multi-stage problems.

In an sGA a chromosome is usually represented as a set of substrings. It also uses conventional genetic operators and the 'survival of the fittest' principle. However, it differs considerably from the Simple Genetic Algorithms in encoding genetic information in the chromosome and in its phenotypic interpretation. The fundamental differences are as follows:

- Structured Genetic Algorithms utilise chromosomes with a multi-level genetic structure (a directed graph). As an example, sGA's having a two-level structure of genes are shown in Figure 8.3a, and chromosomal representations of these structures are shown in Figure 8.3b.
- Genes at any level can be either active or passive.
- 'High level' genes activate or deactivate the lower level genes.

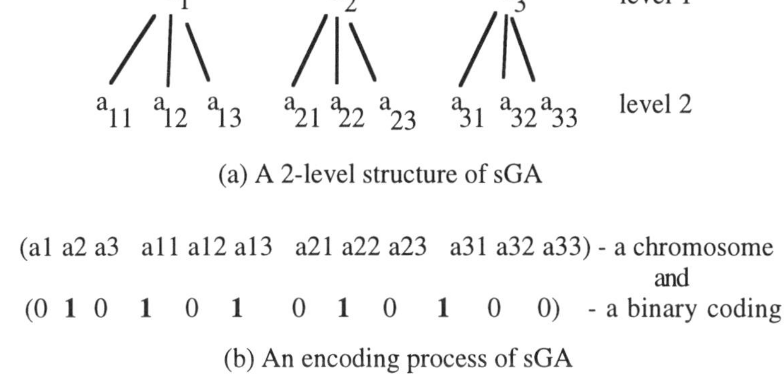

(a) A 2-level structure of sGA

(b) An encoding process of sGA

Figure 8.3: A simple representation of a two-level sGA.

While applying to the field of neural networks, the model can define the network structure and its connection weights in its chromosome, and these parameter sets can be optimized, in parallel, as a single unified process. In each generation, while some members of the population are engaged in searching for the feasible topology; others, which already have feasible structures, are searching for a set of optimal weights, and the process continues until a fully trained network evolves which can solve the task. The details of the model and its application for full designing of neural nets were explained in our previous work [6][7].

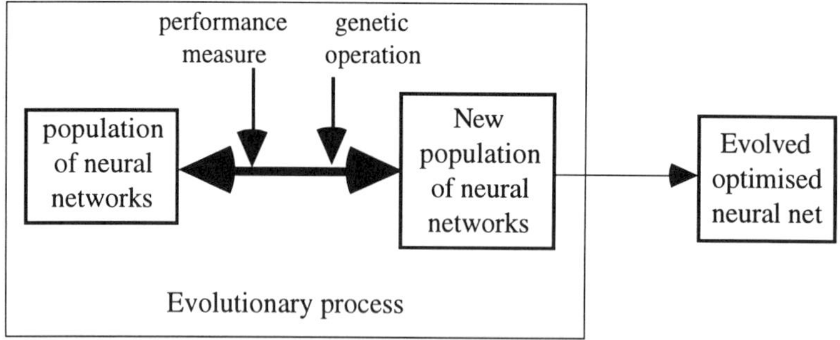


Figure 8.4: Genetic process of evolving neural networks.

Figure 8.4 shows the working principle of the sGA for designing an application specific neural network architectures.

## 8.10 Evolving Neuro-Controllers Using sGA

For this empirical study we adopted a two-level sGA for encoding the complete neural network. Each individual (chromosome) has a two-level genomic structure in Figure 8.5c. The higher level defines the network configuration, while the lower level encodes the connection weights and biases. As mentioned before the high-level of the sGA searches the connectivity space of N units (to evolve an efficient network structure), while the low-level searches for an optimal set of weights to control the system. The fitness of each individual is determined by the combined performance of these two components [6]. A set of individuals (population) is generated randomly to initialise the evolution-learning process.

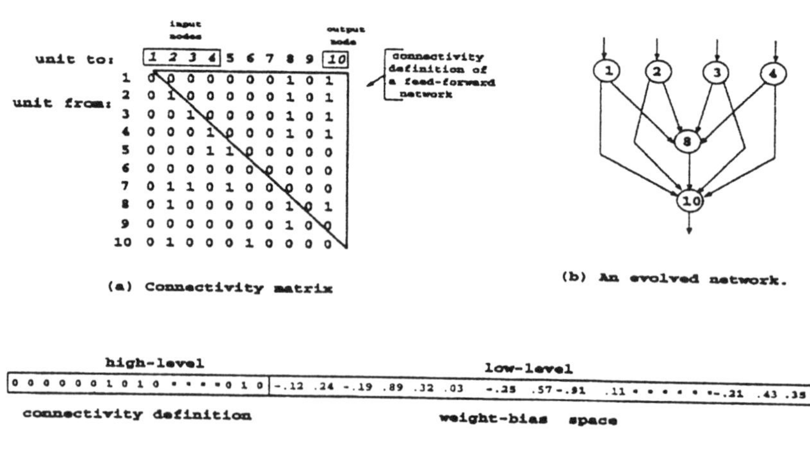


Figure 8.5: A two-level sGA representing neuro-controller.

Here we considered two vertical angles ($12^{\circ}$ *or* $35^{\circ}$) for balancing. As in previous sections, the initial starting position of the cart is randomly set between ±0.1 meters, the starting pole angle between $\pm 6^{\circ}$; the cart velocity and pole's angular velocity are set to 0.0 at the start of each training phase [18]. These values are considered as the initial inputs for each individual neural net at every generation. The input state vector is normalised so that the values lie in the range 0 and 1. The algorithm terminates if at least one evolved net holds the pole for 120,000 time steps (i.e., 40 minutes of simulated time) or the allowed number of iterations are used up (2000 generations).

## 8.11 Fitness Measure and Reward scheme

In every generation, each chromosome is decoded into its phenotype (a network structure with its weights), and its fitness is evaluated by taking into account the feasibility of the structure and its ability to learn the control task. More specifically, since a sGA is used to find both an architecture and the synapse weights, the evaluation function must include not only a measure the learnability of a net, but also a feasibility measure of network structure and its complexity (i.e., number of nodes and their connectivities).

In a randomly-generated initial population there are likely to be a large number of individuals which show poor performance due to two reasons:

1. They have a infeasible network structure i.e., improper connectivity pattern;

2. Arbitrary values of weight-bias parameters which may be far from optimum (even though the structure is feasible).

A network structure is infeasible

- If there exists no path from input nodes, and/or to output nodes,
- If there is fan-in to a hidden node but no fan-out or vice versa,
- If there is any unreachable substructure, etc.

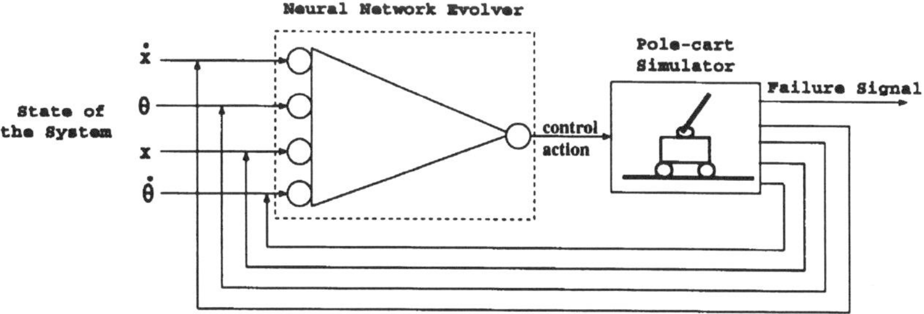


Figure 8.6: Reinforcement learning of neuro-controllers using sGA.

The infeasibility measures quantify the amount by which an individual structure exhibits 'congenital defects' (deformation).

It is necessary to avoid destructive interference between the two searches in their different spaces. For this reason, if an individual decodes to a feasible structure, it is rewarded by keeping its high-level portion stable (i.e., no changes are subsequently allowed to occur), and only the weight-bias space is then explored. However, while training a feasible net, if no improvement is noticed in balancing the pole for 50 successive generations, the individual then loses its structural stability and is downgraded or eliminated. Feasible individuals which have fewer nodes and links also get a selection advantage for reproduction relative to the competing feasible individuals with more complex structures. Since we are rewarding only the feasible structures, there is no chance of an individual structure getting reward by pruning all its connections and nodes so as to become infeasible.

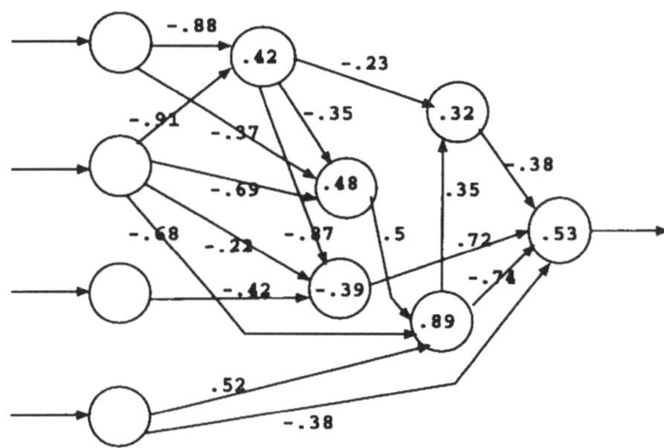


**For 35° case**

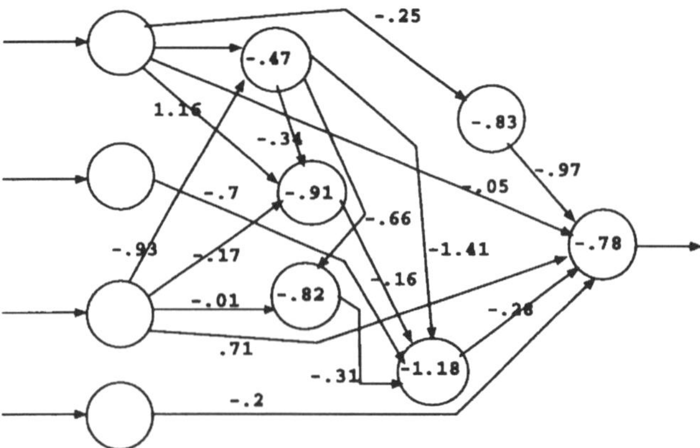


Figure 8.7: The evolved neural net controllers and their weights.

Thus each feasible net (individual) is trained through genetic reinforcement learning [26], where the learning process is also an object of evolution. Figure 8.6 shows the functional blocks involved during learning phase of feasible nets (controller's). The approach considers two 'black boxes' communicating with each other, neither knowing the internal dynamics of the other. The first 'black box' designs the controller to adapt the environment of the other through an evolutionary process. The second responds to the control action of the first, and feedback the system response at each time step. The only information for evaluating performance is a failure signal which indicates that the pole-cart system is/is not out of control.

The learning process of a feasible net starts by providing the initial state of the system to the net and the net's output response is applied to the simulated system. The output of the net is either 0.0 (push left) or 1.0 (push right) representing the direction of a bang-bang control force. The output of the system is a new state vector which is then reintroduced as new inputs to the net. This continues until failure occurs or successful control is performed for the prescribed maximal period of time. The balancing time is the measure of what has been learned by each feasible net.

The individuals decoding to infeasible structures are penalised according to their deformation and undergo a higher rate of mutation in their high level, structural portion of their genome when being selected for reproduction. They thus have the chance to reproduce by changing their connectivity pattern (which may result in feasible offspring) and thus becoming stable members of the population. Exploration of new feasible structures and evolution of weights of the existing stable networks continues until a near optimised network architecture evolves or the whole population converges to a feasible network architecture.

## 8.12 Simulation Results

Following our general methodology for neural network design and training (Section 5.3), in this experiment we used a mixed encoding technique, where the high-level portion of the chromosome is binary-coded representing the topology of a neural net. The low-level is real-valued encoding the weight-bias space in the range of (-1.0, +1.0) and crossover points are allowed to occur only between the weights. A simple bit mutation is applied on the high level and a floating point mutation is used on the low level such that a random value within ±0.1 is added to the existing active weight space rather than replacing it. The mutation rate is varied between 5 and 10% adaptively in two levels of sGA. Different GA parameters were tested in a number of trial runs of the experiment. The reported results used a population size of 80 and a two-point crossover operator with a probability of 75% along with a ranking selection scheme for reproduction. The size of connectivity matrix used is 10 by 10 along with a logistic transfer function for all the nodes.

In most trial runs, the algorithm takes less than 1000 generations to evolve a net that could successfully perform the balancing task. It is found that during different runs many feasible network structures evolved, but ones which could learn quickly proliferate in the population in the later stages. When no restriction is imposed on evolving nets, most of the rapidly-learned structures are highly irregular and have direct links between inputs and the output (these are, however, fully effective controllers) [10]. Our preliminary results [9] also showed that regular structures may be evolved by modifying the evaluation function. Figure 8.7 shows two network structures which evolved in two different runs for balancing the pole at different cut-off angles. Figure 8.8 shows the displacement of the pole and the cart. The performance of the best evolved net in one typical run shown in Figure 8.9.

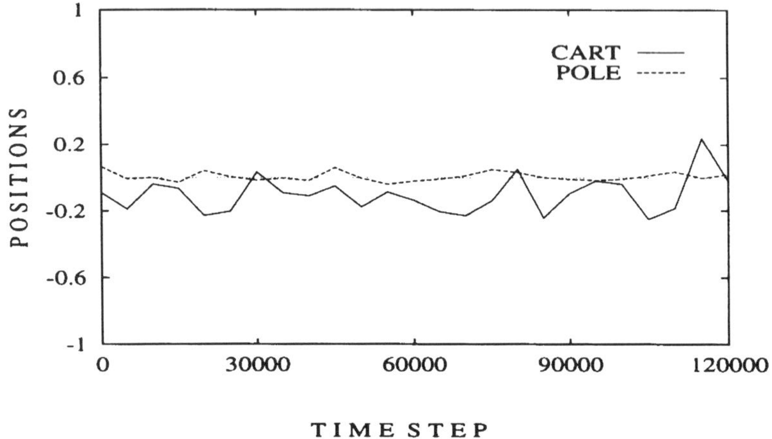


Figure 8.8: Graphs shown the position of the pole and cart with the evolved neuro-controller at different time step (in 12° case).

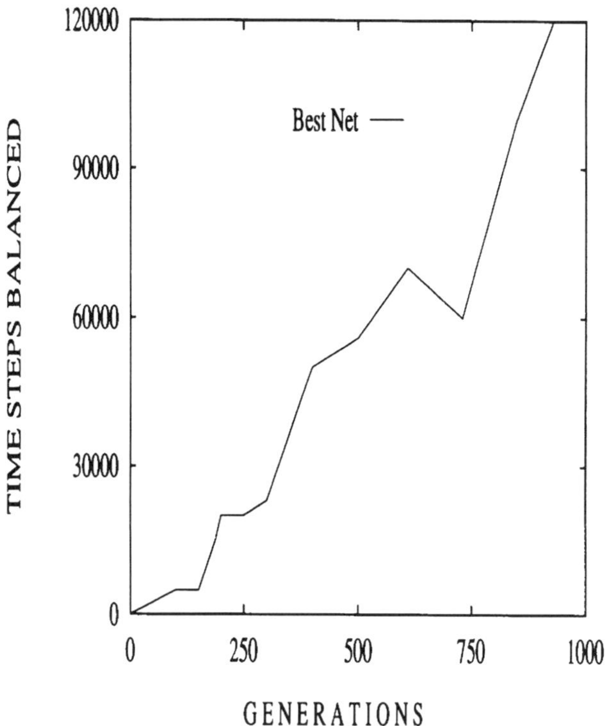


Figure 8.9: The best individual net's performance for balancing task (in 12$^{o}$ case).

## 8.13 Discussion

We applied sGA for evolving neuro-controllers which could learn a mapping between a dynamic system's state space and the space of possible actions. The significance of the sGA result is considerable. It makes possible to automatically design neuro-controller for a complex dynamic control task, by the expenditure of a relatively small amount of computational resource. The structured GA approach offers the following advantages:

1. It can evolve network structures and their weights in a single evolutionary process;
2. Each individual net can be trained using genetic reinforccment learning;
3. The method does not require partitioning of the state space of the problem;
4. No supervisory training data is required for performing the balancing task;
5. It uses global search rather than local search;
6. It can be implemented in parallel to improve the speed of convergence.

Since the results are encouraging, further work should examine generalising this evolutionary neuro-controller to enable it to operate over all possible initial input states of the system, in a similar way to that reported by [26].

## References

[1] C. W. Anderson. Strategy learning with multilayer connectionist representations. In *Proceedings of the Fourth International Workshop on Machine Learning,* pages 103-114. Morgan Kaufmann, Los Altos, 1987.

[2] Andrew G. Barto, Richard S. Sutton, and Charles W. Anderson. Neuron-like adaptive elements that can solve difficult learning control problems. *IEEE Transactions on Systems, Man and Cybernetics,* Smc-13(5):834-846, Sept/Oct 1983.

[3] Hamid R. Berenji and Pratap Khedkar. Learning and tuning fuzzy logic controllers through reinforcements. *IEEE Transaction on Neural Networks,* 3(5):724-740, September 1992.

[4] Ka C. Cheok and K. Loh. A ball-balancing demonstration of optimal and disturbance-accommodating control. *IEEE Control Systems Magazine,* pages 54-57, 1987.

[5] Margaret E. Connell and Paul E. Utgoff. Learning to control a dynamic physical system. In *Proceedings AAAI-87 Sixth National Conference on Artificial Intelligence,* pages 456-460, 1987.

[6] Dipankar Dasgupta and D. R. McGregor. Designing Application-Specific Neural Networks using the Structured Genetic Algorithm. In *Proceedings of the International workshop on Combination of Genetic Algorithms and Neural Networks (COGANN-92),* pages 87-96. IEEE Computer Society Press, June 6, U.S.A 1992.

[7] Dipankar Dasgupta and D. R. McGregor. Designing Neural Networks using the Structured Genetic Algorithm. In *Proceedings of the International Conference on Artifical Neural Networks (ICANN),* pages 263-268, Brighton, U.K., 4-7 September 1992.

[8] Dipankar Dasgupta and D. R. McGregor. Nonstationary function optimization using the Structured Genetic Algorithm. In *Proceedings of Parallel Problem Solving From Nature (PPSN-2),* Brussels, 28-30 September, pages 145-154, 1992.

[9] Dipankar Dasgupta and D. R. McGregor. Evolving Neurocontrollers for Pole Balancing. In *Proceedings of the International Conference on Artificial Neural Networks (ICANN),* pages 834-837, Amesterdam, The Netherlands, 13-16 September 1993.

[10] Dipankar Dasgupta and D. R. McGregor. Genetically Designing Neuro-controllers for a Dynamic System. In *Proceedings of the International Joint Conference on Neural Networks (IJCNN)*, pages 2951-2955, Nagoya, Japan, 25-29 October 1993.

[11] Dipankar Dasgupta and Douglas R. McGregor. A More Biologically Motivated Genetic Algorithm: The Model and some Results. *To appear in Cybernatics and Systems: An International Journal,* 25(3), May 1994.

[12] David E. Goldberg. *Genetic Algorithms in Search, Optimisation and Machine Learning.* Addison-Wesley, first edition, 1989.

[13] E. Grant and Bing Zhang. A neural-net approach to supervised learning of pole balancing. In *Proceedings of IEEE International Symposium on Intelligent Control*, pages 123-129, Albany, New York, 25-26 September 1989.

[14] John H. Holland. *Adaptation in Natural and Artificial Systems.* University of Michigan Press, Ann Arbor, 1975.

[15] N. Karunanithi, D. Whitley and R. Das. Genetic Cascade Learning for Neural Networks. In *Proceedings of International Workshop on Combinations of Genetic Agorithms and Neural Networks,* pages 134-145. IEEE Computer Society Press, 1992.

[16] A. Makarovic. *A qualitative way of solving the pole balancing problem,* volume 12, chapter 16, pages 241-258. Oxford University Press, 1988.

[17] M.L. Mauldin. Maintaining diversity in genetic search. In *Proceedings of the National Conference on Artificial Intelligence,* pages 247-250, 1984.

[18] D. R. McGregor, M.O. Odeytayo, and D. Dasgupta Adaptive control of a dynamic system using genetic-based methods. In *IEEE International Symposium on Intelligent Control*, August 11-13 1992. Glasgow, U.K.

[19] D. Miche and R.A. Chambers. Boxes: An experiment in adaptive control. *Machine Intelligence,* 2:137-152, 1968.

[20] N.J. Hallman, N. Woodcock, and P. D. Picton. Fuzzy boxes as an alternative to neural networks for difficult problems. In G. Rzevski and R. A. Adey, editors, *Application of Artificial Intelligence in Engineering VI (AIENG/91)*, pages 903-919, 1991.

[21] M.O. Odetayo and D. R. McGregor. Genetic algorithm for control rules for a dynamic system. In *Proceedings of ICGA-89*, pages 177-181, 1989.

[22] Michael Omoniyi Odetayo. *On Genetic Algorithms in Machine Learning and Optimisation.* PhD thesis, Department of Computer Science, University of Strathclyde, Glasgow, U. K., December 1990.

[23] Claude Sammut. Experimental results from an evaluation of algorithms that learn to control dynamic systems. *Proceedings of the Fifth International Conference on Machine Learning*, 1988.

[24] Oliver G. Selfridge and Richard S. Sutton. Training and tracking in robotics. In *Proceedings of the Ninth International Conference on Artificial Intelligence (IJCAI).* Morgan Kaufmann, Los Altos, 1985.

[25] Dirk Thierens and Leo Vercauteren. A topology exploiting genetic algorithm to control dynamic systems. In G. Goos and Hartmanis, editors, *Lecture Notes in Computer Science,* pages 104-108. Springer-Verlag, 1991. Proceedings of PPSN-I, 1990.

[26] D. Whitley, Stephen Dominic, and R. Das. Genetic reinforcement learning with multilayer neural networks. In *4th International Conference on Genetic Algorithms*, pages 562-569, 1991.

Chapter 9

**Luis Rabelo**
ISE Department
Ohio University
Athens, OH 45701

**Albert Jones**
National Institute of
Standards and Technology
Gaithersburg, MD 20899

**Yuehwern Yih**
School of Industrial Eng
Purdue University
W. Lafayette, IN 47907

# A Hybrid Approach Using Neural Networks, Simulation, Genetic Algorithms, and Machine Learning for Real-Time Sequencing and Scheduling Problems



**Abstract**
A hybrid approach for sequencing and scheduling is described which integrates neural networks, real-time simulation, genetic algorithms, and machine learning. This approach has been used to solve both single machine sequencing and multi-machine scheduling problems. Neural networks are used to quickly evaluate and select a small set of candidate sequencing or scheduling rules from some larger set of heuristics. This evaluation is necessary to generate a ranking which specifies how each rule performs against the performance measures. Genetic algorithms are applied to this remaining set of rules to generate a single "best" schedule using simulation to capture the system dynamics. A trace-driven knowledge acquisition technique (symbolic learning) is used to generate rules to describe the knowledge contained in that schedule. The derived rules (in English-like terms) are then added to the original set of heuristics for future use. In this chapter, we describe how this integrated approach works, and provide an example.

0-8493-2529-3/95/$0.00 + $.50

## 9.1 Introduction

Sequencing and scheduling are two of the most important decisions made by any shop floor control system. But, while there has been an enormous research effort expended over the years in these areas, it has had little effect in the marketplace. The reason is simple, the research has led to the development of very few software tools that can solve real problems. The tools that do exist are typically 1) too slow and cannot react to changing shop floor conditions, 2) based on simplistic formulations which ignore important constraints like material handling, 3) based on a single objective function or simplistic trade-offs like goal programming, and 4) difficult to install and integrate into pre-existing commercial shop floor control systems.

In this chapter, we describe a methodology which integrates neural networks, simulation, genetic algorithms, and machine learning technniques. It determines the start and finish times of the jobs assigned to any module in the hierarchical shop floor control architecture proposed in (JONES and SALEH, 1990). Because this hierarchy decomposes the global scheduling problem into multiple levels, this methodology 1) never needs to solve very large problems, and 2) can react to delays on the shop floor in a manner which is not disruptive to the rest of the system. Moreover, by exploiting the parallel processing and modeling capabilities of neural networks, simulation, and genetic algorithms it has the potential to be extremely fast and highly adaptable to customer needs. Finally, the use of a learning technique provides the additional capability to learn what works and what does not work in a variety of situations and utilize that knowledge at a later time. For these reasons, we believe that this technique has the potential to solve real-world sequencing and scheduling problems in real-time.

The chapter is organized as follows. In section 9.2, we describe the generic controller and shop floor hierarchy. In section 9.3, we describe the method for generating start and finish times which is applicable at every level in that hierarchy and the learning technique. In section 9.4, we provide an example.

## 9.2 Hierarchical Generic Controller

The foundation of this research is the generic controller developed in (DAVIS et al., 1992) and the hierarchical shop floor control system described in (JONES and SALEH, 1990). This hierarchy is based on a decomposition of the global planning and scheduling problems, rather than the traditional partitioning of the physical shop floor equipment. This new approach to building hierarchies led to the fundamental contribution of (DAVIS et al., 1992) — that every controller in this hierarchy performs the exact same four production management functions — assessment, optimization, execution, and monitoring. We now give a brief description of these functions (see Figure 9.1).

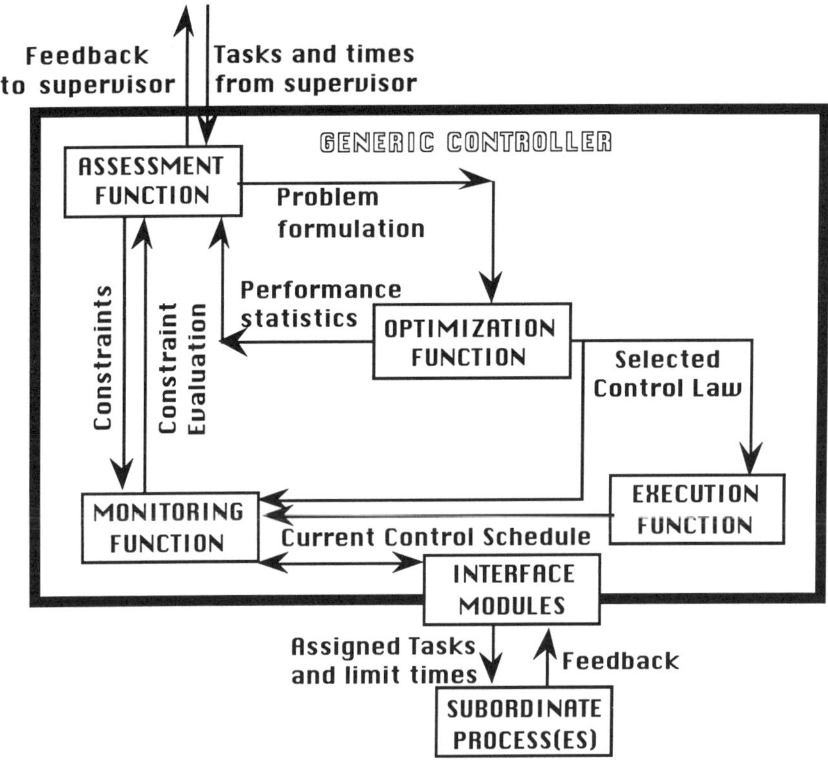


**Figure 9.1 Generic controller module.**

The Assessment Function formulates the real-time decision-making problems for each control module. The exact nature of those decisions depends on the hierarchical level at which the module resides (JACKSON and JONES, 1987). These decisions can be thought of as optimization problems. This means that the Assessment Function must specify both the constraints and the performance measures. Two types of constraints are allowed: hard and soft. Hard constraints are those that cannot be violated either by the other functions in the same module or by modules in subordinate levels in the hierarchy. These constraints come from three sources: supervisors, process plans, and the physical limitations of the system. A supervisor may impose hard constraints such as due dates, priorities, and maintenance schedules. The process planner may impose hard constraints in the form of precedence relationships, tools, and fixturing requirements. Finally, physical limits such as transfer times and queue sizes also result in hard constraints.

The Assessment Function can also specify soft constraints to further control the evolution and behavior of its subordinates. Typically, minor violations of these constraints will be tolerated, but major violations indicate that the system may be getting into trouble. For example start and finish times for the individual tasks that make up a job can be viewed as soft constraints. As long as the job is on time, delays in the start and finish times of some tasks can be tolerated. However, as more and more of these tasks are delayed, the on-time completion of the job is jeopardized. Other commonly imposed soft constraints are utilization rates for subordinates and inventory policies. Unlike the hard constraints, these soft constraints may be unknown to subordinates.

As noted above, the Assessment Function also specifies the performance criteria for each optimization problem. There are typically several, possibly conflicting, criteria to be considered simultaneously, which combine the "performance" of subordinates and the "performance" of jobs. Examples of subordinate performance include utilization and throughput. Examples of job performance include lateness, tardiness, and makespan. Priorities can be assigned to particular jobs and weights to particular performance criteria. All of these can be changed to reflect the current state of the system.

The Optimization Function is responsible for solving these decision-making problems posed by the Assessment Function. The solution consists of a) selecting a run-time production plan for each job and b) selecting the start and finish times for each of the tasks in that plan. The production plan identifies the tasks and the resources needed to complete each job. It also includes all precedence relations that exist among those tasks. This run-time plan is selected from the set of feasible process plans passed down by the Assessment Function. Selecting the start and finish times for these tasks may involve the solution of a single machine sequencing problem, a multi-machine scheduling problem, a multi-cell routing problem, or a resource (tools, fixtures, transporters) allocation problem. All selections are made to optimize the current set of performance measures. In addition to making these initial selections, the Optimization Function must deal with violations of the constraints imposed by the Assessment Function. This may involve the selection of new sequences, schedules, or plans.

The Execution Function implements the decisions selected by the Optimization Function. Using the current state of the system, it does a single pass simulation to compute the start and finish times for each task to be assigned to one of its subordinate modules. In addition, when minor deviations from these times are either reported or projected by the subordinates, the Execution Function attempts to restore feasibility using techniques such as perturbation analysis or match-up scheduling (JONES and SALEH, 1990).

Lastly, the Monitoring Function updates the system state using feedback from subordinates, and evaluates proposed subordinate responses against the current set of imposed (from the Assessment Function) and computed (by the Optimization Function) constraints. It determines when violations occur, their severity, and who should deal with them.

## 9.3 Implementing the Optimization Function

We now describe a methodology for solving the real-time sequencing and scheduling (s/s) problems faced by the Optimization Function. This method consists of a three step refinement process. The first step is to generate a set of candidate s/s rules from a much larger set of heuristics. We have used single-performance, neural networks as discussed in Section 9.3.1. We then evaluate these candidates against all of the performance measures dictated by the

Assessment Function. This ranking is based on a real-time simulation approach as discussed in Section 9.3.2. The last step is to use the top candidates from that ranking as input to a genetic algorithm to determine the "best" sequence or schedule. This is discussed in Section 9.3.3. In Section 9.3.4, we describe a technique for extracting the knowledge contained in that schedule for future use.

### 9.3.1 Candidate Rule Selection

The first step in this process is to select a small list of candidate rules from a larger list of available rules. For example, we might want to find the best five dispatching rules from the list of all known dispatching rules so that each one maximizes (or minimizes) at least one of the performance measures, with no regard to the others. To carry out this part of the analysis, we have used neural networks. This approach extends earlier efforts by (RABELO, 1990) and (CHRYSSOLOURIS et al., 1991).

Neural networks have shown good promise for solving some classic, textbook job shop scheduling problems. (FOO and TAKEFUJI, 1988) and (ZHOU et al., 1990) have applied stochastic Hopfield networks to solve 4-job 3-machine and 10-job 10-machine job shop scheduling problems, respectively. These approaches tend to be computationally inefficient and frequently generate infeasible solutions. (LO and BAVARIAN, 1991) extended the two-dimensional Hopfield network to 3 dimensions to represent jobs, machines, and time. Another implementation based on stochastic neural networks applied to scheduling can be found in (ARIZONO et al., 1992). However, they have been unable to solve real scheduling problems optimally because of limitations in both hardware and algorithm development.

These implementations have been based on relaxation models (i.e., pre-assembled systems which relax from input to output along a predefined energy contour). The neural networks are defined by energy functions in these approaches. (LO and BAVARIAN, 1991) formulated the objective function which minimizes makespan as

$$E_t = (1/2) \Sigma_{j=1} \Sigma_{i=1} \Sigma_{l=1} (v_{ijl}/C_k) (l + T_{ij} - 1)$$

where $C_k$ is a scaling factor; $v_{ijl}$ is the output of neuron $ijl$, and $T_{ij}$ is the time required by $j^{th}$ machine to complete the $i^{th}$ job.

However, due to a large number of variables involved in generating a feasible schedule, it has been difficult for these approaches to solve realistic job shop scheduling problems with multiple objectives. It is even difficult to get a good suboptimal solution when attempting to solve problems in real-time.

There are four reasons to select neural networks as candidate rule selectors. First, because of the decomposition that results from the hierarchical control architecture we are using, we never have to solve the global shop floor scheduling problem all at once. Since it is decomposed into several scheduling and sequencing problems (of smaller size and complexity), we don't anticipate the kinds of problems described above. Second, it is no longer necessary to resolve the global problem each time a minor delay occurs or a new job is put into the system. Local rescheduling and resequencing can be done with little impact on the overall shop floor schedule. Third, as discussed below, each neural network is designed (i.e., they are presented with training sets of representative scheduling instances and they learn to recognize these and other scheduling instances) to optimize a single objective (e.g., minimization of work-in process inventory). Neural networks in our approach are utilized as pattern recognition machines. Neural networks assign a given shop floor status to a specific rule with some degree. Finally, the solution from the neural networks is just the beginning of this methodology, not the end. Therefore, a very fast technique is needed. Neural networks are a proven real-time technique with speed (inherent from their distributed/parallel processing nature), timeliness, responsiveness, and graceful degradation capabilities.

In this research, we will focus our initial efforts on backpropagation neural networks, because they are more developed and much faster than the relaxation models described above (RUMELHART et al., 1988). Backpropagation applies the gradient-descent technique in a feed-forward network to change a collection of weights so that the cost function can be minimized. The cost function, which is only dependent on weights and training patterns, is defined by:

$$\mathbf{C(W) = (1/2)\ \Sigma(T_{ip} - O_{ip})^2}$$

where the $\mathbf{T}$ is the target value, $\mathbf{O}$ is the output of network, $\mathbf{i}$ is the output nodes, and $\mathbf{p}$ is the number of training patterns.

After the network propagates the input values to the output layer, the error between the desired output and actual output will be "back-propagated" to the previous layer. In the hidden layers, the error for each node is computed by the weighted sum of errors in the next layer's nodes. In a three-layered network (see Figure 9.2), the next layer means the output layer. If the activation function is sigmoid, the weights are modified according to

$$\Delta \mathbf{W_{ij} = h\ X_j\ (1 - X_j)(T_j - X_j)\ X_i} \tag{9.1}$$

or

$$\Delta \mathbf{W_{ij} = h\ X_j\ (1 - X_j)\ (S\ d_k W_{jk})\ X_i} \tag{9.2}$$

where $\mathbf{W_{ij}}$ is weight from node $\mathbf{i}$ to node $\mathbf{j}$, $\mathbf{h}$ is the learning rate, $\mathbf{X_j}$ is the output of node $\mathbf{j}$, $\mathbf{T_j}$ is the target value of node $\mathbf{j}$, $\mathbf{d_k}$ is the error function of node $\mathbf{k}$.

If $\mathbf{j}$ is in the output layer, Relation (9.1) is used. Relation (9.2) is for the nodes in the hidden layers. The weights are updated to reduce the cost function at each step.

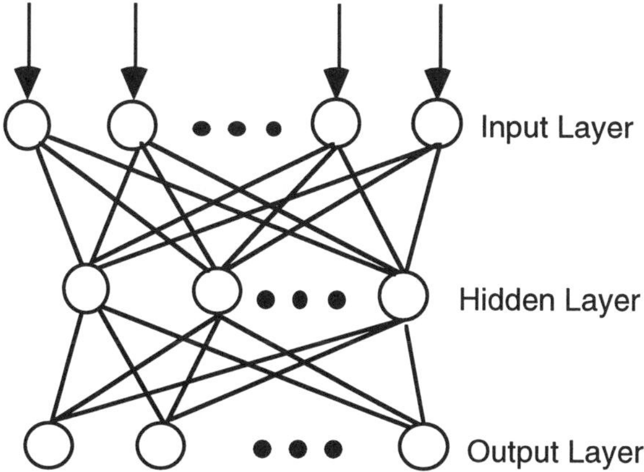


**Figure 9.2 An example of a three-layer feed-forward neural network.**

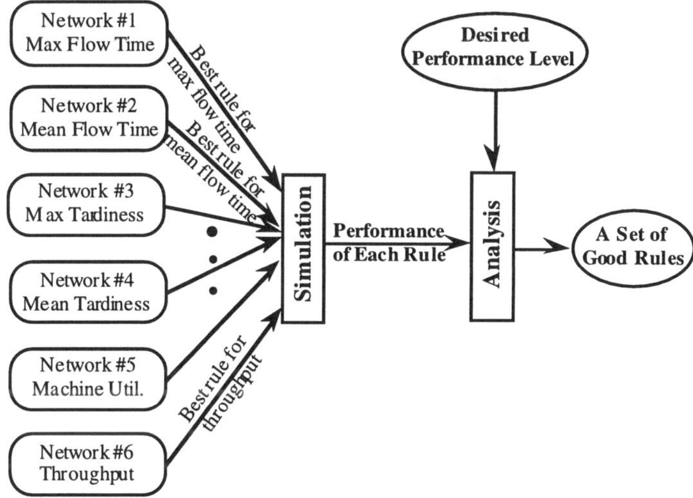


**Figure 9.3 Architecture of the intelligent scheduling aid.**

As indicated in Figure 9.3 our approach to developing the actual rule selector is to have backpropagation neural network trained to rank the available rules for each individual performance measure of interest (multiple performance evaluation comes in the next section). The weights for each of these networks are selected after a thorough training analysis. To carry out this training, we used two methodologies: 1) off-line training and 2) on-line training.

**Off-Line Training**

In off-line training, it is needed to generate training data sets for each of these performance measures from simulation studies. Suppose we wanted to train a neural net to minimize the maximum tardiness and we wanted to consider the following dispatching rules: SPT, LPT, FIFO, LIFO, SST, LST, CR, etc. (see Figure 9.4). After simulating each of these rules off-line under a variety of input conditions, we would be able to rank them to determine the best rule for this measure (The example in Section 9.4 gives some of these simulation results.). We would then use these results to train (i.e., choose weights) a neural net.

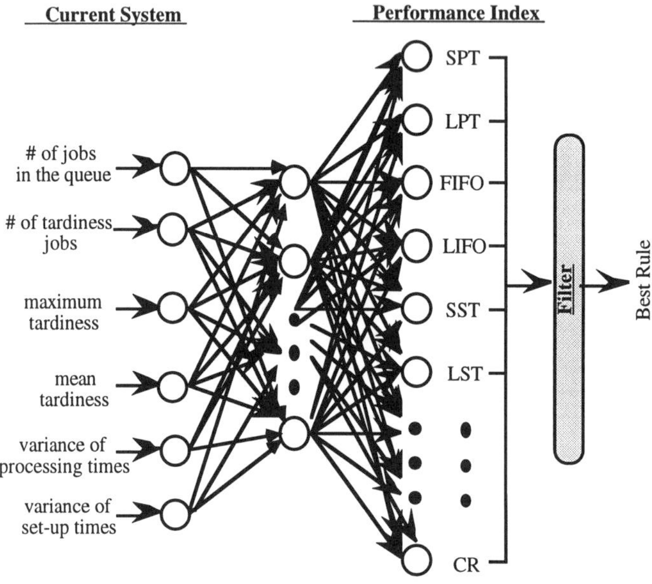


**Figure 9.4 Neural network training for maximum tardiness.**

**On-Line Training**

In on-line training, adaptive critics concepts are utilized to train in real-time the neural network structures (BARTO, 1992, WERBOS, 1992). Q-learning (a derivation of adaptive critics) (WATKINS, 1989) is used to predict a scheduling policy to meet the required performance criterion for a given queue status and undefined period of operation and therefore accomplish an effective candidate rule selector. The key idea of Q-learning is to assign values to state (shop floor status)-action (scheduling policy) pairs. Q-learning does not need an explicit model of the dynamic system underlying the decision problem. It directly estimates the optimal Q values (i.e., ranking) for pairs of states and admissible actions. The optimal Q value for state **i** (shop floor status) and action **u** (a scheduling heuristic) is a cost of executing action **u** in state **i**. Any policy selecting actions that are greater with respect to the optimal Q values is an optimal policy. Actions are ranked based on the Q values. On the other hand, ranking through an evaluation function requires more information like immediate costs of state action pairs and state transition probabilities. Instead of state transition probabilities Q-learning requires a random function to generate successor states. The Q-value of the successful action is updated with learning parameters, although with the other admissible actions, Q values remain the same. Q-learning learns to accurately model the evaluation function. For a given state **x**, the system (e.g., a neural network) chooses the action **a**, where the utility **util(x,a)** is maximal. Q-learning consists of two parts: a utility function and a stochastic action selector. The utility function implemented using a neural network based on backpropagation works as both evaluator and policy maker. It tries to model the system by assigning values to action-state pairs. The neural network has multiple outputs, one for each action (as depicted in Figures 9.3 and 9.4).

We have initiated this training for a wide range of performance measures and dispatching rules for a single machine sequencing and multiple machine scheduling problems. We anticipate using these results for robots, machine tools, material handling devices and inspection devices. They will form the lower level of the two-level scheduling system which we are developing. Preliminary training results are described in (RABELO et al., 1993).

The output from the rule selector phase will be a collection of R matched pairs — {(performance measure, best rule)$_1$, ..., (performance measure, best rule)$_R$}. These pairs form the candidates which are passed on to the next phase for more detailed analysis.

### 9.3.2 Real-Time Simulation

After these R candidates have been determined, each of the rules must be evaluated to determine the impact that each rule will have on the future evolution of the system as measured from the current state of the system. In other words, we must predict how it does against all of the performance measures simultaneously. To carry out this analysis, we intend to use the technique developed by (DAVIS/JONES 1989) termed real-time Monte Carlo simulation. Since **R** rules must be considered, we plan to run **R** real-time simulations concurrently to avoid unacceptable timing delays in the analysis.

This form of Monte Carlo simulation differs considerably from traditional discrete-event simulation in two ways. First, each simulation trial is initialized to the current system state as updated in real-time by the Monitoring Function. This results in the second important distinction — these types of simulations are neither terminating nor steady state. They are not terminating because the initial conditions may change from one trial to the next. Furthermore, they are not steady state because we are specifically interested in analyzing the transient phenomena associated with the near-term system response. The question is, from a statistical perspective, does this really matter. Early experiments conducted by (DAVIS et al. 1991) indicate that the answer varies. That is, the inclusion or exclusion of new events corresponding to changes in the initial state can bias the statistical estimates of certain, but not all, performance measures.

The outputs from these simulation trials yield the projected schedule of events under each scheduling rule. These schedules are then used to compute the values of the various performance measures and constraints imposed by the Assessment Function. The computed statistics are used to select and develop the rule which provides the best statistical compromise among the performance criteria. In the next section, we discuss a new approach to this compromise analysis, genetic algorithms.

### 9.3.3 Genetic Algorithms

No matter how the utility function described above is constructed, only one rule from the candidate list can be selected. This causes an undesirable situation whenever there are negatively correlated performance measures, because no one rule can optimize all objectives simultaneously. Conceptually, one would like to create a new "rule" which 1) combines the best features of the most attractive rules, 2) eliminates the worst features of those rules, and 3) simultaneously achieves satisfactory levels of performance for all objectives. Our approach does not deal with the rules themselves, but rather the actual schedules that result from applying those rules. Consequently, we seek to generate a new schedule from these candidate schedules. To do this, we propose to use a genetic algorithm approach. Presently, we give a brief description of how genetic algorithms (GAs) work. In the next section, we give some results from our preliminary experimentation which demonstrates that this approach can actually generate new and better schedules from existing ones.

#### 9.3.3.1 Genetic Algorithms and Scheduling

GAs have been utilized in job shop scheduling. GAs could be utilized using the following schemes:

**(1)** GAs with blind recombination operators have been utilized by GOLDBERG and LINGLE (1985), DAVIS (1985), SYSWERDA (1990), and WHITLEY et al. (1989). Their emphasis on relative ordering schema, absolute ordering schema, cycles, and edges in the offspring will arise from differences in such blind recombination operators.

**(2)** Sequencing problems have been also addressed by the mapping of their constraints to a Boolean satisfiability problem (DE JONG and SPEARS, 1989) using partial payoff schemes. This scheme has produced good results for simple problems. However, this scheme needs more research.

**(3)** Heuristic genetic algorithms have been applied to job shop scheduling (BAGCHI et al., 1991). In these GAs, problem specific heuristics are incorporated in the recombination operators (such as optimization operators).

**Example: Using a simple genetic algorithm for sequencing**

This example illustrates the utilization of a simple genetic algorithm based on a blind recombination operator for sequencing problems. The partially mapped crossover (PMX) operator developed by GOLDBERG and LINGLE (1985) will be utilized. Consider a single machine sequencing problem with 7 types of jobs. Each job-type has its own arrival time, due date, and processing time distributions. The set-up time is sequence dependent as shown in Table 9.1. The objective is to determine a sequence that minimizes Maximum Tardiness for the 10-job problem described in Table 9.2.

| | Current job-type | | | | | | |
|---|---|---|---|---|---|---|---|
| Previous job-type | 1 | 2 | 3 | 4 | 5 | 6 | 7 |
| 1 | 0 | 1 | 2 | 2 | 3 | 2 | 2 |
| 2 | 1 | 0 | 2 | 3 | 4 | 3 | 2 |
| 3 | 2 | 2 | 0 | 3 | 4 | 3 | 2 |
| 4 | 1 | 2 | 3 | 0 | 4 | 4 | 2 |
| 5 | 1 | 2 | 2 | 3 | 0 | 3 | 2 |
| 6 | 1 | 2 | 2 | 3 | 3 | 0 | 3 |
| 7 | 1 | 2 | 2 | 2 | 2 | 2 | 0 |

**Table 9.1 Set-up times.**

| Job # | Job Type | Mean Processing Time | Arrival Time | Due Date |
|---|---|---|---|---|
| 1 | 6 | 8 | 789 | 890 |
| 2 | 6 | 8 | 805 | 911 |
| 3 | 5 | 10 | 809 | 910 |
| 4 | 1 | 4 | 826 | 886 |
| 5 | 2 | 6 | 830 | 905 |
| 6 | 7 | 15 | 832 | 1009 |
| 7 | 6 | 8 | 847 | 956 |
| 8 | 3 | 5 | 848 | 919 |
| 9 | 1 | 4 | 855 | 919 |
| 10 | 4 | 3 | 860 | 920 |

**Current Time:** 863
**Previous Job Type executed:** 3

**Table 9.2 10-Job problem description.**

The simple genetic algorithm procedure developed (different possible procedures could be developed, the one demonstrated is only for illustration purposes) for this sequence problem could be described as follows:

**1.** Randomly generate n legal sequences (n is the population size, e.g., n = 50).

**2.** Evaluate each sequence using a fitness function (for this problem: Minimization of Maximum Tardiness — the sequences will be ranked according to their Maximum Tardiness — the lower the better)

**3.** Choose the best sequences (m sequences with the lower values for Maximum Tardiness, m < n, e.g., 25).

**4.** Reproduction (e.g., duplicate them, stop when you have a population of n). This reproduction could be in function of the fitness value (sequences with the best fitness values will have higher probability to reproduce).

**5.** Crossover. Select randomly pairs of sequences. (Crossover could be applied to the best sequences. However, the offspring do not replace their parents, but rather a low ranking individual in the population (WHITLEY and STARKWEATHER, 1990). Apply PMX:

The PMX operator produces legal solutions by choosing a swapping interval between two crossover points selected randomly. The offspring will inherit the elements from the interval of one of the parents. Then, it is necessary to detect and fix the illegal situations by mapping and exchanging. For example, consider two sequences (A and B):

| **Position:** | **1** | **2** | **3** | **4** | **5** | **6** | **7** | **8** | **9** | **10** |
|---|---|---|---|---|---|---|---|---|---|---|
| A(Job Numbers): | 9 | 8 | 4 | 5 | 6 | 7 | 1 | 3 | 2 | 10 |
| B(Job Numbers): | 8 | 7 | 1 | 2 | 3 | 10 | 9 | 5 | 4 | 6 |

**Swapping interval (randomly generated):** 4 to 6.

| **Position:** | **1** | **2** | **3** | **4** | **5** | **6** | **7** | **8** | **9** | **10** |
|---|---|---|---|---|---|---|---|---|---|---|
| A: | 9 | 8 | 4 \| | 5 | 6 | 7 \| | 1 | 3 | 2 | 10 |
| B: | 8 | 7 | 1 \| | 2 | 3 | 10 \| | 9 | 5 | 4 | 6 |

**Exchanging:**

| **Position:** | **1** | **2** | **3** | **4** | **5** | **6** | **7** | **8** | **9** | **10** |
|---|---|---|---|---|---|---|---|---|---|---|
| A': | 9 | 8 | 4 \| | 2 | 3 | 10 \| | 1 | 3 | 2 | 10 |
| B': | 8 | 7 | 1 \| | 5 | 6 | 7 \| | 9 | 5 | 4 | 6 |

**Mapping and Exchanging to create legal sequences:**

| **Position:** | 1 | 2 | 3 | 4 | 5 | 6 | 7 | 8 | 9 | 10 |
|---|---|---|---|---|---|---|---|---|---|---|
| A": | 9 | 8 | 4 | 2 | 3 | 10 | 1 | 6 | 5 | 7 |
| B": | 8 | 10 | 1 | 5 | 6 | 7 | 9 | 2 | 4 | 3 |

**6.** Mutation (with a low probability, e.g., 0.1, exchange two arbitrary jobs' position). (Mutation might not be applied to some schedules.)

Example: we have the following sequence **9 8 4 5 6 7 1 3 2 10,**
applying mutation, the sequence could become **9 8 6 5 4 7 1 3 2 10,**

Jobs 4 and 6 exchanged positions in the sequence.

**7.** Repeat (2) to (6), until no more improvements are possible.

After 17 iterations (approximately: 17 * 50 = 850 sequences were generated and tested), the genetic algorithm produces the following sequence for this simple problem (this took less than 500 milliseconds in a PC 486 @ 33MHz):

**8 5 4 1 2 3 9 10 6 7** with a Maximum Tardiness of **2** (Fitness Function).

This is an optimal sequence. Studies of all possible combinations (**10! = 3628800**) produced Table 9.3 for the same sequencing problem. In addition,

Table 9.4 indicates some of the solutions por the same problem using dispatching rules.

| **Maximum Tardiness (MT)** (Range) | **2.<= MT < 6** | **6<=MT<10** | **10<=MT<13** | **13<=MT<=75** |
|---|---|---|---|---|
| **Number of Possible Sequences** | 68 | 1344 | 5568 | 3621820 |

**Table 9.3 Evaluating 3628800 Sequences (exhaustive search)**

| **Dispatching Rule** | **Sequence** | **MT** |
|---|---|---|
| **SPT** Shortest Processing Time | 10 4 9 8 5 1 2 7 3 6 | 23 |
| **FIFO** First-In/First-Out | 1 2 3 4 5 6 7 8 9 10 | 32 |
| **EDD** Earliest Due Date | 4 1 5 3 2 8 9 10 7 6 | 10 |
| **LIFO** Last-In/First-Out | 10 9 8 7 6 5 4 3 2 1 | 65 |

**Table 9.4 Using dispatching rules.**

#### 9.3.3.2 Genetic Algorithms for Compromise Analysis

The compromise analysis process carried out by a genetic algorithm can be thought of as a complex hierarchical "generate and test" process. The generator produces building blocks which are combined into schedules. At various points in the procedure, tests are made that help weed out poor building blocks and promote the use of good ones. To reduce and support uncertainty management of the search space, the previous two steps (candidate rules selection and parallel simulation with statistical analysis) provide partial solutions to the problem of compromise analysis. Reducing uncertainty makes the search process more effective, with the complexity of the scheduling problem becoming more manageable in the process (see Figure 9.5).

### 9.3.4 Inductive Learning Algorithm — TDKA

Now that this new schedule has been generated, we want to extract the knowledge contained in that schedule for future use. To do this, we must derive a "new rule" which can be used to regenerate schedules in the same way that other dispatching rules like SPT (Shortest Processing Time first) are used. This new rule will not, however, be a simple dispatching rule. To do this, we will use a technique developed in (YIH, 1988 and YIH, 1990) called Trace-Driven Knowledge Acquisition (TDKA). TDKA is a method that extracts knowledge from the actual results of decisions rather than statements or explanations of the presumed effects of decisions. There are three steps in the process of trace-driven knowledge acquisition: *data collection*, *data analysis*, and *rule evaluation*.

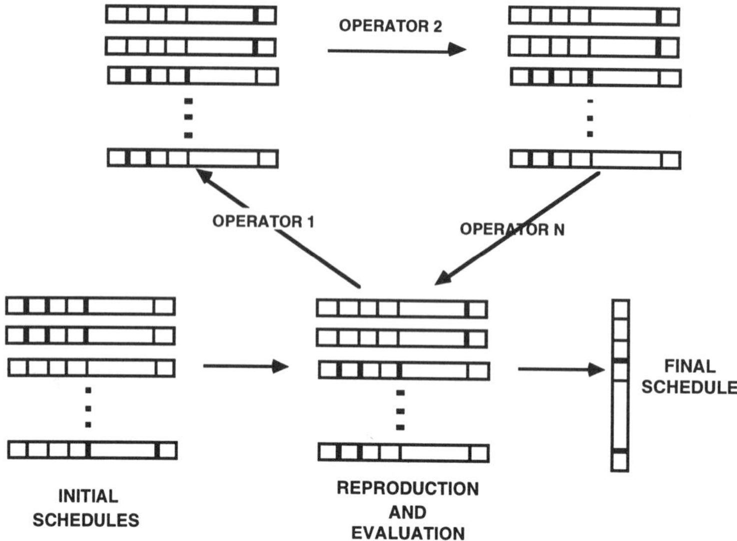


**FIGURE 9.5 Genetic algorithm for compromise analysis.**

In step one, data is collected through simulation, historical records, human inputs or actual physical experiments. (THESEN et al., 1987 and YIH, 1992) describe their efforts to use human experts. In this application, the data (trace) is simply the schedule generated from the Genetic Algorithm. In step two, the schedule is analyzed to extract a set of production rules (If-Then rules) which could be used to regenerate the same schedule. In step three, simulation is used to compare the generated schedule with the original schedule. The process returns to step two to refine the rules if the comparison is unacceptable.

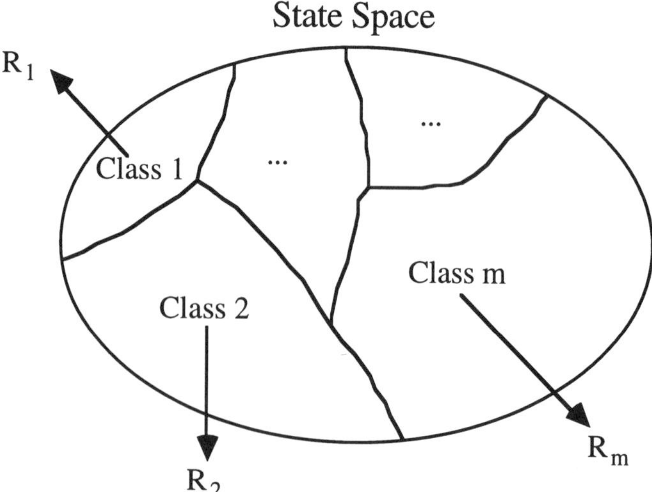


**Figure 9.6 State space hyper plane($R_m$ is the decision rule applied in class m).**

The core of the TDKA lies in the step two-data analysis. There are two types of rules involved — decision rules and class assignment rules. If the state space is viewed as a hyper plane, class assignment rules draw the boundaries on the hyper plane to define the areas of classes as shown in Figure 9.6. For each class, a single production rule is used to determine what to do next. These rules are in the form of

**"If [state $\in$ class i] then apply $R_i$* "**

In data analysis, the records collected are used to define classes and to select a single rule for each class. The rule formation algorithm can be described by three steps as follows.

*I.* ***Initialize***

Determine state variables
Determine decision rules
Set initial value for acceptance level (**L**)
Determine initial class $\mathbf{C_1}$
Obtain a trace (a set of records) from simulation

*II.* ***Vote***

Each record in this class ($\mathbf{C_i}$) votes for all the decision rules that will result in the same decision as in the record
Summarize the vote results in percentage (No. of votes / No. of records)

*III.* ***Form rules***

The decision rule ($\mathbf{R_k}^*$)with the highest percentage wins.
If the percentage is higher than L, then form the following rule

**If [state $\in$ $C_i$] then apply $R_k^*$**
else
Split class $C_i$ into two classes based on selected state variable ($\mathbf{V_p*}$) and threshold ($\mathbf{Th_i}^*$).

Add two class assignment rules.

**IF [state $\in$ $C_i$] and [$V_p* < Th_i^*$] THEN [state $\in$ $C_{i1}$]**

**IF [state $\in$ $C_i$] and [$V_p* \geq Th_i^*$] THEN [state $\in$ $C_{i2}$]**

Repeat Steps II and III for classes $\mathbf{C_{i1}}$ and $\mathbf{C_{i2}}$.

*IV.* ***Stop***

This iterative process will stop whenever the rule formation is completed. That is, the state space has been properly divided into several classes and a single production rule is selected for each class. If the performance of the extracted rules is worse than the trace, then it is necessary to return to data analysis and refine the rules by increasing the value of acceptance level (**L**). Otherwise, the process stops.

## 9.4 An Example

Consider a single machine sequencing problem with 7 types of jobs. Each job-type has its own arrival time, due date, and processing time distributions. The set-up time is sequence dependent as shown in Table 9.1. The objective is to determine a sequence which minimizes the summation of mean flow time and maximum tardiness. Assume that we are to generate a sequence for the 10 jobs in the input queue of the machine as described in Table 9.5.

| Job # | Job Type | Mean Processing Time | Arrival Time | Due Date |
|---|---|---|---|---|
| 1 | 7 | 5 | 154 | 203 |
| 2 | 3 | 5 | 154 | 193 |
| 3 | 4 | 3 | 159 | 194 |
| 4 | 3 | 5 | 160 | 208 |
| 5 | 2 | 4 | 166 | 194 |
| 6 | 1 | 4 | 170 | 202 |
| 7 | 3 | 5 | 185 | 231 |
| 8 | 2 | 4 | 186 | 221 |
| 9 | 7 | 5 | 192 | 243 |
| 10 | 7 | 5 | 200 | 250 |

**Current Time:** 200
**Previous Job Type executed:** 3

**Table 9.5 Job description for the example problem.**

The candidate rule selector implemented, using backpropagation neural networks (previously trained by off-line training), uses the system status and the performance criteria (mean flow time and maximum tardiness) in order to select a small set of candidates from the 13 rules available (see Table 9.6). Each neural network has six inputs (as shown in Figure 9.4), 12 hidden units in the hidden layer, and 13 outputs (one for each dispatching rule). Each neural network (one for mean flow time and the other for mean tardiness) in parallel ranks all rules. The networks developed in the C programming language take on average less than 10 mS (486 PC compatible @ 33 MHz) to give an answer to the problem. The neural network for mean flow time ranks higher SPST (Shortest Processing and Shortest Set-upTime first) and SST (Set-up Time first). On the other hand, the neural network for maximum tardiness ranks higher EDD (Earliest Due Date first) and mSLACK (Smallest Slack first).

A genetic algorithm is utilized having as initial populations the selected schedules and some randomly generated schedules. The fitting function is a combination of all performance measures with coefficients reflecting the "importance" of each one according to the imposed criteria. The crossover mechanism utilized is based on "order Crossover" as developed by Syswerda (1990). The genetic algorithm takes on average eleven iterations with an average time of less than 700 mS on a 486 PC @ 33 MHz. The new schedule compromises both measures with an acceptable degree of success (Mean Flow Time = 73.5 and Maximum Tardiness = 6). In order to verify the answers, a program that generates all possible solutions to the scheduling problem is generated (10!), taking on average 2 hours of cpu time of the 486 PC @ 33 MHz. The "new" schedule generated by the genetic algorithm is superior (based on the combined performance criteria) to the initial schedules selected by the neural networks — (see Table 9.6). The output, shown at the bottom of Table 9.6, is (**2 3 6 5 1 4 7 8 10 9**). This sequence performs better than EDD for maximum tardiness while maintaining good performance in mean flow time (better than the third ranked rule-SPT). However, it is possible to identify some of the relative positions of the dispatching heuristics in the final schedule. This schedule is selected to be applied to the manufacturing system.

| Dispatching Rule | Mean Flow Time | Rank | Maximum Tardiness | Rank | Job Sequence |
|---|---|---|---|---|---|
| SPT | 62.4 | 3 | 46 | 7 | 3 5 6 8 1 2 4 7 9 10 |
| LPT | 63.4 | 5 | 67 | 10 | 1 2 4 7 9 10 5 6 8 3 |
| FIFO | 66.0 | 11 | 42 | 5 | 1 2 3 4 5 6 7 8 9 10 |
| LIFO | 64.4 | 7 | 73 | 13 | 10 9 8 7 6 5 4 3 1 2 |
| SST | 59.2 | 2 | 61 | 8 | 2 4 7 1 9 10 6 5 8 3 |
| LST | 66.2 | 12 | 40 | 4 | 3 1 2 5 4 6 7 8 9 10 |
| SPST | 58.5 | 1 | 42 | 6 | 2 4 7 3 6 5 8 1 9 10 |
| LPST | 67.6 | 13 | 63 | 9 | 1 2 9 4 10 7 3 5 6 8 |
| EDD | 63.2 | 4 | 31 | 1 | 2 3 5 6 1 4 8 7 9 10 |
| LDD | 64.8 | 8 | 72 | 11 | 10 9 7 8 4 1 6 3 5 2 |
| mSLACK | 63.4 | 6 | 31 | 2 | 2 3 5 1 6 4 8 7 9 10 |
| MSLACK | 64.8 | 9 | 72 | 12 | 10 9 7 8 4 1 6 3 5 2 |
| CR | 65.7 | 10 | 34 | 3 | 3 5 2 6 1 4 8 7 9 10 |
| **Genetic Algorithm*** | 61.7 | - | 30 | - | 2 3 6 5 1 4 7 8 10 9 |

**Note:**

SPT - Shortest Processing Time
LPT - Longest Processing Time
FIFO - First In First Out
LIFO - Last In First Out
SST - Shortest Set-up Time
LST - Longest Set-up Time
SPST - Shortest Pro and Set-up Time
LPST - Longest Proc and Set-up Time
EDD - Earliest Due Date
LDD - Latest Due Date
mSLACK - Smallest slack
MSLACK - Largest slack
CR - Critical Ratio (slack/processing time)

**Table 9.6 Results from commonly used heuristics and genetic algorithm.**

We now show how to use TDKA to generate a new sequencing rule. Using the GA sequence (**2 3 6 5 1 4 7 8 10 9**), we first obtain the trace from simulation in the following format.

(Q, NT, MT, AT, VP, VS) ---> Action where
Q: number of jobs in the queue
NT: number of tardy jobs
MT: maximum tardiness

AT: mean tardiness
VP: variance of processing times
VS: variance of set-up times
Action: job identification number to be selected next

For instance, the trace from simulation is:

(10, 4, 10, 7.25, 0.50, 1.17) --> 2
(9, 4, 14, 9.75, 0.53, 1.00) --> 3
(8, 4, 17, 11.3, 0.27, 0.13) --> 6
(7, 3, 14, 13.0, 0.24, 0.24) --> 5
(6, 2, 16, 16.0, 0.17, 0.67) --> 1
(5, 2, 16, 9.5, 0.20, 1.20) --> 4
(4, 1, 8, 8.0, 0.25, 1.00) --> 7
(3, 1, 13, 13.0, 0.33, 0.00) --> 8
(2, 0, 0, 0.0, 0.00, 0.00) -->10

Each record in the trace will vote for the rules that could yield the same decision as in the sequence. The summary of votes is listed in the following table.

We start with one class, called Class 1. If we are satisfied with the accuracy of 67%, that is 67% is higher than the acceptance level (L), we may arbitrarily choose SST, SPST, EDD or mSLACK and form the following rule.

**"If [state $\in$ Class 1] then apply SST"** (9.1)

If we would like to obtain higher accuracy, the variable MT (maximum tardiness) can be used to split Class 1 into two subclasses, Class 11 and Class 12. Class 11 includes the states with MT $\geq$ 10, and Class 12 has the remainder. The following class assignment rules will be formed.

| Rule | Votes | % |
|---|---|---|
| SPT | 5 | 55% |
| LPT | 5 | 55% |
| FIFO | 5 | 55% |
| LIFO | 1 | 11% |
| SST | 6 | 67% |
| LST | 5 | 55% |
| SPST | 6 | 67% |
| LPST | 3 | 33% |
| EDD | 6 | 67% |
| LDD | 1 | 11% |
| mSLACK | 6 | 67% |
| MSLACK | 1 | 11% |
| CR | 5 | 55% |

**Number of records = 9**

**Table 9.7 Summary of the record votes (Class 1).**

$$\text{"If [MT} \geq 10\text{] then state} \in \text{Class 11"} \quad (9.2)$$

$$\text{"If [MT} < 10\text{] then state} \in \text{Class 12"} \quad (9.3)$$

After splitting into two classes, the voting process repeats within each class. The results are listed in Tables 9.8 and 9.9 in Appendix 1.

The following rule for Class 12 could be formed with 100% accuracy with LPT, SST, or SPST.

$$\text{"If [state} \in \text{Class 12] then apply SPST"} \quad (9.4)$$

In Class 11, if we are satisfied with the accuracy of 86%, we may form the following rule with EDD or mSLACK.

$$\text{"If [state} \in \text{Class 11] then apply EDD"} \quad (9.5)$$

However, if we would like to achieve higher accuracy, MT is used again to split Class 11 into Class 111 and Class 112 with threshold of 17. The following rules are formed.

$$\text{"If [state} \in \text{Class 11] and [MT} \geq 17\text{] then state} \in \text{Class 111"} \quad (9.6)$$

$$\text{"If [state} \in \text{Class 11] and [MT} < 17\text{] then state} \in \text{Class 112"} \quad (9.7)$$

After splitting, the voting results are summarized in Tables 9.10 and 9.11 in the Appendix included at the end of this chapter.

Two rules may be formed with 100% accuracy as follows.

$$\text{"If [state} \in \text{Class 111] then apply SPT"} \quad (9.8)$$

$$\text{"If [state} \in \text{Class 112] then apply EDD"} \quad (9.9)$$

Finally, rules (9.2), (9.3), (9.4), (9.6), (9.7), (9.8), and (9.9) will be included in the rule base and this set of rules is able to generate sequences based on the strategy embedded in the sequence derived from the GA.

## 9.5 Remarks

In this chapter, we have described a methodology to solve sequencing and scheduling problems which integrate neural networks, simulation, genetic algorithms, and machine learning techniques. We also described a small example which demonstrates the methodology. By exploiting the parallel processing and modeling capabilities of the neural ncts, simulation, and genetic algorithms it has the potential to be extremely fast and highly adaptable to customer needs. Finally, we described a learning technique which extracts the knowledge from the derived schedules and creates new rules. This provides the additional capability to learn what works and what does not work in a variety of situations and utilize that knowledge at a later time. For these reasons, we believe that this technique has the potential to solve real-world sequencing and scheduling problems in real-time.

**References**

ARIZONO I., YAMAMOTO A., and OHTA, H., "Scheduling for Minimizing Total Actual Flow Time by Neural Networks," *International Journal of Production Research*, 1992, Vol. 30, No. 3, pp. 503-511.

BAGCHI, S., UCKUN, S., MIYABE, Y., and KAWAMURA, K., "Exploring Problem-Specific Recombination Operators for Job Shop Scheduling," *Proceedings of the Fourth International Conference on Genetic Algorithms and Their Applications*, pp. 10-17, 1991.

BARTO, A., "*Reinforcement Learning and Adaptive Critic Methods,*" Handbook of Intelligent Control: Neural, Fuzzy, and Adaptive approaches. D.A. White and D.A. Sofge (Ed.) Van Nostrand Reinhold Publication, 1992.

CHRYSSOLOURIS, G., LEE, M., and DOMROESE, M., "The Use of Neural Networks in Determining Operational Policies for Manufacturing Systems," *Journal of Manufacturing Systems*, 10, pp. 166-175, 1991.

DAVIS L., "Job Shop Scheduling with Genetic Algorithms," *Proceedings on an International Conference on Genetic Algorithms and Their Applications*, Carnegie-Mellon University, pp. 136-140, 1985.

DAVIS W. and JONES A., "Issues in real-time simulation for flexible manufacturing systems", *Proceedings of the European Simulation Multiconference*, Rome, Italy, June 7-9, 1989.

DAVIS, W., WANG, H., and HSIEH, C., "Experimental Studies in Real-time, Monte Carlo Simulation", *IEEE Transactions on Systems, Man, and Cybernetics*, Vol. 21, No. 4, pp. 802-814, 1991.

DAVIS W., JONES A., and SALEH A., "A Generic Architecture for Intelligent Control Systems", *Computer Integrated Manufacturing Systems*, Vol. 5, No. 2, pp. 105-113, 1992.

DE JONG, K. and SPEARS, W., "Using Genetic Algorithms to solve NP-Complete Problems," *Proceedings of the Third International Conference on Genetic Algorithms*, pp. 124-132, 1989.

JACKSON, R. and JONES, A., "An Architecture for Decision Making in the Factory of the Future, *INTERFACES*, Vol. 17, NO. 6, pp. 15-28, 1987.

JONES, A. and SALEH, A., "A Multi-level/Multi-layer Architecture for Intelligent Shop Floor Control," *International Journal of Computer Integrated Manufacturing Special Issue on Intelligent Control*, 3, 1, pp. 60-70, 1990.

FOO Y. and TAKEFUJI Y., "Stochastic Neural Networks for solving job shop Scheduling", *Proceedings of the IEEE international Conference on Neural Networks*, published by IEEE TAB, 1988, pp. II275-II290.

GOLDBERG D., *Genetic Algorithms in Machine Learning*, Addison-Wesley, Menlo Park, California, 1988.

GOLDBERG, D., and LINGLE, R., "Alleles, loci, and the Traveling Salesman Problem," *Proceedings of the of the International Conference on Genetic Algorithms and Their Applications*, 1985.

LO, Z. and BAVARIAN, B., "Scheduling with Neural Networks for Flexible Manufacturing Systems," *Proceedings of the 1991 IEEE International Conference on Robotics and Automation*, Sacramento, California, pp. 818-823, 1991.

RABELO, L., "A hybrid artificial neural network and expert system approach to flexible manufacturing system scheduling", PhD Thesis, University of Missouri-Rolla, 1990.

RABELO, L, YIH, Y., JONES, A., and WITZGALL, G., "Intelligent FMS Scheduling using Modular Neural Netwroks", *Proceedings of ICNN'93*, pp. 1224-1229, 1993.

RUMELHART, D., McCLELLAND, J., and the PDP Research Group, *Parallel Distributed Processing: Explorations in the Microstructure of Cognition*, Vol. 1: Foundations, Cambridge, MA: MIT Press/Bradford Books, 1988.

SYSWERDA, G., "*Scheduling Optimization using Generic Algorithms,*" Handbook of Genetic Algorithms, pp. 332-349, 1990.

WATKINS, C., Learning From Delayed Rewards, PhD Thesis, Cambridge University, Cambridge, England, 1989.

WERBOS, P., "Approximate Dynamic Programming for Real Time Control Neural Modelling," *Handbook of Intelligent Control: Neural, Fuzzy, and Adaptive Approaches.* White and Sofge (eds.) Van Nostrand Reinhold Publication, pp. 493-525, 1992.

WHITLEY, D. and STARKWEATHER, T., "*GENITOR II: A Distributed Genetic Algorithm,*" Journal of Experimental and Theoretical Artificial Intelligence, Vol. 2, pp. 189-214, 1990.

WHITLEY D., STARKWEATHER T., and FUQUAY D., "Scheduling Problems and the Traveling Salesman: the genetic edge recombination operator," *Proceedings of the Third International Conference on Genetic Algorithms,* pp. 133-140, 1989.

YIH, Y., Trace Driven Knowledge Acquisition for Expert Scheduling System, Ph.D. dissertation, University of Wisconsin-Madison, December, 1988.

YIH, Y., "Trace-Driven Knowledge Acquisition (TDKA) for Rule-Based Real-Time Scheduling Systems," *Journal of Intelligent Manufacturing*, 1, 4, pp. 217-230, 1990.

YIH, Y., "Learning Real-Time Scheduling Rules from Optimal Policy of Semi-Markov Decision Processes," *International Journal of Computer Integrated Manufacturing*, Vol. 5, No. 3, pp. 171-181, 1992.

ZHOU D., CHERKASSKY V., BALDWIN T., and HONG D., "Scaling Neural Network for Job Shop Scheduling," *Proceedings of the International Conference on Neural Networks*, Vol. 3, pp. 889-894, 1990.

**APPENDIX**

| Rule | Votes | % |
|---|---|---|
| SPT | 4 | 57% |
| LPT | 3 | 43% |
| FIFO | 4 | 57% |
| LIFO | 0 | 0% |
| SST | 4 | 57% |
| LST | 4 | 57% |
| SPST | 4 | 57% |
| LPST | 2 | 29% |
| EDD | 6 | 86% |
| LDD | 0 | 0% |
| mSLACK | 6 | 86% |
| MSLACK | 0 | 0% |
| CR | 5 | 71% |

**Number of records = 7**

**Table 9.8 Summary of the record votes (Class 11).**

| Rule | Votes | % |
|---|---|---|
| SPT | 1 | 50% |
| LPT | 2 | 100% |
| FIFO | 1 | 50% |
| LIFO | 1 | 50% |
| SST | 2 | 100% |
| LST | 1 | 50% |
| SPST | 2 | 100% |
| LPST | 1 | 50% |
| EDD | 0 | 0% |
| LDD | 1 | 50% |
| mSLACK | 0 | 0% |
| MSLACK | 1 | 50% |
| CR | 0 | 0% |

**Number of records = 2**

**Table 9.9 Summary of the record votes (Class 12).**

| Rule Candidate | # of Votes | Percentage |
|---|---|---|
| SPT | 1 | 100% |
| LPT | 0 | 0% |
| FIFO | 0 | 0% |
| LIFO | 0 | 0% |
| SST | 1 | 100% |
| LST | 0 | 0% |
| SPST | 1 | 100% |
| LPST | 0 | 0% |
| EDD | 0 | 0% |
| LDD | 0 | 0% |
| mSLACK | 0 | 0% |
| MSLACK | 0 | 0% |
| CR | 0 | 0% |

**Number of records = 1**

**Table 9.10 Summary of the record votes (Class 111).**

| Rule Candidate | # of Votes | Percentage |
|---|---|---|
| SPT | 4 | 57% |
| LPT | 3 | 43% |
| FIFO | 4 | 57% |
| LIFO | 0 | 0% |
| SST | 4 | 57% |
| LST | 4 | 57% |
| SPST | 4 | 57% |
| LPST | 2 | 29% |
| EDD | 6 | 86% |
| LDD | 0 | 0% |
| mSLACK | 6 | 86% |
| MSLACK | 0 | 0% |
| CR | 5 | 71% |

**Number of records = 6**

**Table 9.11 Summary of the record votes (Class 112).**

## Chapter 10

**Vijaykumar Hanagandi**[1] **and Michael Nikolaou**[2]
Texas A&M University
College Station, TX 77843-3122

mOn2431@venus.tamu.edu

# Chemical Engineering



**Abstract**

Genetic Algorithms (GAs) are emerging as powerful alternatives to traditional optimization methods which are too restrictive and CPU intensive. The field of chemical engineering offers challenging optimization and search problems. We show that, apart from being able to efficiently handle highly nonlinear, multimodal and nonlinear objective functions, GAs can also handle cases where the objective function is not clearly defined. Three case studies are presented to illustrate the effectiveness of GAs in solving complex optimization problems. The FORTRAN code along with the templates for the case studies can be obtained from the authors.

---

[1] Present address: MS C312, Los Alamos National Laboratory Los Alamos, NM 87545.

[2] Author to whom all correspondence should be addressed. For software enquiries please send e-mail to: vmh@killdeer.lanl.gov or vmh0648@sigma.tamu.edu.

0-8493-2529-3/95/$0.00 + $.50

## 10.1 Introduction

Genetic Algorithms (GAs) are emerging as powerful alternatives to traditional optimization methods which are too restrictive and CPU intensive. Genetic Algorithms (GAs) accomplish the task of optimization by starting with a random "population" of values for the parameters of an optimization problem, and thereafter producing new "generations" of improved values that combine the best "parts" of values from previous populations.

The field of chemical engineering offers challenging optimization and search problems. In this paper we use GAs to solve representative problems in design and control of chemical processes and to save a complex problem in transport phenomena In the light of some unique problems encountered in chemical engineering applications the following points summarize the advantages of using GAs to solve them:

- As a GA proceeds randomly (yet systematically) in its search, it does not require smoothness, derivability, continuity etc. in the objective function. The only requirement is that for a set of values for the optimization parameters, one must assign a fitness value. This feature makes it possible to solve complex optimization/search problems where the concept of an objective function is itself fuzzy and is more qualitative than quantitative.

- Through computer experimentation and some heuristic analysis GAs have been found to arrive at or close to the global optimum of an objective function.

- GAs can efficiently handle highly nonlinear and noisy objective functions as encountered in stochastic processes where traditional gradient based methods are inefficient.

- GAs are amenable to parallel processing. Unlike in gradient search algorithms, in GAs the objective function evaluation for one parameter set is independent of that for all others in the same generation. This facilitates the use of parallel computers for the search procedure.

We refer the reader to standard literature (Goldberg, 1989, Davis, 1991) for details of the theory and implementation of GAs. Here we discuss three case studies which represent a cross section of chemical engineering applications. The FORTRAN software and manual can be obtained from the authors. Parameters of GAs employed in the three case studies are grouped at the end of the third case study in Table. 10.4.

## 10.2 Case Study 1: Best Controller Synthesis using Qualitative Criteria

In this section we demonstrate the use of GAs to accomplish controller design task with quantitative goals to be accomplished. The goals are defined by production engineers and operators. First, we illustrate our methodology on a textbook example and then solve the linear controller design problem for a more complex nonlinear multi-input multi-output (MIMO) plant.

### 10.2.1 A Textbook Example

Controller:

The following example illustrates the use of GAs for qualitative optimization:

Plant:

$$P(s) = \frac{1}{(s+1)(s+2)}$$

Controller:

$$C(s) = K_c(1 + \frac{1}{\tau_1 s} + \tau_D s)$$

Problem:
Find the values of $K_C$, $\tau_1$ and $\tau_D$ so that the response of the closed loop for a step change in the set-point has the following qualitative features:

- it has *fast* rise;
- it has *smooth* rise;
- it has a *stable* response.

The optimization procedure is initialized by ten randomly generated sets of values for $K_C$, $\tau_1$ and $\tau_D$ in the intervals [0, 10], [0, 10] and [0, 10], respectively. This set of parameters is the first generation. Using these values, closed loop simulations are conducted and the responses are ranked by the designer/operator, who decides the ranking by assigning a numerical value to each response. These fitness values do not correspond to integral square error (ISE) or any quantitative characteristic of the responses but they reflect the quality of the responses as seen by the expert with the above mentioned qualitative goals in mind. In the absence of such a fitness value only a mere ranking is enough for the GA to proceed, but convergence might be slow. Once the rankings and/or fitness values are assigned, the GA produces the parameter set values corresponding to the second generation and the above evaluation procedure is repeated. In this example problem runs up to ten generations were performed. Figures 10.1.1 through 10.10.10 give the responses of the dosed loop to step changes in the set-point. These figures are arranged in their decreasing order of fitness in each generation. As we note from the figures, the GA has managed to recognize our qualitative objective and is giving more weight to those candidates (in the parameter set) that have desirable features thereby successively improving the responses.

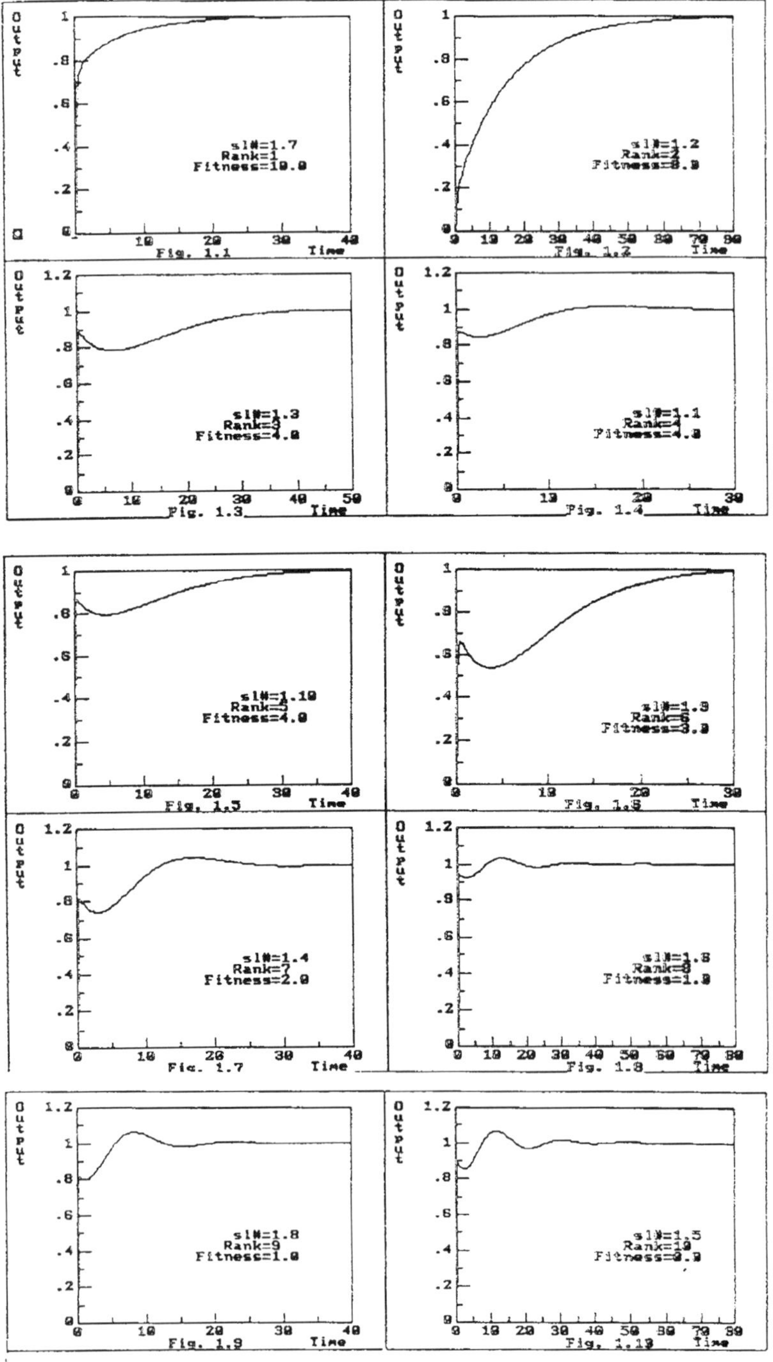
Output
Time
sl#=1.7
Rank=1
Fitness=10.0
Fig. 1.1
sl#=1.2
Rank=2
Fitness=8.0
Fig. 1.2
sl#=1.3
Rank=3
Fitness=4.0
Fig. 1.3
sl#=1.1
Rank=4
Fitness=4.0
Fig. 1.4
sl#=1.10
Rank=5
Fitness=4.0
Fig. 1.5
sl#=1.9
Rank=6
Fitness=3.0
Fig. 1.6
sl#=1.4
Rank=7
Fitness=2.0
Fig. 1.7
sl#=1.6
Rank=8
Fitness=1.0
Fig. 1.8
sl#=1.8
Rank=9
Fitness=1.0
Fig. 1.9
sl#=1.5
Rank=10
Fitness=0.0
Fig. 1.10

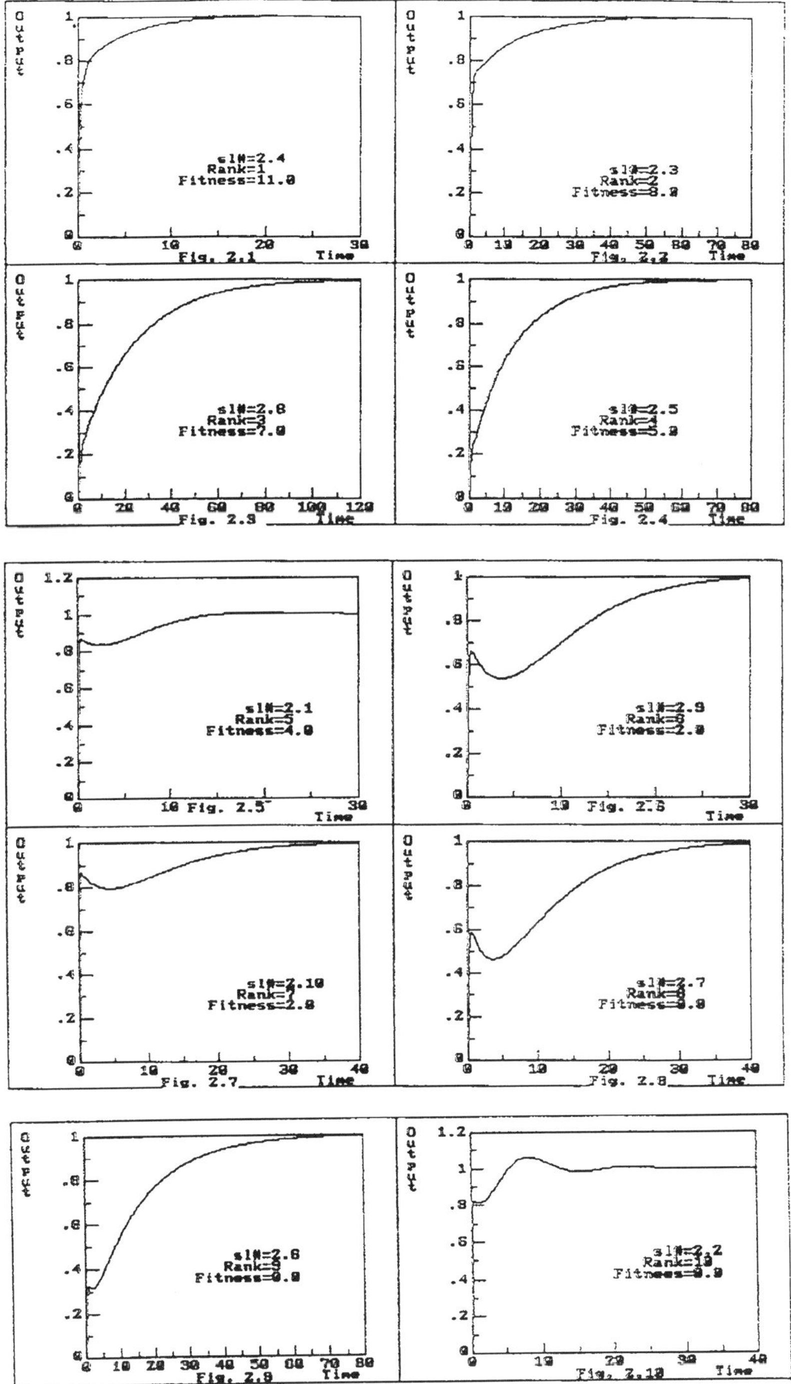

Output
sl#=2.4
Rank=1
Fitness=11.0
Time
Fig. 2.1

Output
sl#=2.3
Rank=2
Fitness=8.0
Time
Fig. 2.2

Output
sl#=2.8
Rank=3
Fitness=7.0
Time
Fig. 2.3

Output
sl#=2.5
Rank=4
Fitness=5.0
Time
Fig. 2.4

Output
sl#=2.1
Rank=5
Fitness=4.0
Time
Fig. 2.5

Output
sl#=2.9
Rank=6
Fitness=2.0
Time
Fig. 2.6

Output
sl#=2.10
Rank=7
Fitness=2.0
Time
Fig. 2.7

Output
sl#=2.7
Rank=8
Fitness=0.0
Time
Fig. 2.8

Output
sl#=2.6
Rank=9
Fitness=0.0
Time
Fig. 2.9

Output
sl#=2.2
Rank=10
Fitness=0.0
Time
Fig. 2.10

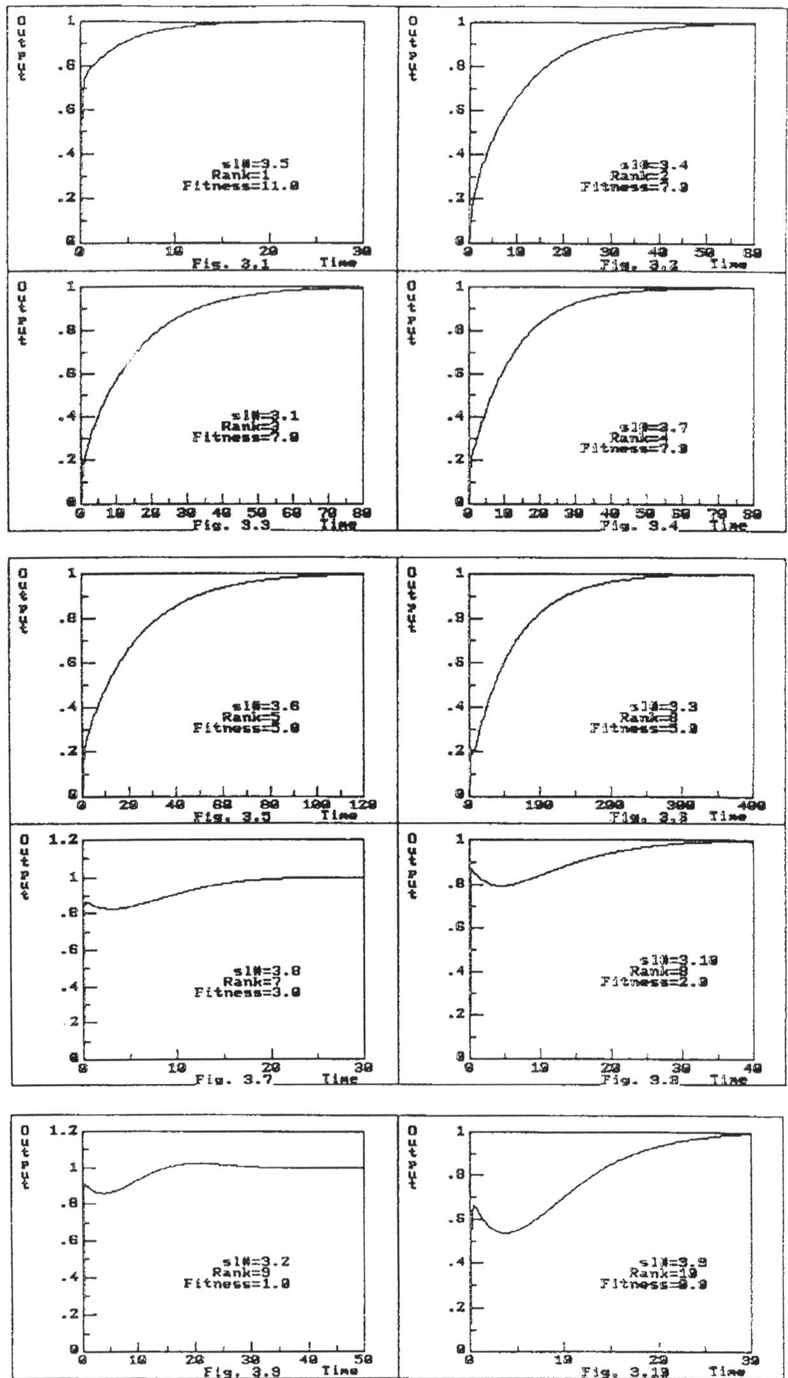

Output
sl#=3.5
Rank=1
Fitness=11.0
Time
Fig. 3.1

Output
sl#=3.4
Rank=2
Fitness=7.0
Time
Fig. 3.2

Output
sl#=3.1
Rank=3
Fitness=7.0
Time
Fig. 3.3

Output
sl#=3.7
Rank=4
Fitness=7.0
Time
Fig. 3.4

Output
sl#=3.6
Rank=5
Fitness=5.0
Time
Fig. 3.5

Output
sl#=3.3
Rank=6
Fitness=5.0
Time
Fig. 3.6

Output
sl#=3.8
Rank=7
Fitness=3.0
Time
Fig. 3.7

Output
sl#=3.10
Rank=8
Fitness=2.0
Time
Fig. 3.8

Output
sl#=3.2
Rank=9
Fitness=1.0
Time
Fig. 3.9

Output
sl#=3.9
Rank=10
Fitness=0.0
Time
Fig. 3.10

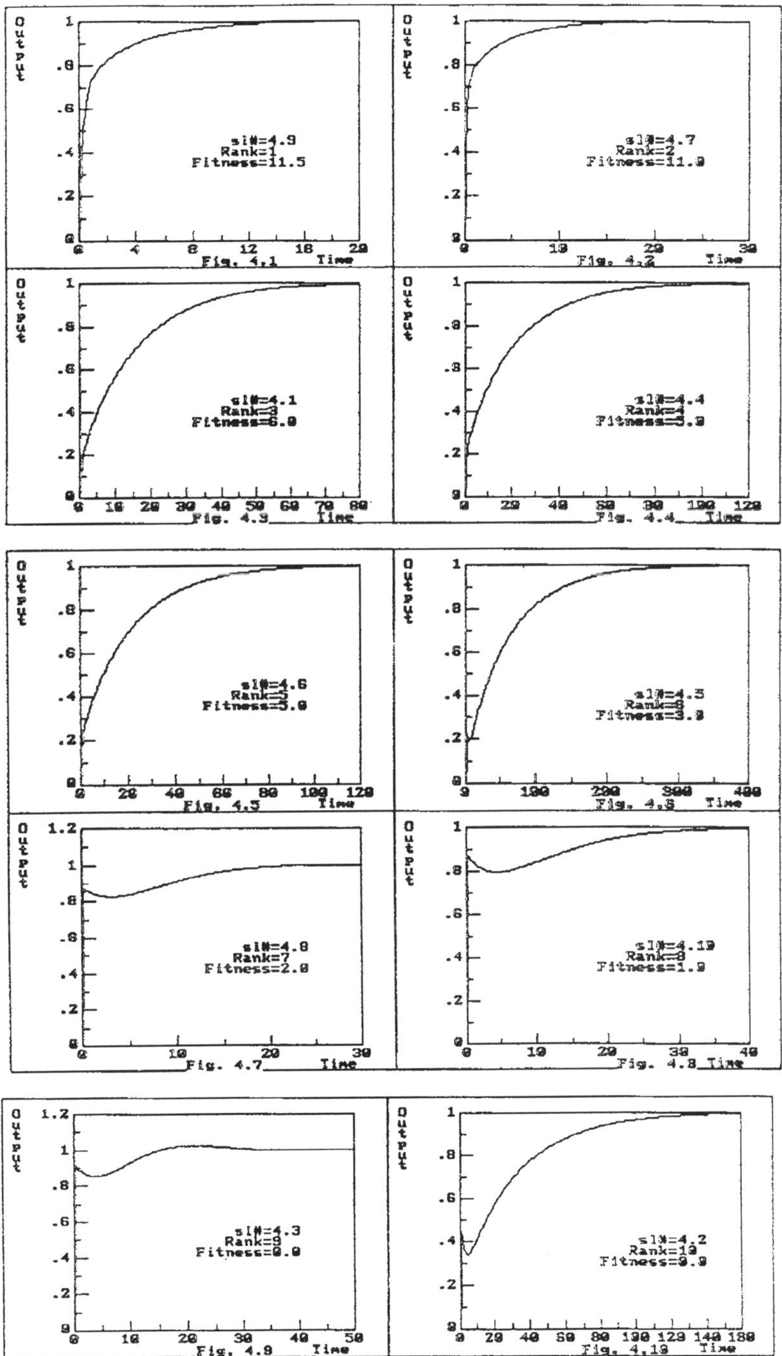
Output
sl#=4.9
Rank=1
Fitness=11.5
Fig. 4.1 Time
Output
sl#=4.7
Rank=2
Fitness=11.0
Fig. 4.2 Time
Output
sl#=4.1
Rank=3
Fitness=6.0
Fig. 4.3 Time
Output
sl#=4.4
Rank=4
Fitness=5.0
Fig. 4.4 Time
Output
sl#=4.6
Rank=5
Fitness=5.0
Fig. 4.5 Time
Output
sl#=4.5
Rank=6
Fitness=3.0
Fig. 4.6 Time
Output
sl#=4.8
Rank=7
Fitness=2.0
Fig. 4.7 Time
Output
sl#=4.10
Rank=8
Fitness=1.0
Fig. 4.8 Time
Output
sl#=4.3
Rank=9
Fitness=0.0
Fig. 4.9 Time
Output
sl#=4.2
Rank=10
Fitness=0.0
Fig. 4.10 Time

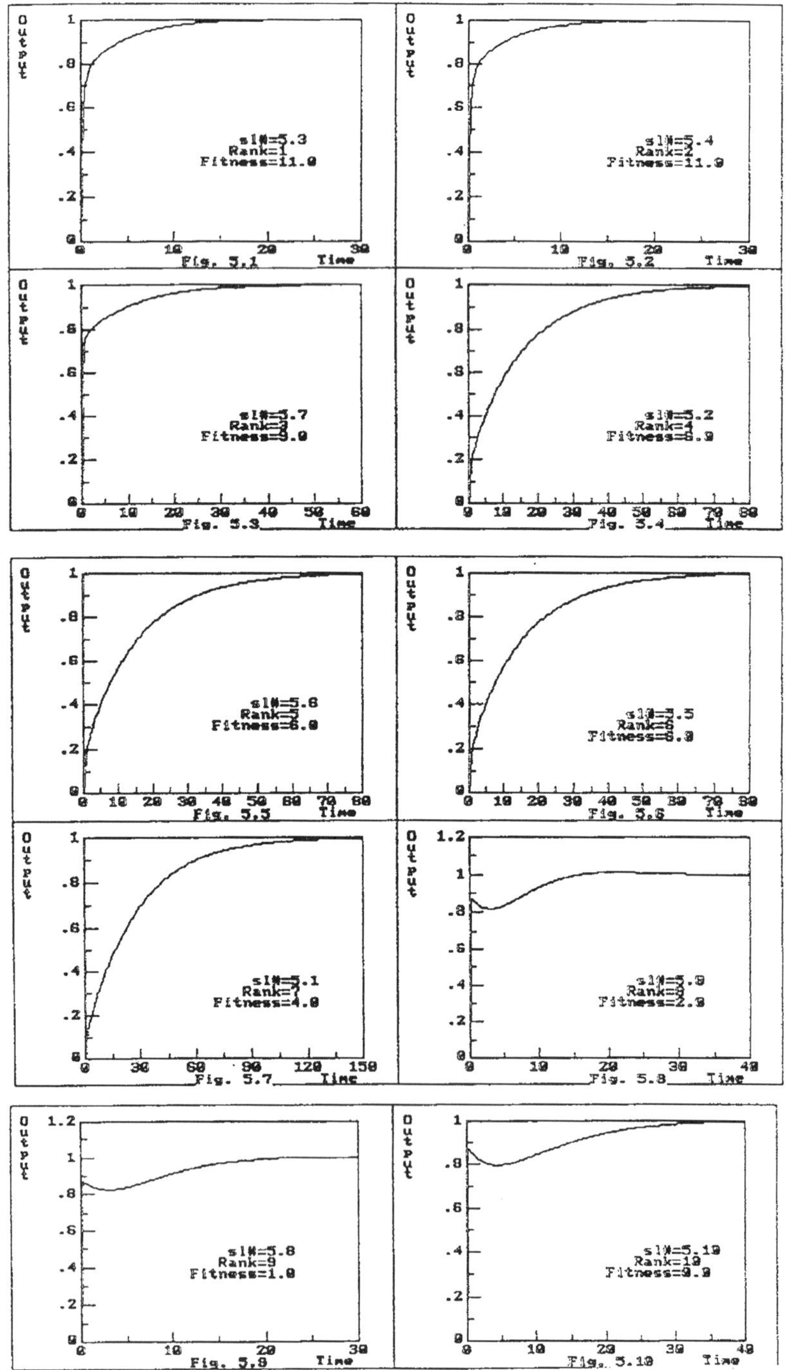


Fig. 5.1 Fig. 5.2 Fig. 5.3 Fig. 5.4 Fig. 5.5 Fig. 5.6 Fig. 5.7 Fig. 5.8 Fig. 5.9 Fig. 5.10

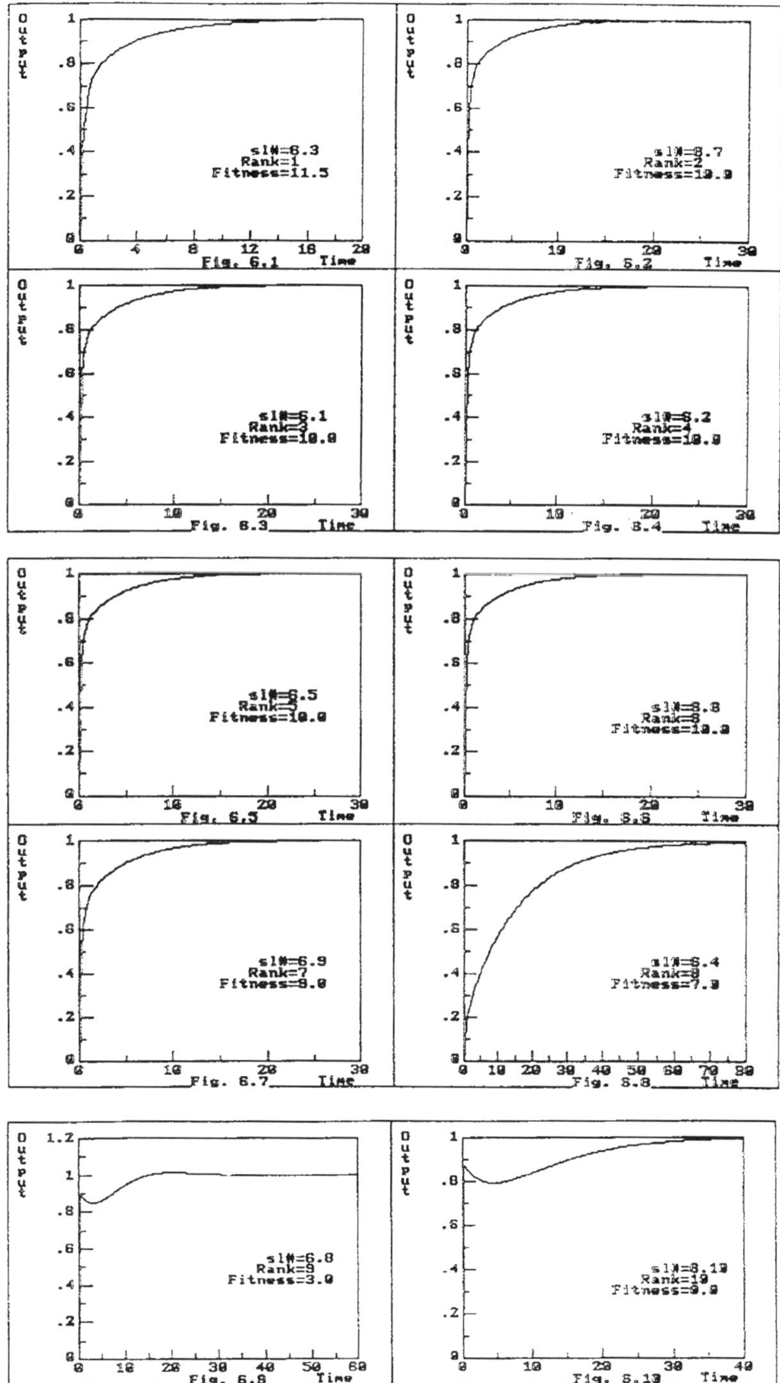
Output
sl#=6.3
Rank=1
Fitness=11.5
Fig. 6.1 Time
Output
sl#=6.7
Rank=2
Fitness=10.0
Fig. 6.2 Time
Output
sl#=6.1
Rank=3
Fitness=10.0
Fig. 6.3 Time
Output
sl#=6.2
Rank=4
Fitness=10.0
Fig. 6.4 Time
Output
sl#=6.5
Rank=5
Fitness=10.0
Fig. 6.5 Time
Output
sl#=6.6
Rank=6
Fitness=10.0
Fig. 6.6 Time
Output
sl#=6.9
Rank=7
Fitness=8.0
Fig. 6.7 Time
Output
sl#=6.4
Rank=8
Fitness=7.0
Fig. 6.8 Time
Output
sl#=6.8
Rank=9
Fitness=3.0
Fig. 6.9 Time
Output
sl#=6.10
Rank=10
Fitness=0.0
Fig. 6.10 Time

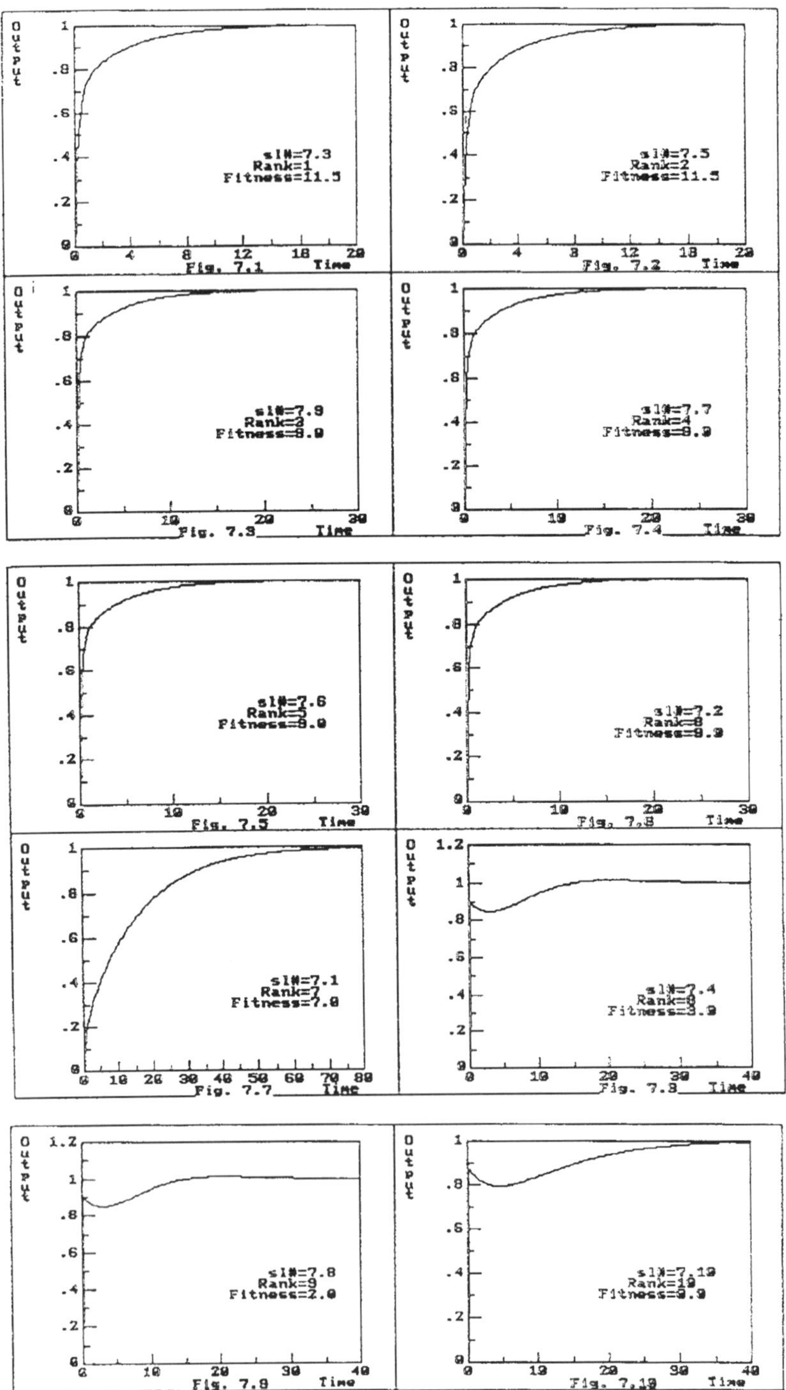
Output
Time
sl#=7.3
Rank=1
Fitness=11.5
Fig. 7.1
sl#=7.5
Rank=2
Fitness=11.5
Fig. 7.2
sl#=7.9
Rank=3
Fitness=8.0
Fig. 7.3
sl#=7.7
Rank=4
Fitness=8.0
Fig. 7.4
sl#=7.6
Rank=5
Fitness=8.0
Fig. 7.5
sl#=7.2
Rank=6
Fitness=8.0
Fig. 7.6
sl#=7.1
Rank=7
Fitness=7.0
Fig. 7.7
sl#=7.4
Rank=8
Fitness=3.0
Fig. 7.8
sl#=7.8
Rank=9
Fitness=2.0
Fig. 7.9
sl#=7.10
Rank=10
Fitness=0.0
Fig. 7.10

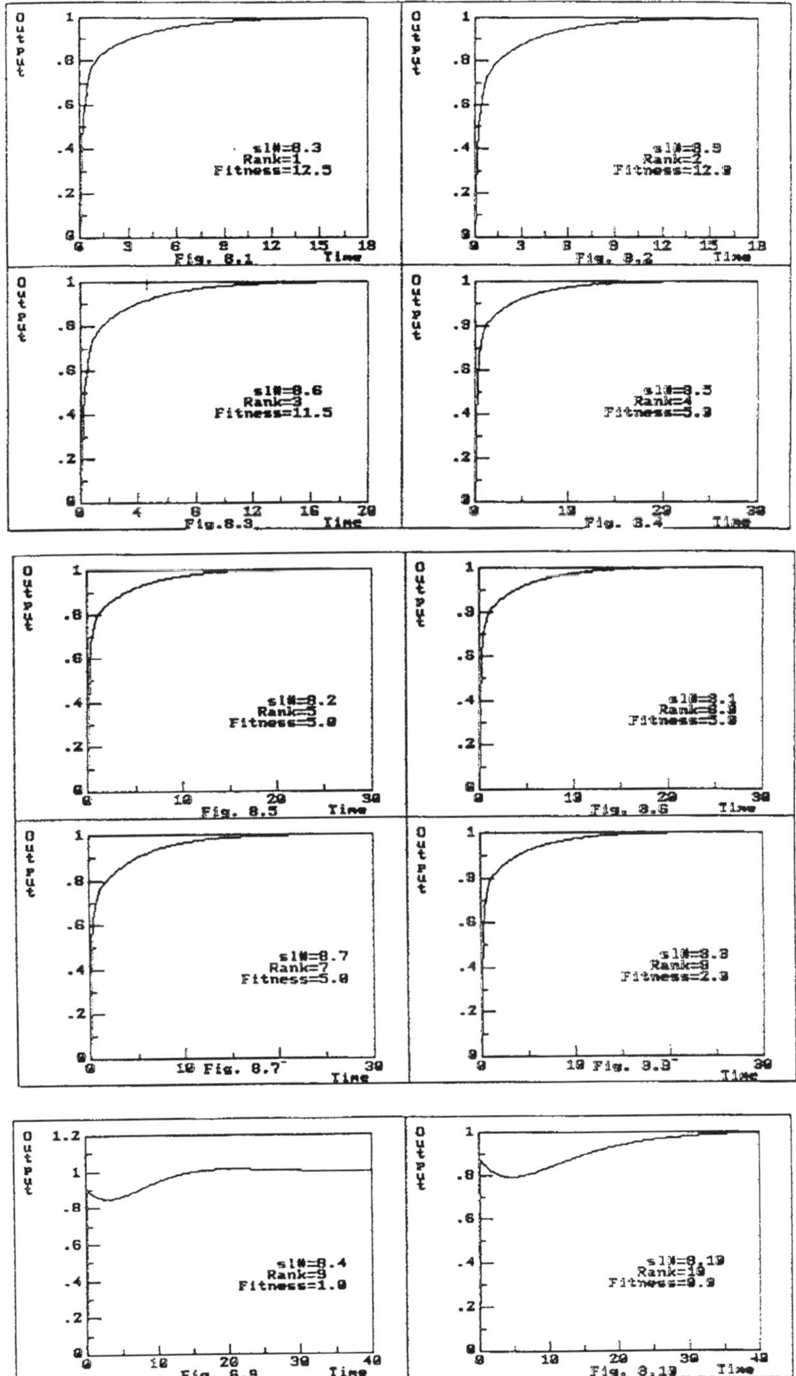
Output
Time
sl#=8.3
Rank=1
Fitness=12.5
Fig. 8.1
sl#=8.9
Rank=2
Fitness=12.0
Fig. 8.2
sl#=8.6
Rank=3
Fitness=11.5
Fig. 8.3
sl#=8.5
Rank=4
Fitness=5.0
Fig. 8.4
sl#=8.2
Rank=5
Fitness=5.0
Fig. 8.5
sl#=8.1
Fitness=5.0
Fig. 8.6
sl#=8.7
Rank=7
Fitness=5.0
Fig. 8.7
sl#=8.8
Rank=8
Fitness=2.0
Fig. 8.8
sl#=8.4
Rank=9
Fitness=1.0
Fig. 8.9
sl#=8.10
Rank=10
Fitness=0.0
Fig. 8.10

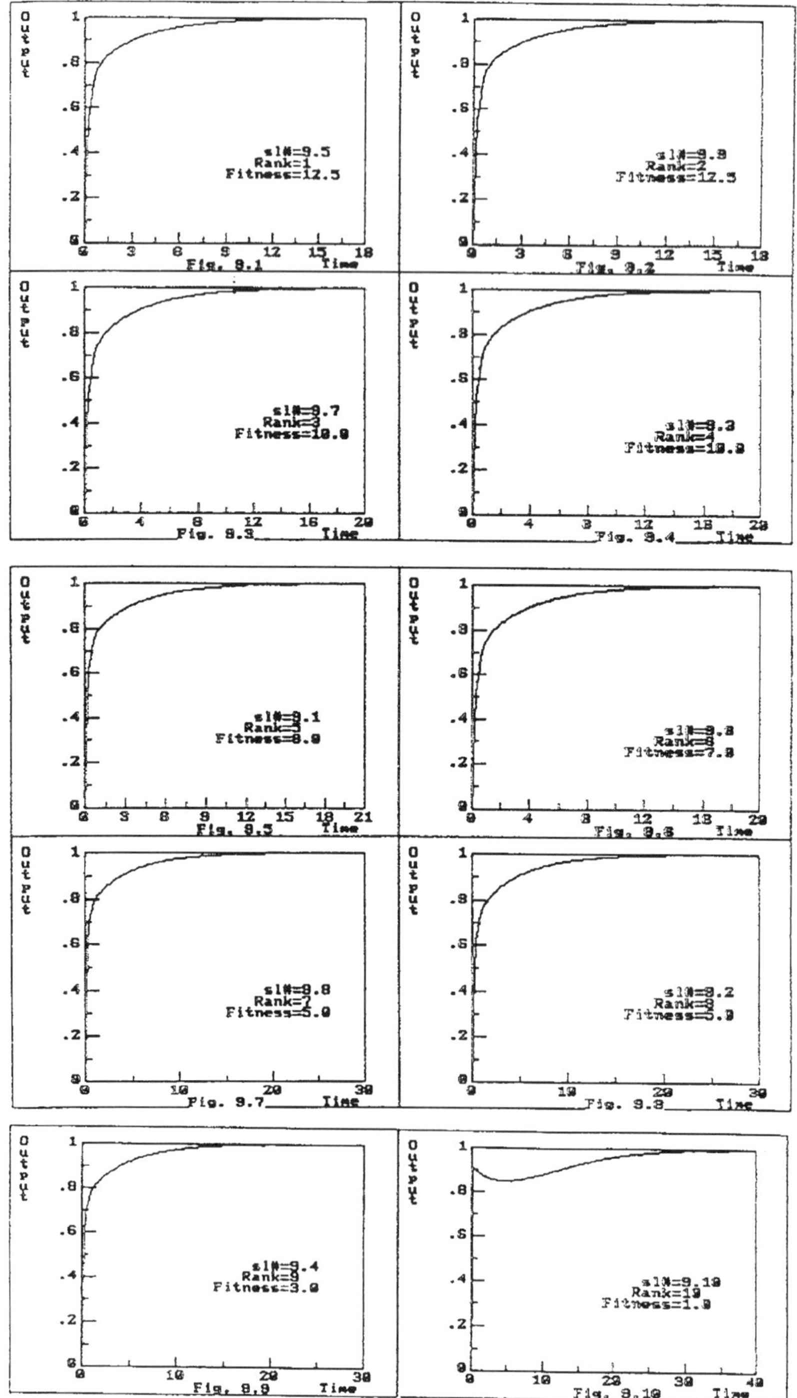
Output
Time
sl#=9.5
Rank=1
Fitness=12.5
Fig. 9.1
sl#=9.9
Rank=2
Fitness=12.5
Fig. 9.2
sl#=9.7
Rank=3
Fitness=10.0
Fig. 9.3
sl#=9.3
Rank=4
Fitness=10.0
Fig. 9.4
sl#=9.1
Rank=5
Fitness=8.0
Fig. 9.5
sl#=9.8
Rank=6
Fitness=7.0
Fig. 9.6
sl#=9.6
Rank=7
Fitness=5.0
Fig. 9.7
sl#=9.2
Rank=8
Fitness=5.0
Fig. 9.8
sl#=9.4
Rank=9
Fitness=3.0
Fig. 9.9
sl#=9.10
Rank=10
Fitness=1.0
Fig. 9.10

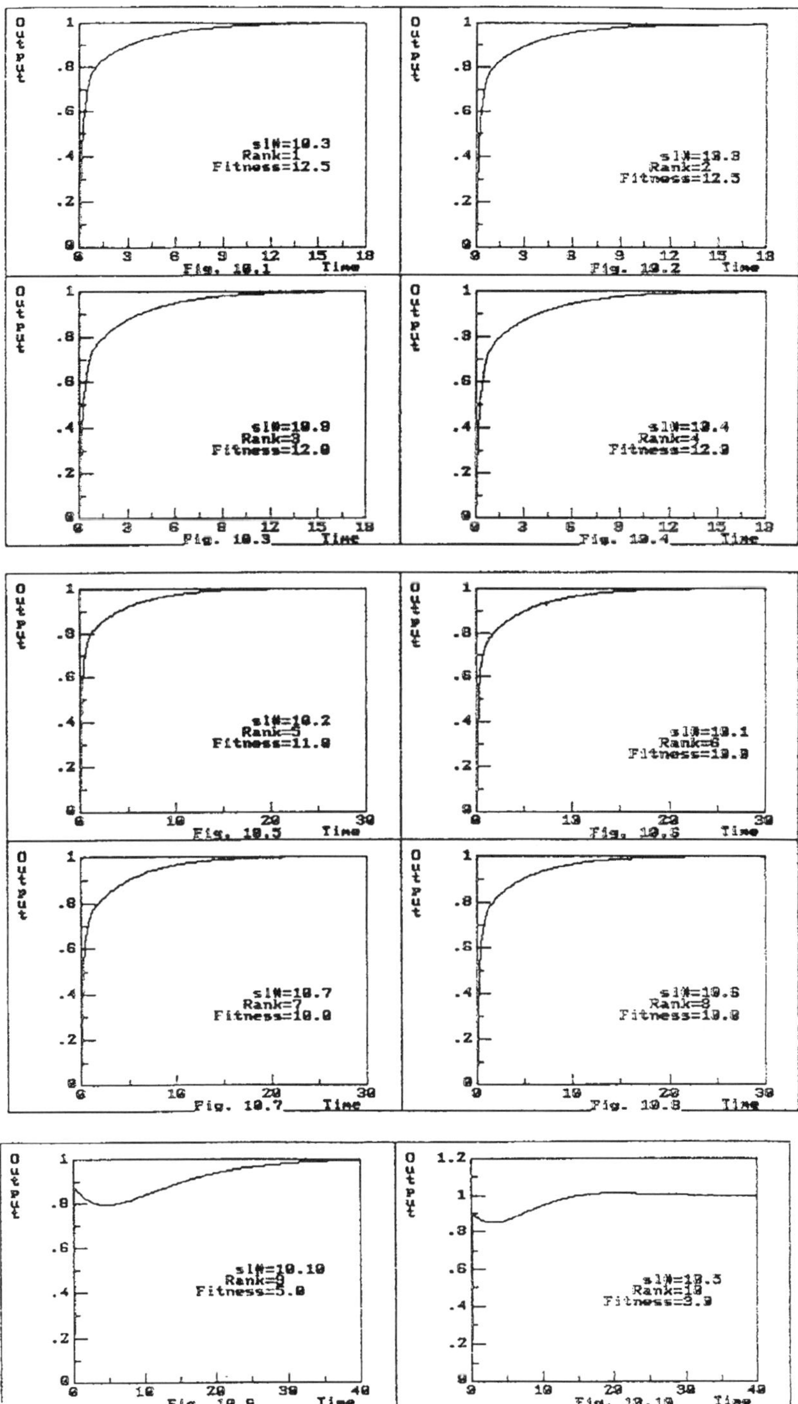


Fig. 10.1
Fig. 10.2
Fig. 10.3
Fig. 10.4
Fig. 10.5
Fig. 10.6
Fig. 10.7
Fig. 10.8
Fig. 10.9
Fig. 10.10

For comparison purposes, the optimal solution for this problem (according to the internal model control (IMC) principle) is

$$K_c = 10 \ (\textit{ideally } \infty) \quad \tau_I = 1.5 \quad \tau_D = 0.333$$

and the corresponding step-change response is shown in Figure 10.11.

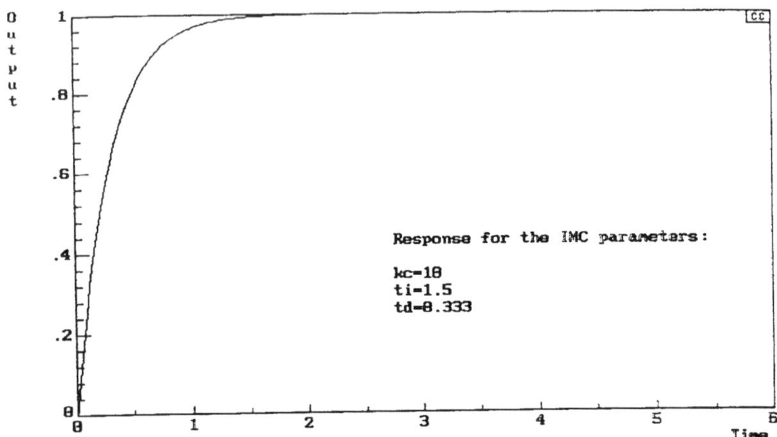


Figure 10.11 Best linear controller synthesis for a nonlinear plant.

Chien and Ogunnaike (1992) give details of the high-purity distillation column we attempt to control. For control purposes, this is a two-input ($y_1$, $y_2$) two-output ($u_1$, $u_2$) dynamical system. The open loop responses to various step changes in the inputs indicate that the system is highly nonlinear. The controller structure comprises two proportional-integral (PI) loops as indicated in Figure 10.12. $kc_1$,$\tau_{I,1}$,$kc2$ and $\tau_{I,2}$ are the parameters to be tuned. We follow the steps exactly as outlined in the previous example to arrive at the best PI settings. The set-point changes in the simulations are [0, 0] $\rightarrow$ [0.05, 0]. In many plant operations, set point changes are known beforehand (e.g., in startup, shutdown, etc.) so the optimal linear controller can be tuned using all those set-point changes in the simulations. The criteria used to evaluate controller fitness were:

- *fast* rise;
- *smooth* rise;
- *stable* responses with *less* oscillations.
- *realizable* control actions (both move sizes and move velocities).

Figures 10.13.1 through 10.13.10 show the simulations for the first generation controller performances. As one can imagine, this set has a wide range of random controller settings and performances. Figures 10.14.1 through 10.14.10 give the performances of the controller settings in the tenth generation. The best setting of this set is shown in Figure 10.14.1 which shows an improvement in performance over the best setting of the first generation shown by Figure 10.13.1. Also, unlike the first generation performances the tenth generation settings give more stable and good responses which satisfy the qualitative criteria showing that the GA is converging toward better and better settings.

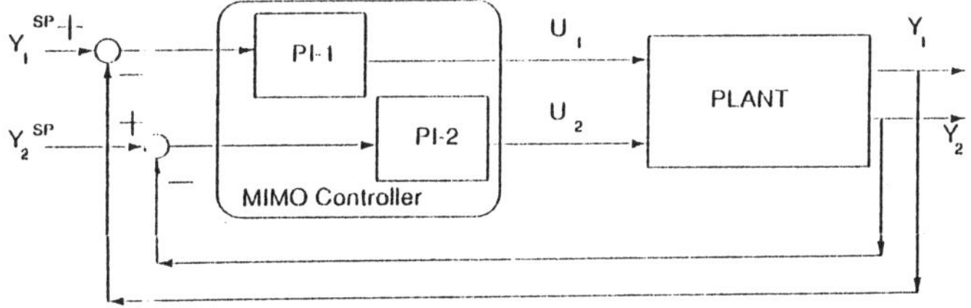


Figure. 12. The Closed loop using Linear MIMO Controller.

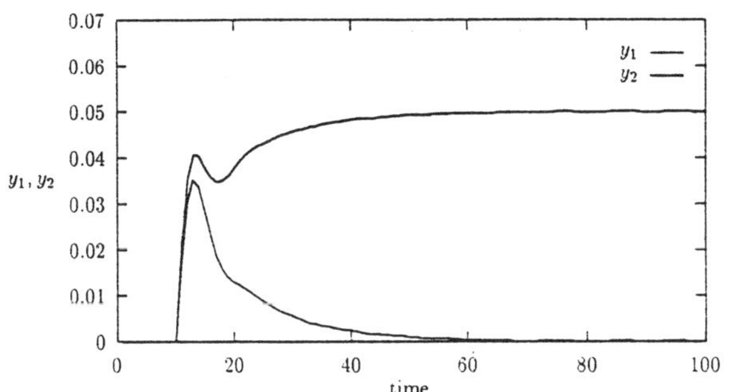


Figure. 13.1. Rank=1, Fitness=10.

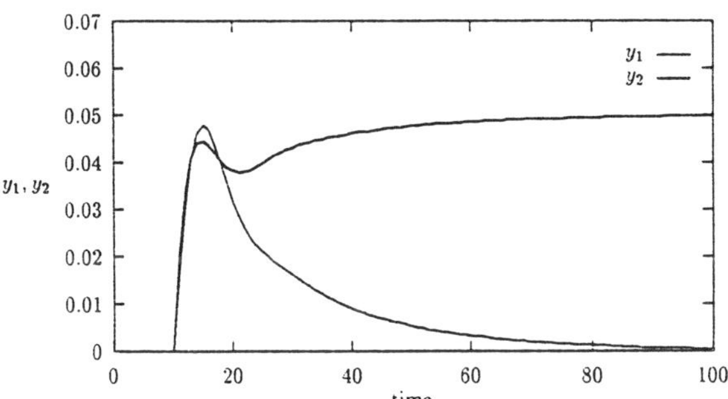


Figure. 13.2. Rank=2, Fitness=9.5

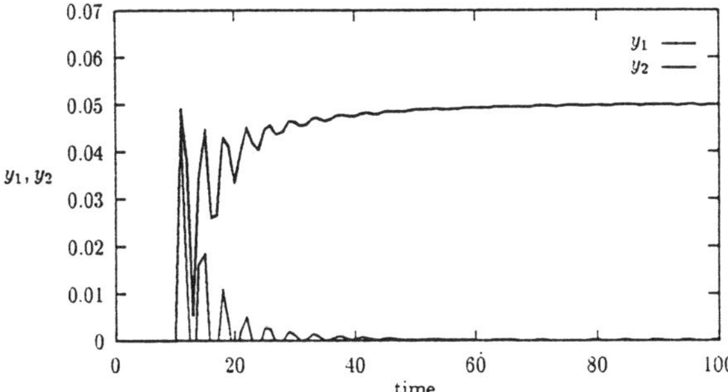


Figure. 13.3. Rank=3, Fitness=8.0

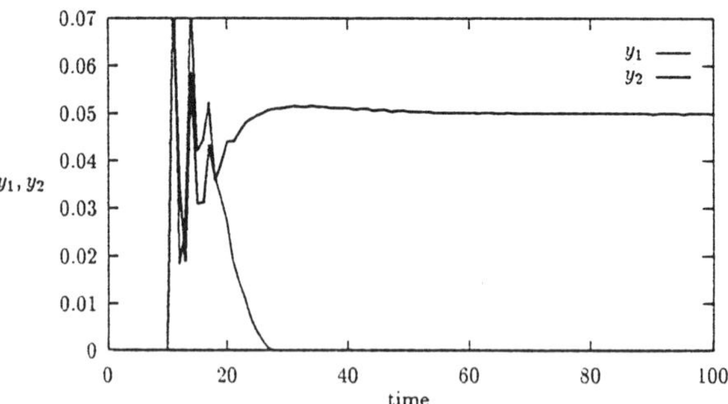


Figure. 13.4. Rank=4, Fitness=7.0

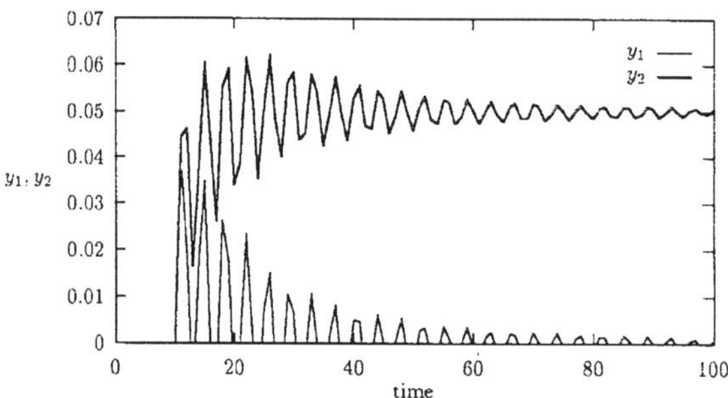


Figure. 13.5. Rank=5, Fitness=6.0

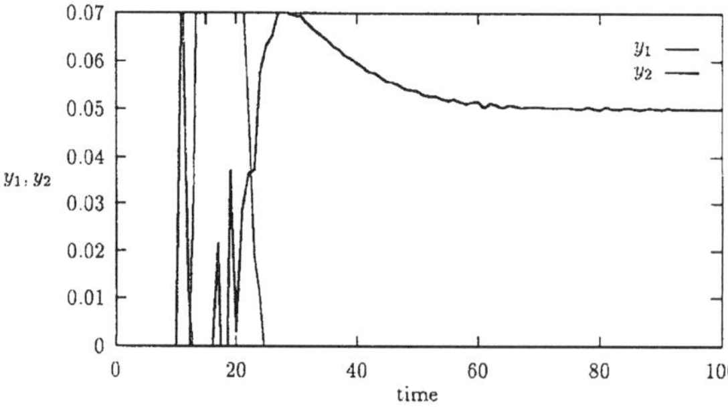


Figure. 13.6. Rank=6, Fitness=5.5

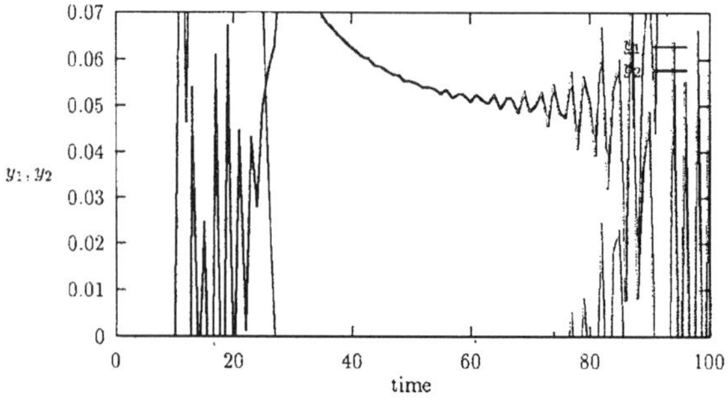

Figure. 13.7. Rank=7, Fitness=3.0

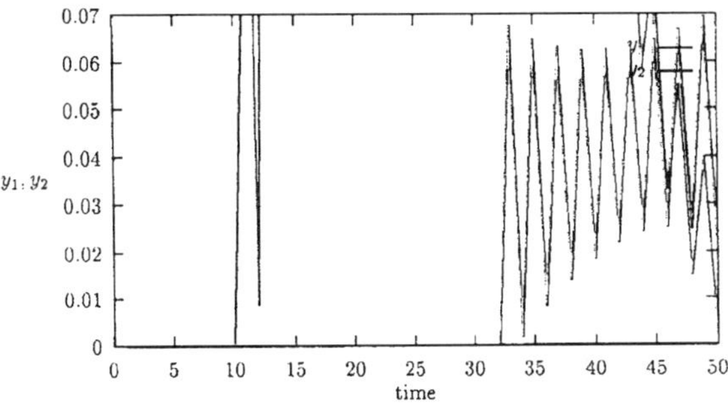


Figure. 13.8. Rank=8, Fitness=2.0

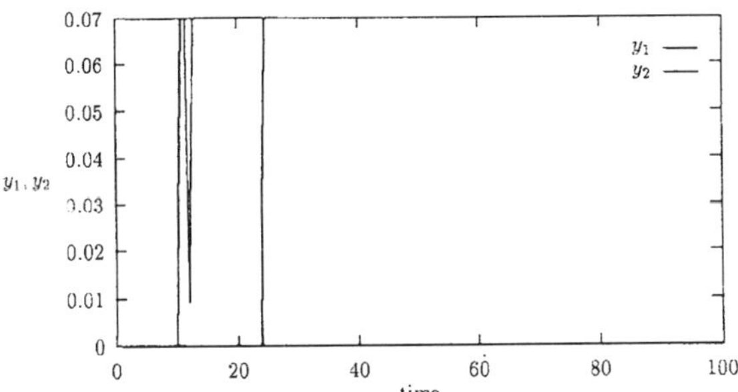


Figure. 13.9. Rank=9, Fitness=0.0

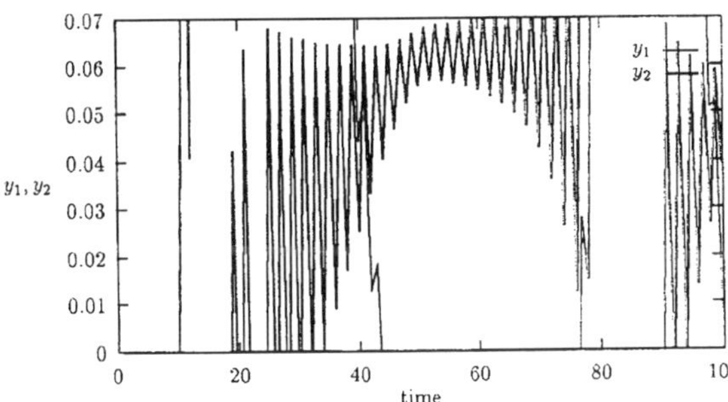


Figure. 13.10. Rank=10, Fitness=0.0

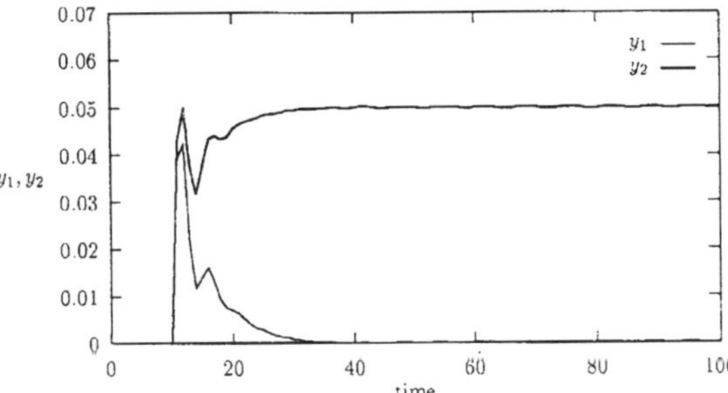


Figure. 14.1. Rank=1, Fitness=12.0

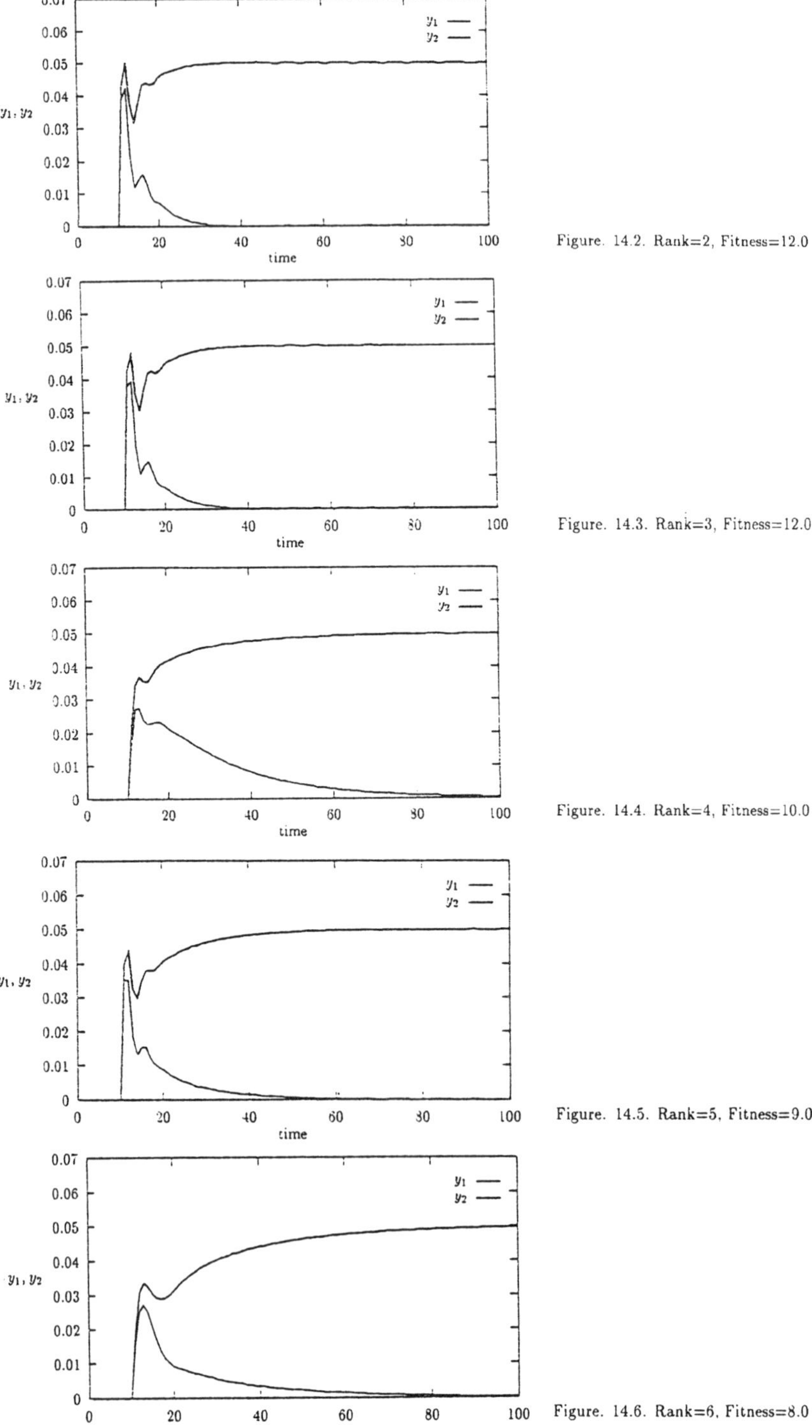


Figure. 14.2. Rank=2, Fitness=12.0

Figure. 14.3. Rank=3, Fitness=12.0

Figure. 14.4. Rank=4, Fitness=10.0

Figure. 14.5. Rank=5, Fitness=9.0

Figure. 14.6. Rank=6, Fitness=8.0

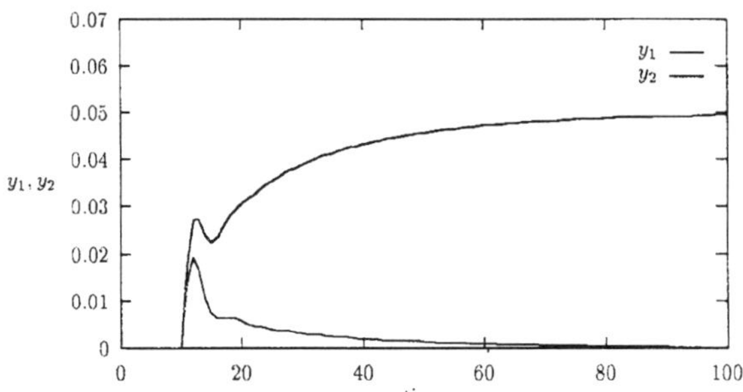


Figure. 14.7. Rank=7, Fitness=8.0

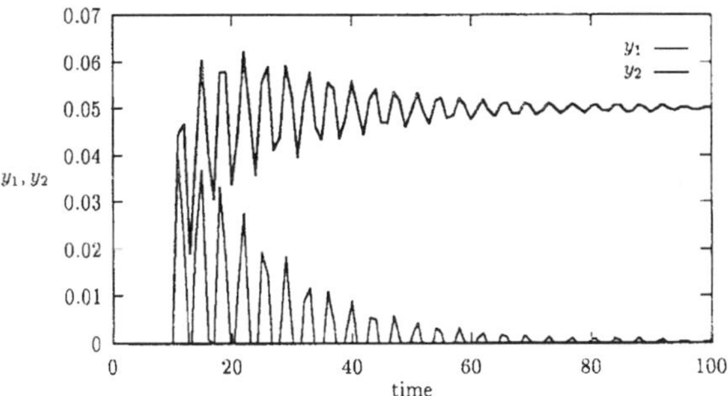


Figure. 14.8. Rank=8, Fitness=7.0

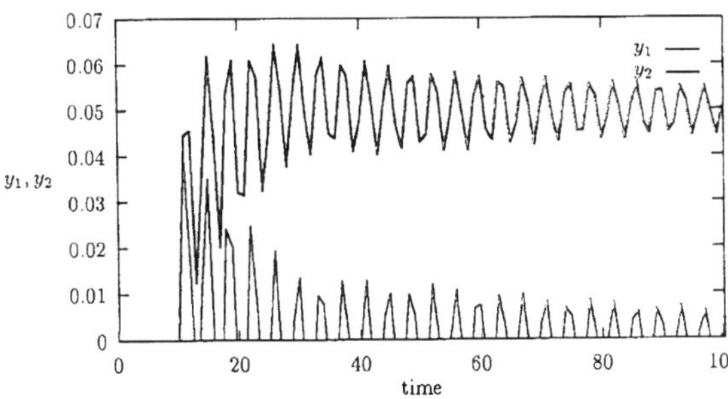


Figure. 14.9. Rank=9, Fitness=6.0

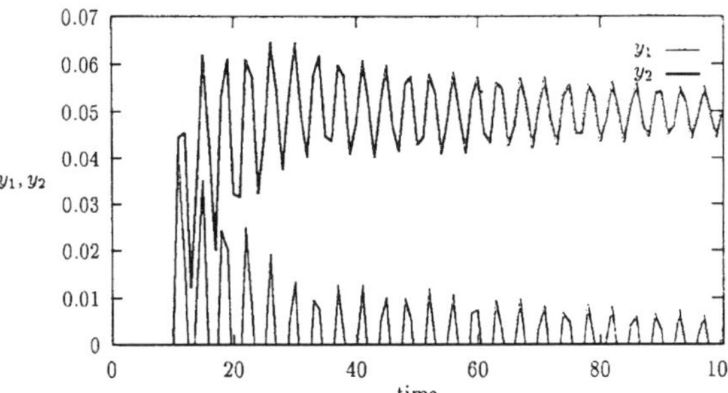


Figure. 14.10. Rank=10, Fitness=6.0

This example illustrates that GAs can successfully converge to controller settings to meet qualitative goals. An advantage of this technique is that in deciding which is the best response the experience of the operator or the engineer is used. This allows us to incorporate qualitative criteria which cannot be captured by quantitative measures like ISE, percent overshoot, etc. Also, in our study we found that this technique is robust in the sense that some (but inevitable) lack of consistency in fitness assignments is tolerable.

## 10.3 Case Study 2: Optimization of back mix reactors in series

In this section we examine the classic problem of optimally designing a continuously stirred tank reactor (CSTR) train. In the past this problem has been studied by various researchers (Szèpe and Levenspiel (1964), Wood and Stevens (1964), Luss (1965), Crooks (1966) and Edgar and Himmelblau (1988)). In a train of four CSTRs the problem is to design CSTR volumes to achieve maximum conversion. The sum of the individual CSTR volumes is constrained by a prespecified value. This problem is successfully solved for an arbitrary order, irreversible, single reaction power law kinetics by previous researchers. For comparison purposes, we solve the same problem by a GA. The reaction

$$A \rightarrow B \qquad r = kA^n$$

takes place under isothermal conditions in a series of four CSTRs whose dynamics are given by

$$\frac{d(V_i c_i)}{dt} = Fc_{i-1} - Fc_i - r_i V_i, \quad i = 1,\ldots,4 \qquad (10.1)$$

The L.H.S. of the above equation is set to zero and the exit concentrations $c_i$, i = 1,...,4, of the four CSTRs are solely determined by inlet flow rate $F$, reaction constraint $k$ and feed concentration $c_0$. Parameter values and variable nomenclature are given in Table 10.1. Since GAs are insensitive to the analytic properties of the objective function, they can handle any general kinetic expression. Here a GA is used as a function optimizer to solve the following:

$$\min_{V_1,\ldots,V_4} \; [c_4]$$

subject to

$$V_1 + V_2 + V_3 + V_4 = 20\text{m}^3$$

In each fitness evaluation, the routine FZERO (Kahaner, Moler and Nash, 1988) is used to solve for the steady state algebraic equation yielding c4 and the fitness is set equal to $-c_4$. When the constraints are violated the fitness is set equal to that of the minimum fitness encountered in that generation. Figure 10.15 shows the evolution of the solution. The fact that the average population minima approaches that of the best member in each population indicates that the minimum is indeed global. Table 10.2 shows a comparison between this study and that of Edgar and Himmelblau (1988).

In gradient based solution of this problem, as $c_4$ cannot be solved explicitly as a function of $c_0$, $c_4$ is held constant and $c_0$ is maximized w.r.t. $V_i$, i = 1, ..., 4 in each cycle. If the maximum $c_0$ does not match with 20 kgmol/m$^3$, $c_4$ is changed using linear interpolation and the optimization is done again. This procedure is continued until $c_0$ matches the given inlet concentration of 20 kgmol/m$^3$. To demonstrate the complexity of this technique a sequential quadratic programming (SQP) technique (Zhou and Tits, 1989) was used to arrive at optimal $V_i$s for $c_4$ = 0.3961 kgmol/m$^3$ and 1190 function evaluations were needed to converge to the maximum inlet concentration of 20 kgmol/m$^3$ (see Figure 10.16 for the convergence profile). Obviously many more function evaluations are needed if we start from an arbitrary initial value of $c_4$ and change $c_4$ after each cycle. Clearly this is a very cumbersome method.

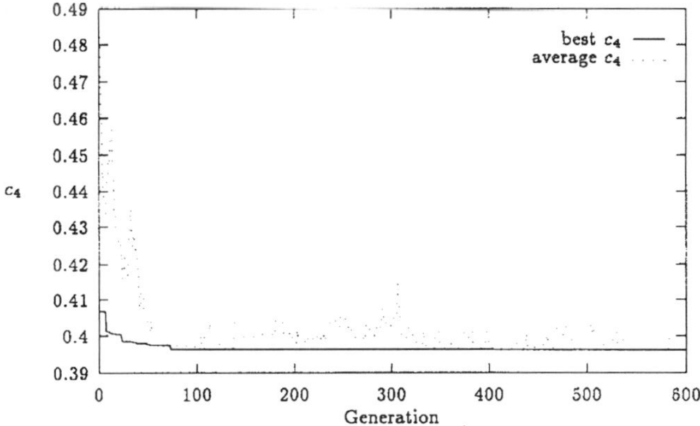


Figure. 15. The value of $c_4$ achieved by the best volume ratio and the average of $c_4$ achieved by all settings of volume ratios in each generation vs. the number of generations elapsed.

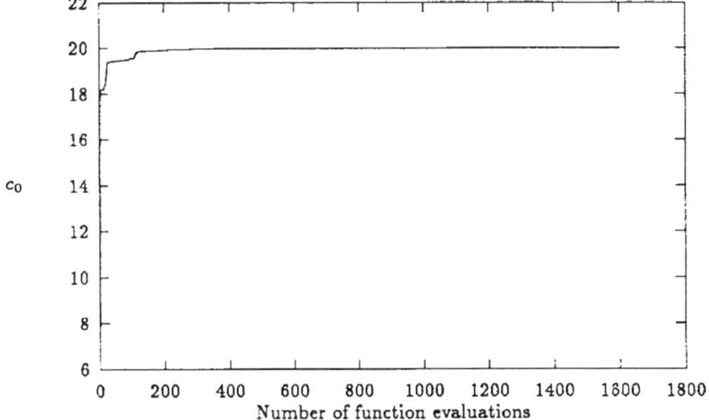


Figure. 16. Plot showing convergence of the SQP routine to $c_4 = 20\frac{kgmol}{m^3}$.

| Symbol | meaning | value |
|---|---|---|
| F | feedrate | 71 $m^3$/hr |
| $V_i$, i = 1,2, 3, 4 | volume of $i^{th}$ reactor | variable |
| $c_i$ ,i = 1,2,3,4 | concentration of *A* in $i^{th}$ CSTR | variable |
| $c_o$ | concentration of species *A* in the feed to the first reactor | 20 kgmol/$m^3$ |
| $r_i$, i = 1,2,3,4 | rate of dissipation of *A* in CSTR *i* | $= kc_i^n$ |
| $n$ | reaction order | 2.5 |
| $k$ | reaction constant | 0.00625 $\left[\frac{m^3}{kgmol}\right]^{1.5} s^{-1}$ |

Table. 10.1. Parameter values and nomenclature for the CSTR train.

| *Variable value at the minimum* | *This Study* | *Edgar and Himmelblau (1988)* |
|---|---|---|
| $c_4$ | 0.39627 kgmol/$m^3$ | 0.3961 kgmol/$m^3$ |
| $V_1$ | 2.234$m^3$ | 2.242$m^3$ |
| $V_2$ | 3.698$m^3$ | 3.884$m^3$ |
| $V_3$ | 6.163$m^3$ | 5.849$m^3$ |
| $V_4$ | 7.905$m^3$ | 8.025$m^3$ |

Table. 10.2. Results of Case Study 2

## 10.4 Case Study 3: Solution of lattice model to predict the adsorption of polymer molecules[3]

In this section we discuss a method of solving the lattice model by GAs and compare it with a classical technique. Next, we state the lattice model and pose the optimization problem. We will solve the optimization problem using GAs. We will also suggest ways to improve the convergence properties of GAs and compare the results of these modifications with Levenberg-Marquardt technique.

### 10.4.1 The lattice model

The lattice model aims at computing the structure of adsorbed polymer molecules near the surface and into the bulk. It aims at computing probabilities of various polymer chain conformations by using physical properties. The structure is described by two parameters viz. $\phi_i$ the segment volume fraction in layer *i* and $P_i$ the free segment probability in layer *i*. As we see below, the physics of the problem is such that computation of $P_i$ and $\phi_i$ is not straightforward since both

[3] The authors wish to thank Dr. Harry J. Ploehn of Texas A&M University for his assistance in this case study

depend on one another implicitly. The probability of finding a free segment (monomer) in layer $i$ is defined by

$$\ln P_i = \chi_s \delta_{1,i} + \chi\left(\langle\phi_i\rangle - \langle\phi_i^0\rangle\right) + \ln\phi_i^0 \qquad (10.2)$$

where

$$\langle\phi_i\rangle = \sum_{j=1}^{M} \lambda_{j-i}\phi_j \qquad (10.3)$$

and

$$\langle\phi_i^0\rangle = \sum_{j=1}^{M} \lambda_{j-i}\phi_j^0 \qquad (10.4)$$

The free segment probability $P*$ for a segment in the bulk solution is defined by

$$\ln P_* = \chi(\phi_* - \phi_*^0) + \ln\phi_*^0 \qquad (10.5)$$

Now the free segment probability $p_i$ with respect to the bulk solution is

$$p_i = \frac{P_i}{P_*} \qquad (10.6)$$

which with

$$\phi_* = \frac{nr}{Lp(r)} \qquad (10.7)$$

gives

$$\phi_i = \frac{\phi_*}{r}\frac{1}{p_i}\sum_{s=1}^{r} p(i,s)p(i,r-s+1) \qquad (10.8)$$

where $p_{i,s}$ is end segment probability, i.e. the probability that the end segment of an $s$-mer is in layer $i$ and

$$p(i,s) = \frac{P_{i,s}}{P_*^s} = p_i\lambda_1 p(i-1,s-1) + \lambda_0 p(i,s-1) + \lambda_1 p(i+1,s-1) \qquad (10.9)$$

Table. 10.3 gives the parameter values, for more details see Scheutjens et al. (1979).

| *symbol* | *meaning* | *value* |
|---|---|---|
| $M$ | number of lattice layers | 20 |
| $r$ | number of chains per segment | 1000 |
| $\phi_*$ | segment volume fraction in bulk | 0.01 |
| $\lambda_0, \lambda_1$ | fraction of nearest neighbors in the same layer, and in the adjacent layer | 0.5,0.25 |
| $\chi$ | Flory-Huggins polymer solvent interaction parameter | in Case 1.: 0.0<br>in Case 2.: 0.5 |
| $\chi_s$ | differential adsorption energy parameter | 1.0 |
| $\phi_\iota$ | segment volume fraction | variable |
| $\phi_i^0$ | solvent volume fraction in layer $i$ | = 1 - $\phi_\iota$ |
| $\phi_*^0$ | solvent volume fraction in the bulk solution | = 1 - $\phi_\iota$ |
| $p_i$ | free segment probability w.r.t. the bulk solution | variable |

Table. 10.3. Lattice model parameters and variables.

### 10.4.2 A General Framework to Solve the Lattice Model

The lattice model can be solved in an iterative manner as follows:

- Step 1. Guess $\phi_i$, $i = 1,2, ...M$
- Step 2. Solve for $pi_,$ $i = 1,2, ...M,$ using Eqs. 10.7, 10.8, 10.9, 10.10 and 10.11
- Step 3. Get $\phi_i'$, $i = 1,2,...M$ using Eqs. 10.12, 10.13 and 10.14
- Step 4. Use the difference $J = \sum_{i=1}^{M} (\phi_i - \phi_i')^2$ to correct $\phi_i$, $i = 1,2 ...M$
- Step 5. Stop if the difference is satisfactorily small, else goto step 2.

So the problem of solving the lattice model can be cast as a constrained optimization problem (the constraint is $0.0 < \phi_i < 1.0$).

### 10.4.3 Solution of the Lattice Model by GAs

We solve the lattice model by solving the optimization problem stated earlier. Observing results in the existing literature (Scheutjens, 1979) we note that the profiles of $Pi$ and $\phi_i$ are not varying everywhere so it will be easier if we approximate the solution by the following polynomial form to reduce dimensionality, i.e.

$$\phi_i = \sum_{k=1}^{N} a_k i^{k-1}, \quad i = 1,2,...,M \text{ and } N < M \qquad (10.10)$$

and estimate $a_k$, $k = 1, 2, ..., N$ that minimize the objective function $J = \sum_{i=1}^{M} (\phi_i - \phi_i')^2$. When $N = M$ this approximation approaches to the original problem where no such approximation was made. $N$ can be increased or decreased depending on problem complexity. We start with a set of several subsets, each subset containing the coefficients of the polynomial approximation and proceed as outlined in the previous section. In order to enhance the convergence properties we can augment the GA by a gradient search technique either at the end or between each generation. The tuning parameters for the GA are the population size, length of the vector representations, *mutation* and *crossover* rates and the number of *children* in each generation. Results of the optimization procedure are shown in Figures 10.17 and 10.18 for $\chi = 0.0$ and $\chi = 0.5$, respectively. Figures 10.19 and 10.20 show the plots of $\phi_i$ and $p_i$ vs. $i$ for the last two iteration cycles of the optimization procedure indicating convergence.

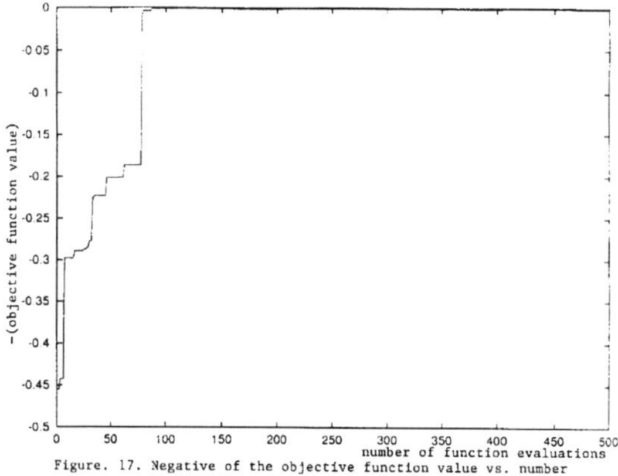

Figure. 17. Negative of the objective function value vs. number of function evaluations for Lattice Model solution when GA was used. (for x=0.0)

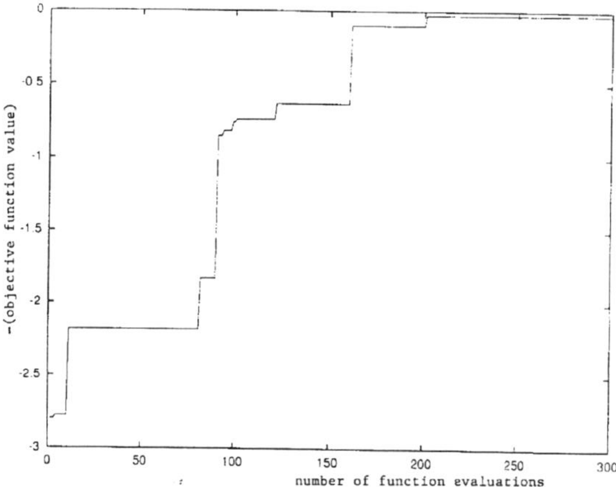

Figure. 18. Negative of the objective function value vs. number of function evaluations for Lattice Model solution

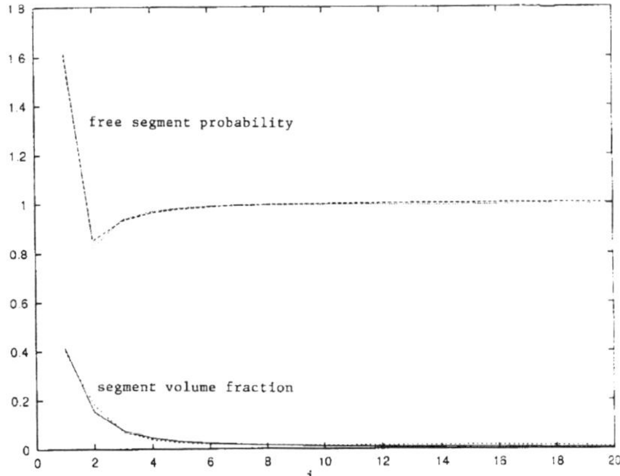

Figure. 19. Segment volume fraction and free segment probability vs. i for polynomial approximation when GA was used.

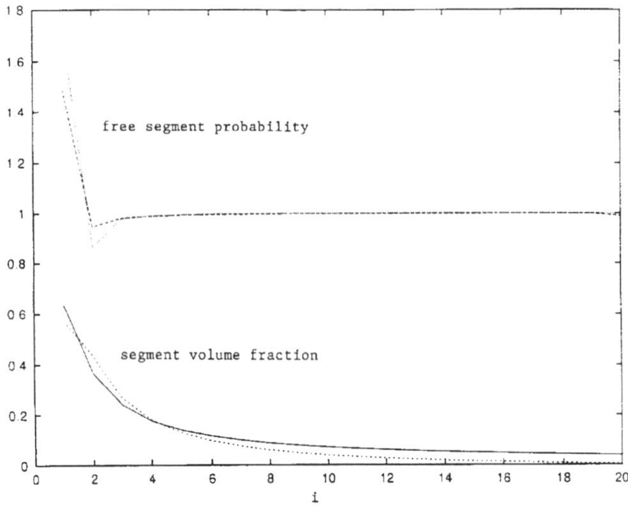


Figure. 20. Segment volume fraction and free segment probability

## 10.5 Comparison with other Techniques

For purposes of comparison we solved the lattice model using a constrained Levenberg-Marquardt (LM) method using finite difference gradients. Figures 10.21, 10.22, 10.23 and 10.24 show the results when we did not use the polynomial approximation of the solution but optimized using the whole vector $[\phi_1, \phi_2, ..., \phi_M]^T$. Figures 10.25, 10.26, 10.27 and 10.28 show the corresponding results when we used the LM method along with polynomial approximation of the solution. Figures 10.29, 10.30, 10.31 and 10.32 show the results of the modified GA technique (GA+LM) with the LM method implemented after 10 generations. In the last method we let the GA run for the first ten generations and from the tenth generation we picked the best candidate and did gradient search by a constrained LM technique. For all the LM trials, we observed that the solution was very sensitive to the initial guesses and convergence was not always guaranteed. For the method employing a GA, we observed that convergence was guaranteed from any random initial population but was not sharp. The GA+LM method not only ensured convergence but gave sharp convergence in relatively less number of total function evaluations. It should be noted that in this study in order to develop a general solution strategy Roe's approximation (Scheutjens, 1979) was not used as an initial guess for optimization. Instead, the initial solution was randomly picked in all cases. Contrary to the experience of Scheutjens (1979) constrained nonlinear optimization works and might work even better when Roe's approximation is used as a starting point. Polynomial approximation pays when we have extended and complex cases of the same problem. Our experience in this case shows that GA+LM is the safest approach for guaranteed convergence.

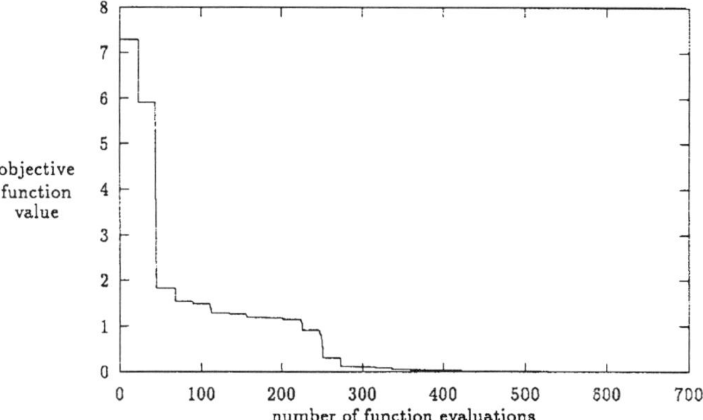


Figure. 21. Objective Function Value vs. Number of Function Evaluations for Non-Polynomial Approximation of the Lattice Model Solution. ($\chi = 0.0$)

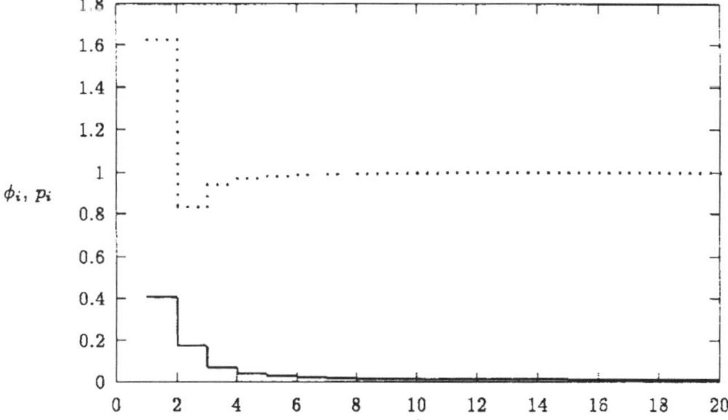


Figure. 22. $\phi_i$ vs. $i$ and $p_i$ vs. $i$ for Non-Polynomial Approximation of the Lattice Model Solution. ($\chi = 0.0$)

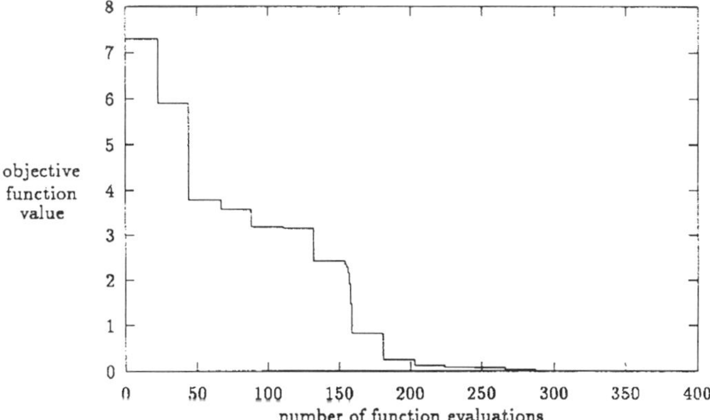


Figure. 23. Objective Function Value vs. Number of Function Evaluations for Non-Polynomial Approximation of the Lattice Model Solution. ($\chi = 0.5$)

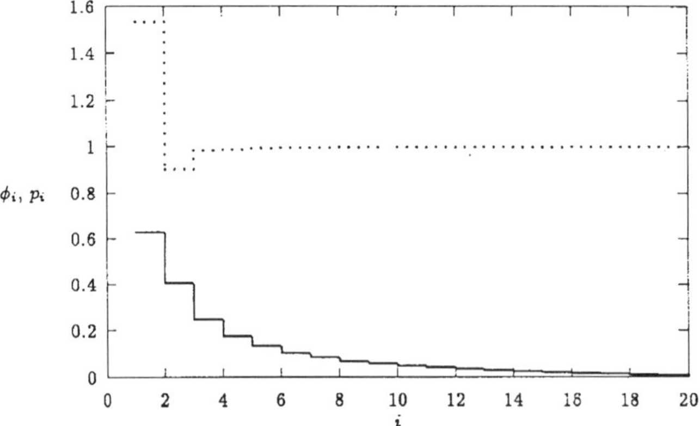


Figure. 24. $\phi_i$ vs. $i$ and $p_i$ vs. $i$ for Non-Polynomial Approximation of the Lattice Model Solution. ($\chi = 0.5$)

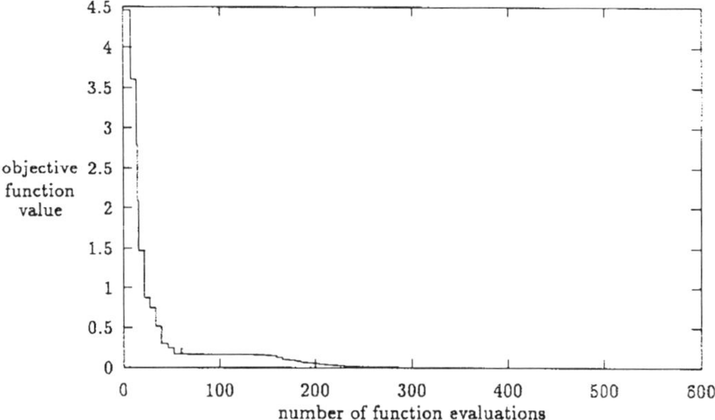


Figure. 25. Objective Function Value vs. Number of Function Evaluations for Polynomial Approximation of the Lattice Model Solution. ($\chi = 0.0$)

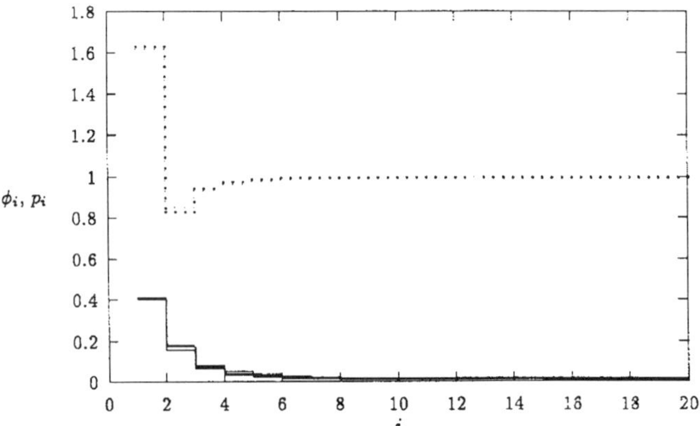


Figure. 26. $\phi_i$ vs. $i$ and $p_i$ vs. $i$ for Polynomial Approximation of the Lattice Model Solution. ($\chi = 0.0$)

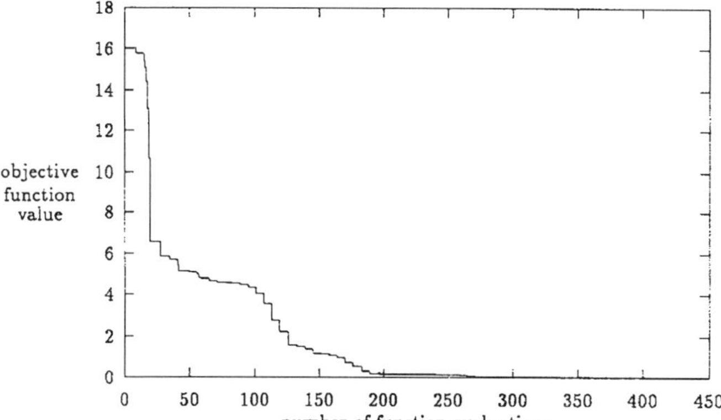


Figure. 27. Objective Function Value vs. Number of Function Evaluations for Polynomial Approximation of the Lattice Model Solution. ($\chi = 0.5$)

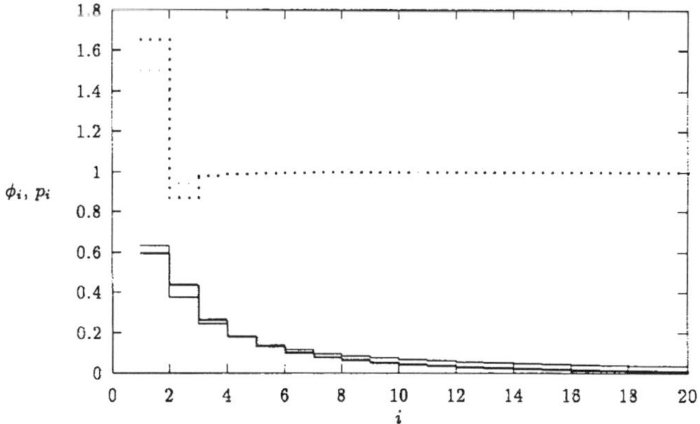


Figure. 28. $\phi_i$ vs. $i$ and $p_i$ vs. $i$ for Polynomial Approximation of the Lattice Model Solution. ($\chi = 0.5$)

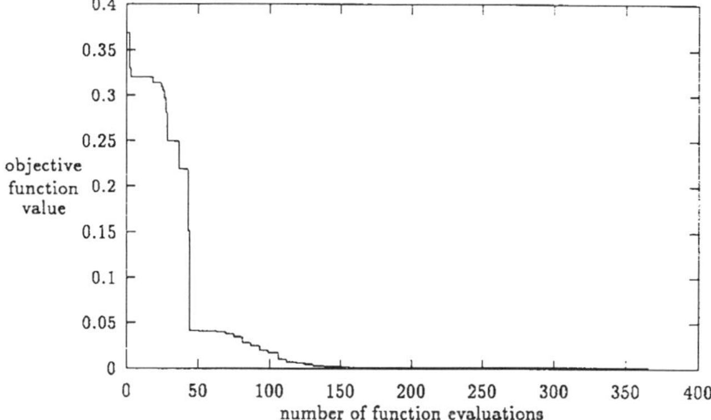


Figure. 29. Objective Function Value vs. Number of Function Evaluations for Polynomial Approximation of the Lattice Model Solution. ($\chi = 0.0$)

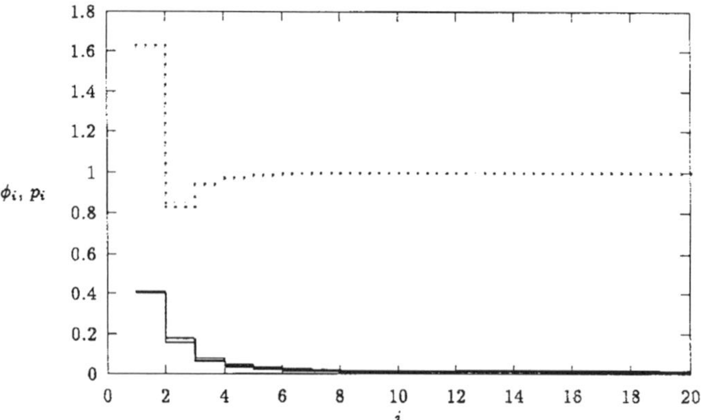


Figure. 30. $\phi_i$ vs. $i$ and $p_i$ vs. $i$ for Polynomial Approximation of the Lattice Model Solution. ($\chi = 0.0$)

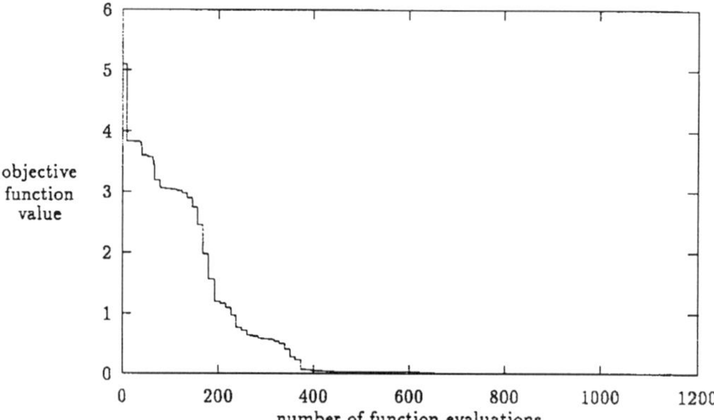


Figure. 31. Objective Function Value vs. Number of Function Evaluations for Polynomial Approximation of the Lattice Model Solution. ($\chi = 0.5$)

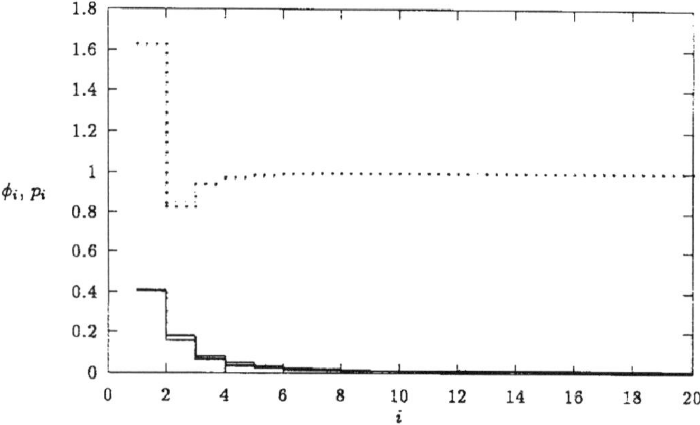


Figure. 32. $\phi_i$ vs. $i$ and $p_i$ vs. $i$ for Polynomial Approximation of the Lattice Model Solution. ($\chi = 0.5$)

Since GAs are blind to the underlying nonlinearities and the analytic properties of the objective function they are suitable for solving complex lattice models. GAs can be effectively combined with traditional constrained gradient techniques to ensure guaranteed and sharp convergence in the solution with very little or no a priori knowledge of the physics of the system.

| | |
|---|---|
| *Case Study 1* (Best controller synthesis for a nonlinear plant) | population size = 10<br>mutation probability = 0.0081<br>crowding factor = 0.3<br>generation gap = 0.4<br>crossover rate = 1 |
| *Case Study 2* (Optimization of back mix reactors in series) | population size = 10<br>mutation probability = 0.0030<br>crowding factor = 0.5<br>generation gap = 0.4<br>crossover rate = 1 |
| *Case Study 3* (Solution of lattice model) | population size = 10<br>mutation probability = 0.0080<br>crowding factor = 0.5<br>generation gap = 0.6<br>crossover rate = 1 |

Table. 10.4. Parameters of GAs employed in the three case studies.[4]

## 10.6 Conclusions

We demonstrated the ability of GAs to solve complex optimization problems in chemical engineering applications. The performance of GAs was compared with the traditional techniques where available.[5] This study shows that GAs can handle abstract and nonlinear objective functions efficiently With a good front-end software (made available by the authors) the application of GAs to any optimization problem is very easy and requires minimal a priori knowledge about the physics of the problem or the mathematical theory behind the optimization technique.

## References

[1] Chien, I.L. and B.A. Ogunnaike, "Modeling and Control of High Purity Distillation Columns", *AIChE* annual meeting, Miami, FL (1992).

[2] Crooks, W.M., "Denbig's 2-tank CSTR system", *British Chemical Engng.*, **11**, 7, 710-712 (1966).

---

[4] A generation gap of 0.4 implies that the number of offspring in each generation equals 4(= 10 x 0.4).

[5] It should be emphasized that the CPU times in case studies 2, 3 and 4 were in the order of seconds on a Sparc station 2. Comparison of CPU times between GAs and other techniques will only be fair if the former were implemented on a parallel machine.

[3] Davis, L., *Handbook of Genetic Algorithms,* Van Nostrand, New York (1991).

[4] Edgar, T.F. and D.N. Himmelblau, *Optimization of Chemical Processes,* McGraw-Hill (1988).

[5] Goldberg, D.E., *Genetic Algorithms in Search Optimization and Machine Learning,* Addison-Wesley, Reading, MA (1989).

[6] Hansen, J.M., C. Lim, and J. Hong, "Optimization of Autocatalytic Reactions", *Chemical Eng. Sci.,* **48**, 13, 2375-2390 (1993).

[7] Kahaner, D., C. Moler, and S. Nash, *Numerical Methods and Software,* Prentice Hall (1988).

[8] Luss, D.,"Optimum Volume Ratios for Minimum Residence Time in CSTR sequences", *Chemical Eng. Sci.,* **20**, 171 (1965)

[9] Scheutjens, J.M.H.M. and G.J. Fleer, "Statistical Theory of the Adsorption of Interacting Chain Molecules. 1. Partion Function, Segment Density Distribution, and Adsorption Isotherms", *J. Phys. Chem.,* **83**, 12, (1979).

[10] Szépe, S. and O. Levenspiel, "Optimization of Backmix Reactors in Series for a Single Reaction", *Ind. Eng. Chem. Proc. Des. and Der.,* **3**, 3, 214-217 (1964).

[11] Wood, R.K., and W.F. Stevens, "Optimum Volume Ratios for Minimum Residence Time in CSTR sequences", *Chemical Engng Sci.,* **19**, 426 (1964)

[12] Zhou, J.L. and A.L. Tits, "User's Guide for FSQP Version 3.3b: A FORTRAN Code for Solving Constrained Nonlinear (Minimax) Optimization Problems, Generating Iterates Satisfying All Inequality and Linear Constraints" (1989).

## Chapter 11

**Sam R. Thangiah**
Artificial Intelligence and Robotics Laboratory
Computer Science Department
Slippery Rock University
Slippery Rock
PA 16057

# Vehicle Routing with Time Windows using Genetic Algorithms



**Abstract**

In vehicle routing problems with time windows (VRPTW), a set of vehicles with limits on capacity and travel time are available to service a set of customers with demands and earliest and latest time for servicing. The objective is to minimize the cost of servicing the set of customers without being tardy or exceeding the capacity or travel time of the vehicles. As finding a feasible solution to the problem is NP-complete, search methods based upon heuristics are most promising for problems of practical size. In this chapter we describe GIDEON, a Genetic Algorithm for heuristically solving the VRPTW. GIDEON has a global customer clustering method and a local post-optimization method. The global customer clustering method uses an adaptive search strategy based upon population genetics, to assign vehicles to customers. The best solution, obtained from the clustering method is improved by a local post-optimization method. The synergy between a global adaptive clustering method and a local route optimization method produce results superior to those obtained by competing heuristic search methods. The results obtained by GIDEON on a standard set of 56 VRPTW problems obtained from the literature were as good as or better than solutions from known competing heuristics.

0-8493-2529-3/95/$0.00 + $.50

## 11.1 Introduction

The problem we address is the Vehicle Routing Problem with Time Windows (VRPTW). The VRPTW involves routing a fleet of vehicles, with limited capacities and travel times, from a central depot to a set of geographically dispersed customers with known demands within specified time windows. The time windows are two-sided, meaning that a customer must be serviced at or after its earliest time and before its latest time. If a vehicle reaches a customer before the earliest time it results in idle or waiting time. A vehicle that reaches a customer after the latest time is tardy. A service time is also associated with servicing each customer. The route cost of a vehicle is the total of the traveling time (proportional to the Euclidean distance), waiting time and service time taken to visit a set of customers.

The VRPTW arises in a wide array of practical decision making problems. Instances of the VRPTW occur in retail distribution, school bus routing, mail and newspaper delivery, municipal waste collection, fuel oil delivery, dial-a-ride service and airline and railway fleet routing and scheduling. Efficient routing and scheduling of vehicles can save government and industry millions of dollars a year. The current survey of vehicle routing methodologies are available in [2] [12][21]. Solomon and Desrosiers [28] provide an excellent survey on vehicle routing with time windows.

In this chapter we describe GIDEON, a Genetic Algorithm system to heuristically solve the VRPTW. GIDEON is a cluster-first route-second method that assigns customers to vehicles by a process we call Genetic Sectoring and improves on the routes using a local post-optimization method. The Genetic Sectoring method uses a genetic algorithm to adaptively search for sector rays that partition the customers into sectors or clusters served by each vehicle. It ensures that each vehicle route begins and ends at the depot and that every customer is serviced by one vehicle. The solutions obtained by the Genetic Sectoring method are not always feasible and are improved using a local post-optimization method that moves customers between clusters.

The chapter is arranged in the following form. Section 11.2 gives a mathematical formulation of the VRPTW. Section 11.3 gives a description of the GIDEON system. Section 11.4 describes the results of computational testing on a standard set of VRPTW problems obtained from the literature. Section 11.5 is the computational analysis of the solutions obtained from the GIDEON system and with respect to competing heuristics. Section 11.6 contains the summary and concluding remarks.

## 11.2 Mathematical Formulation for the VRPTW

The notation and expressions used in the model are useful in explaining the genetic search. We present a mixed-integer formulation of the vehicle routing problem with time window constraints. Our formulation is based upon the model defined by Solomon [30]. The following notations will help in the description of the GIDEON system. In the mixed-integer formulation the indices $i,j$=l,...,$N$ and $k$=$l$,...,$K$.

Parameters:
$K$ = number of vehicles
$N$ = number of customers (0 denotes the central depot)
$T$ = maximum travel time permitted for a vehicle
$C_i$ = customer $i$
$C_0$ = the central depot
$V_k$ = vehicle route $k$
$O_k$ = total overload for vehicle route $k$
$T_k$ = total tardiness for vehicle route $k$
$D_k$ = total distance for a vehicle route $k$
$R_k$ = total route time for a vehicle route $k$
$Q_k$ = total over-route time for a vehicle route $k$
$t_{ij}$ = travel time between customer i and j (proportional to the Euclidean distance)
$v_k$ = maximum capacity of vehicle $k$
$t_i$ = arrival time at customer $i$
$f_i$ = service time at customer $i$
$w_i$ = waiting time before servicing customer $i$
$e_i$ = earliest release time for customer $i$
$l_i$ = latest delivery for customer $i$
$q_{ik}$ = total demand of vehicle k until customer $i$
$r_{ik}$ = travel time of vehicle $k$ until customer $i$ (including service time and waiting time)
$p_i$ = polar coordinate angle of customer $i$
$s_i$ = pseudo polar coordinate angle of customer $i$
$F$ = fixed angle for Genetic Sectoring, Max$[p_i,...,p_n]/2K$, where $n = 1,...,N$
$B$ = length of the bit string in a chromosome representing an offset, $B = 3$
$P$ = population size of the Genetic Algorithm, $P = 50$
$G$ = number of generations the Genetic Algorithm is simulated, G = 1000
$E_k$ = offset of the $k^{th}$ sector, i.e, decimal value of the $k^{th}$ bit string of size $B$
$I$ = a constant value used to increase the range of $E_i$
$Sk$ = seed angle for sector $k$
$S_0$ = initial seed angle for Genetic Sectoring, $S_0 = 0$
$\alpha$ = weight factor for the distance
$\beta$ = weight factor for the route time
$\eta$ = penalty weight factor for an overloaded vehicle
$\gamma$ = penalty weight factor for exceeding maximum route time in a vehicle route
$\kappa$ = penalty weight factor for the total tardy time in a vehicle route

Variables:

$$y_{ik} = \begin{cases} 1, \text{if } i \text{ is serviced by vehicle } k \\ 0, \text{otherwise} \end{cases}$$

$$x_{ijk} = \begin{cases} 1, \text{if the vehicle } k \text{ travels directly from } i \text{ to } j \\ 0, \text{otherwise} \end{cases}$$

The mixed integer formulation for the vehicle routing problem is stated as follows:

$$\text{(VRPTW) Min} \sum_{i=1}^{N}\sum_{j=1}^{N}\sum_{k=1}^{K} c_{ijk} x_{ijk} \qquad (11.2.1)$$

where

$c_{ijk} = t_{ij} + w_i + f_i$

Subject to:

$$\sum_{i=0}^{N} q_{ik} y_{ik} \le v_k, \quad k = 1,\ldots,K \qquad (11.2.2)$$

$$\sum_{i=0}^{N}\sum_{j=0}^{N} y_{ik}\left(t_{ij} + f_i + w_i\right) \le v_k, \; k = 1,\ldots,K \qquad (11.2.3)$$

$$y_{ik} = 0 \text{ or } 1; i = 0,\ldots,1; k = 1,\ldots,K \qquad (11.2.4)$$

$$x_{ijk} = 0 \text{ or } 1; i,j = 1,\ldots,N; k = 1,\ldots,K \qquad (11.2.5)$$

$$\sum_{k=1}^{K} y_{ik} = \begin{cases} K, i = 0 \\ 1, i = 1,\ldots,N \end{cases} \qquad (11.2.6)$$

$$\sum_{ji=0}^{N} x_{ijk} = y_{jk}, \; j = 0,\ldots,N; k = 1,\ldots,N \qquad (11.2.7)$$

$$\sum_{j=0}^{N} x_{ijk} = y_{ik}, \; i = 1,\ldots,N; k = 1,\ldots,N \qquad (11.2.8)$$

$$t_j \ge t_i + s_i + t_{ij} - \left(1 - x_{ijk}\right) \cdot T, \quad i,j = 1,\ldots,N, \quad k = 1,\ldots,K \qquad (11.2.9)$$

$$e_i \le t_i < l_i, \quad i = 1,\ldots,N \qquad (11.2.10)$$

$$t_i \ge 0, \quad i = 1,\ldots,N \qquad (11.2.11)$$

The objective is to minimize the vehicle routing cost $C_{ijk}$ (11.2.1) subject to vehicle capacity, travel time and arrival time feasibility constraints. A feasible solution for the VRPTW services all the customers without the vehicle exceeding the maximum capacity of the vehicle (11.2.2) or the travel time of the vehicle (11.2.3). In addition, each customer can be served by one and only one vehicle (11.2.6). Travel time for a vehicle is the sum total of the distance travelled by the vehicle including the waiting and service time. Waiting time is the amount of time that a vehicle has to wait if it arrives at a customer location before the earliest arrival time for that customer. The time feasibility constraints for the problem are defined in (11.2.9), (11.2.10) and (11.2.11). The constraint (11.2.9) ensures that the arrival times between two customers are compatible. The constraint (11.2.10) enforces the arrival time of a vehicle at a customer site to be within the customers earliest and latest arrival times and (11.2.11) ensures that the arrival time of the vehicle at a customer location is always positive.

The vehicle routing problem (VRP), without time windows, is NP-complete [3] [18]. Solomon [30] and Savelsbergh [25] indicate that the time constrained problem is fundamentally more difficult than a simple VRP even for a fixed fleet of vehicles. Savelsbergh [25] has shown that finding a feasible solution for a VRPTW using a fixed fleet size is NP-complete. Due to the intrinsic difficulty of the problem, search methods based upon heuristics are most promising for solving practical size problems [1] [9] [17] [23] [23] [25] [27] [29]. Heuristic methods often produce optimum or near optimum solutions for large problems in a reasonable amount of computer time. Therefore the development of heuristic algorithms that can obtain near optimal feasible solutions for large VRPTW are of primary interest.

The GIDEON system that we propose to solve the VRPTW is a cluster-first route-second heuristic algorithm that solves an approximation of the mathematical model described in (11.2.1). The algorithm has two phases consisting of a global search strategy to obtain clusters of customers and a local post-optimization method that improves the solution. The clustering of customers is done using a Genetic Algorithm (GA) and the post-optimization method moves and exchanges customers between routes to improve the solution. The two processes are run iteratively a finite number of times to improve the solution quality.

## 11.3 The GIDEON System

The global search strategy for clustering customers in GIDEON is done using a Genetic Algorithm(GA). GA's are a class of heuristic search algorithms based upon population genetics [6] [7] [16]. As they are inherently adaptive, genetic algorithms can converge to near optimal solutions in many applications. They have heen used to solve a number of complex combinatorial problems [4] [5] [15] [19]. The GA is an iterative procedure that maintains a pool of candidates simulated over a number of generations. The population members are referred to as chromosomes. The chromosomes are fixed length strings with a finite number of binary values. Each chromosome has a fitness value assigned to it based upon a fitness function. The fitness value determines the relative ability of the chromosome to survive over the generations. Chromosomes with high fit values have a higher probability of surviving into the next generation compared to chromosomes with low fit values. At each generation, chromosomes are subjected to selection, crossover and mutation. Selection allows chromosomes with high fit values to survive into the next generation. Crossover splices chromosomes at random points and exchanges it with other spliced chromosomes. Mutation changes the bit value of a chromosome to its complementary value. Selection and crossover search a problem space exploiting information present in the chromosomes by selecting and recombining primarily those offspring that have high fitness values. These two processes eventually produce a population of chromosomes with high performance characteristics. The mutate operator is a secondary operator that prevents premature loss of information by randomly mutating bits in a chromosome. For a detailed description of this process refer to [11].

The local post-optimization method in GIDEON improves a solution by shifting or exchanging customers between routes if it results in reduction of the total routing cost. The method shifts and exchanges customers between routes until no more improvements are found [22][36][37]. In the shift procedure, one customer is removed from a route and inserted into a different route. In the exchange procedure, one customer each from two different routes is removed and inserted into the other's route. In both shift exchange procedures, improved solutions are accepted if the insertion results in the reduction of the total cost for routing the vehicles. The shift and exchange heuristics have been implemented successfully in many combinatorial problems [20][22][32][36]. The local post-optimization method for the GIDEON system uses the shift and exchange of one and two customers between routes.

The search space used by GIDEON is a relaxation of the feasible region of the mathematical model proposed in (11.2.1). The mathematical model (11.2.1) is approximated by the GIDEON system by a relaxation of the capacity, route time and time window constraints in a Lagrangian Relaxation fashion. The cost function used by the GIDEON system drives the search for a good feasible solution by penalizing violation of capacity, route or time window constraints. The objective function used by the GIDEON system is stated as:

$$\left(\overline{\text{VRPTW}}\right) \quad \text{Min} \sum_{i=1}^{N}\sum_{j=1}^{N}\sum_{k=1}^{K} c_{ijk} x_{ijk} \qquad (11.2.12)$$

where

$$c_{ijk} = \alpha\, t_{ij} + \beta \cdot \left(t_i + f_i + t_{ij}\right) + \eta \cdot \max\left\{0, \left(q_{ik} - v_k\right)\right\}$$
$$+ \kappa \cdot \max\left\{0, \left(r_{ik} - l_i\right)\right\} + \gamma \cdot \max\left\{0, \left(t_i + f_i + t_{ij} - T\right)\right\}$$

The cost function includes components weighted by coefficients α for distance, β for route time and penalty weighting factors, η for vehicle overload, γ for travel time in excess of the allocated route time for the vehicle and κ for tardiness. The GIDEON system explores for feasible solutions to the VRPTW with weights that drive the model towards feasibility in the VRPTW problem. The weights for GIDEON were derived empirically and set at $\alpha = 0.5$, $\beta = 0.05$, $\eta = 50$, $\kappa = 25$ and γ, = 50. The weights are biased towards finding a feasible solution in comparison to reducing the total distance and route time. The main priority of the cost function (11.2.12) is to obtain a feasible solution. Therefore the coefficients of the cost function (11.2.12) gives higher priority to reducing tardiness and overloading vehicles, followed by vehicles that exceed the maximum allotted route time for a vehicle. If there is no violation of the capacity, time feasibility and route time constraints, then the coefficients of the cost function (11.2.12) are to reduce the total distance followed by the total route time. The weights for the coefficients of the cost function were chosen to first obtain a feasible solution and then minimize the total distance and route time. The cost function (11.2.12) was experimented with other weight values, values that gave higher weights to the cost coefficients α and β and lower weights to η, α and γ, but these resulted in infeasible or solutions of poor quality.

The GIDEON system uses the cluster-first route-second method to solve a VRPTW. That is, given a set of customers and a central depot, the system clusters the customers using the GA, and the customers within each sector are routed using the cheapest insertion method [13]. The clustering of customers using a GA is referred to as Genetic Sectoring. Genetic Sectoring has been successfully used to solve vehicle routing and scheduling problems with complex constraints [31][32][33][34][35]. The GIDEON system allows exploration and exploitation of the search space to find good feasible solutions with the exploration being done by the GA and the exploitation by the local post-optimization procedure.

The GENESIS [14] genetic algorithm software was used in the implementation of the GIDEON system. The chromosomes in GENESIS are represented as bit strings. The sectors (clusters) for the VRPTW are obtained from a chromosome by subdividing it into K divisions of size B bits. Each subdivision is used to compute the size of a sector. The fitness value for the chromosome is the cost function (2.12) for serving all the customers computed with respect to the sector divisions derived from it.

In an $N$ customer problem with the origin at the depot, the GIDEON system replaces the customer angles $p_1,...,p_N$ with pseudo polar coordinate angles $s_1,..,s_N$. The pseudo polar coordinate angles are obtained by normalizing the angles between the customers so that the angular difference between any two adjacent customers is equal. This allows sector boundaries to fall freely between any pair of customers that have adjacent angles, whether the separation is small or large. The customers are divided into $K$ sectors, where K is the number of vehicles, by planting a set of "seed" angles, $S_0,...,S_k$, in the search space and drawing a ray from the origin to each seed angle. The initial number of vehicles, $K$, required to service the customers is obtained using Solomon's insertion heuristic [30]. The initial seed angle $S_0$ is assumed to be $0^\circ$. The first sector will lie between seed angles $S_0$ and $S_1$ the second sector will lie between seed angles $S_1$ and $S_2$, and so on.

The Genetic Sectoring process assigns a customer, $C_i$, to a sector or vehicle route, $V_k$, based on the following equation:

$C_i$ is assigned to $V_k$ if $S_k < s_i <= S_{k+1}$, where $k = 0,...,K\text{-}1$

Customer $C_i$ is assigned to vehicle $V_k$ if the pseudo polar coordinate angle $s_i$ is greater than seed angle $S_k$ but is less than or equal to seed angle $S_{k+1}$. Each seed angle is computed using a fixed angle and an offset from the fixed angle. The fixed angle, $F$, is the minimum angular value for a sector and assures that each sector gets represented in the Genetic Sectoring process. The fixed angle is computed by taking the maximum polar coordinate angle within the set of customers and dividing it by $2K$. The offset is the extra region from the fixed angle that allows the sector to encompass a larger or a smaller sector area. The GA is used to search for the set of offsets that will result in the minimization of the total cost of routing the vehicles. If a fixed angle and its offset exceeds $360^\circ$, then that seed angle is set to 360° thereby allowing the Genetic Sectoring process to consider vehicles less than K to service all its customers. Therefore $K$, the initial number of vehicles with which the GIDEON system is invoked, serves as the upper bound on the number of vehicles that can be used for servicing all the customers.

The bit size representation of an offset in a chromosome B was set at 3 bits. Bit size representations larger than 3 were experimented with and resulted in poor quality solutions. The decimal conversion of 3 bits results in a range of integer values between 0 and 7. The decimal values retrieved from the subset of a chromosome are multiplied by a constant I that increases the range of the offset. The value derived from the decimal conversion of the bit values times the constant value I are mapped proportionately to the offsets with the value 0 as a 0° offset and the bit value 10.5 and 10.5° as the maximum offset. Figure 11.1 describes the chromosome mapping used to obtain the offsets.

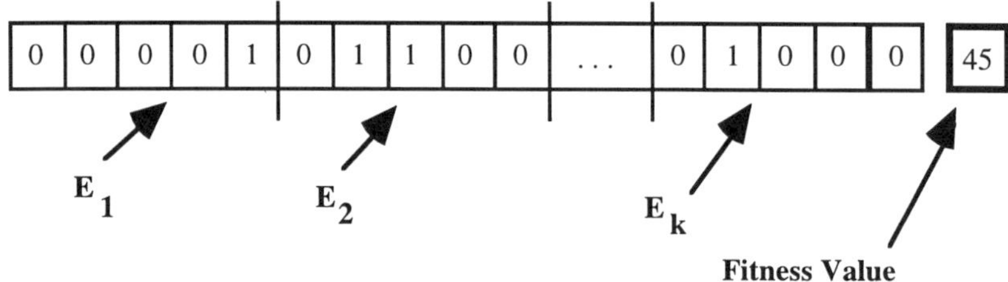


Figure 11.1 Representation of the offsets using a chromosome.

The seed angles are derived from the chromosome using the following equation:

$$S_i = S_{i-1} + F + (E_i \cdot 1) \qquad (11.2.13)$$

The fitness value of a chromosome is the total cost of routing $K$ vehicles for servicing $N$ customers using the sectors formed from the set of seed angles derived from the chromosome. The seed angles are derived using the fixed angle and the offsets from the chromosomes. The customers within the sectors, obtained from the chromosomes, are routed using the cheapest insertion method. The cheapest insertion method takes each unrouted customer in the sector and each edge $\{i, j\}$ in the current tour and computes the cost of inserting the unrouted customer between $i$ and $j$. The unrouted customer that has the least insertion cost at edge $\{i, j\}$ is selected to be inserted between $i$ and $j$. The cost of inserting customer $C_i$ into route $V_k$ using the cheapest insertion method is calculated using the insertion cost function:

$$\text{insertion cost of } C_i = \alpha D_k + \beta R_k + \eta O_k + \kappa T_k + \gamma Q_k \qquad (11.2.14)$$

The insertion cost function (11.2.14) will accept infeasible solutions if the reduction in total distance is high enough to allow either a vehicle to be overloaded or be tardy. Overloading and tardiness in a vehicle route are penalized in the insertion cost function (11.2.14). The weights for insertion cost (11.2.14) were set at $\alpha = 0.5$, $\beta = 0.05$, $\eta = 50$, $\kappa = 50$, and $\gamma = 25$.

In the GIDEON system each chromosome represents a set of offsets for a VRPTW. Therefore, a population of $P$ chromosomes usually has $P$ different solutions for a VRPTW. That is, there may he some chromosomes in the population that are not unique. At each generation, $P$ chromosomes are evaluated for fitness. The chromosomes that have the least cost will have a high probability of surviving into the next generation through the selection process. As the crossover operator exchanges a randomly selected portion of the bit string between the chromosomes, partial information about sector divisions for the VRPTW is exchanged between the chromosomes. New information is generated within the chromosomes by the mutation operator.

The GIDEON system uses selection, crossover and mutation to adaptively explore the search space for the set of sectors that will minimize the total cost of the routes over the simulated generations for the VRPTW. The GIDEON system would utilize more computer time than traditional heuristic algorithms because every time the Genetic Sectoring process is invoked it has to evaluate $P.G$ vehicle routes, where $P$ is the population size and $G$ is number of generations to he simulated. The worst case running time bounds for the Genetic Sectoring process is $O\left(\frac{N^2}{K}\right)$ as there is on the average $\frac{N}{K}$ customers for each route and $\frac{N}{K}$ edges have to be checked for each of the $K$ routes.

The parameter values for the number of generations, population size, crossover and mutation rates for the Genetic Sectoring process were set at 1000, 50, 0.8 and 0.001. During the simulation of the generations, the GIDEON system keeps track of the set of sectors obtained from the genetic search that has the lowest total route cost. The genetic search terminates either when it reaches the number of generations to be simulated or if all the chromosomes have the same fitness value. The best set of sectors obtained after the termination of the genetic search does not always result in a feasible solution. The solution obtained from the GA is improved using the local post-optimization procedure that shifts and exchanges customers between the vehicle routes.

The post-optimization method is similar to the 2-opt method, as it deletes two arcs of a customer and inserts the customer into a location of a different route that has the lowest cost. The worst case running time bounds for the local post-optimization process is $O\left(\left(\frac{N}{K}\right)^2\right)$ as there is on the average $\frac{N}{K}$ customers for each route and on the average $\frac{N}{K}$ edges have to be checked.

The local post-optimization process is carried out until no more improvements can be made to the solution obtained from the GA. At the termination of the local post-optimization procedure, the customers are ranked in order of the sectors, and within the sectors in the sequence in which they are visited by the vehicles. The customer angles, $p_1,...,p_N$, are replaced with pseudo polar coordinate angles $s_1,...,s_N$ in order of the customer rank. The assignment of pseudo polar coordinate angles, using route and customer sequence, clusters together customers with geographical and temporal characteristics that can be serviced by a single vehicle.

The customers with the new pseudo polar coordinate angles are once again used to form new sectors using the GA. The best set of sectors obtained from the GA using the new customer polar coordinate angles is improved using the local post-optimization procedure. This iteration between Genetic Sectoring process and local post-optimization method is carried out a predetermined number of times and was set at 5. The flow of the GIDEON system is described in Figure 11.2

The Genetic Sectoring and local post-optimization procedures are symbiotic as the Genetic Sectoring is a meta-search strategy that forms the sectors and the local post-optimization method gives adjacency information about the customers back to the Genetic Sectoring process. These two methods derive information from each other in order to obtain a good feasible solution.

Step 1: Set the number of cluster-route iterations: *itermax* = 3.
Set the current iteration number: *iter* = 0.
Set the bit string size for the offset: *Bsize* = 5.

Step 2: Sort the customers in order of their polar coordinate angles, and assign pseudo polar coordinate angles to the customers.
Set the lowest global route cost to infinity: $g = \infty$.
Set the lowest local route cost to infinity: $l = \infty$.

Step 3: Increment the number of iterations: *iter* = *iter* + 1.
If *iter* > *itermax*, go to Step 7.

Step 4: If GA has terminated, go to Step 5.
For each chromosome in the population:
    For each bit string of size *BSize*,
        calculate the seed angle,
        sector the customers, and
        route the customers within the sectors using the cheapest insertion method.
    If the cost of the current set of sectors is lower than $l$
        set $l$ to the current route cost, and
        save the set of sectors in *lr*.
    If the cost of the current set of sectors is lower than $g$,
        set $g$ to the current route cost, and
        save the set of sectors in *gr*.
Do Selection, Crossover and Mutation on the chromosomes.
Go to Step 4.

Step 5: Do local post-optimization using the route *lr*.
If no improvements can he made to route *lr*,
go to Step 6.
If the current improved route has lower cost than *l*,
set *l* to the current cost, and
save the set of sectors in *lr*.
If the current improved route has lower cost than *g*,
set *g* to the current cost, and
save the set of sectors in *gr*.
Go to step 5.

Step 6: Rank the customers of route *lr* in order of the sectors, and within the sectors in order of the sequence in which they are visited.
Sort the customers by the rank.
Assign pseudo polar coords to the customers in order of sorted rank.
Go to Step 3.

Step 7: Stop the Genetic Sectoring Heuristic with a local post-optimization solution.

Figure 11.2 Flow of the GIDEON system.

## 11.4 Computational Results

GIDEON was run on a set of 56 VRPTW problems in six data sets denoted R1, C1, RC1, R2, C2, and RC2, developed by Solomon [30]. Solomon generated vehicle routing problems with two time windows using the standard set of vehicle routing test problems from Christofides et al. [3].The vehicle routing problems with two time windows were generated by assigning earliest and latest time windows to each of the customers in addition to the service time required by each of the customers. In terms of time window density (the percentage of customers with time windows), the problems have 25%, 50%, 75%, and 100% time window density. Each of the problems in these data sets has 100 customers. The fleet size to service them varied between 2 and 21 vehicles.

For the R1 data set, without time window constraints, a fleet of 10 vehicles, each with a capacity of 200 units, was required to attain a feasible solution. Each of the customers in the R1 data set required 10 units of service time and a maximum route time of 230 units. In the C1 data set, each customer required 90 units of service time and the vehicles had a capacity of 200 units and a maximum route time of 1236 units. The optimal solution for this problem class requires 10 vehicles and has a distance of 827 units [9]. The RC1 data set was created using data sets, R1 and C1. The vehicle capacity for this problem was set at 200 units with a maximum route time of 240 units. Each of the customers in this problem required 10 units of service time.

The R2 data set was a modification of the R1 data set to allow for servicing of many customers by one vehicle. The maximum route time of the vehicles was set at 1000 units and each vehicle had a capacity of 1000 units. Two vehicles are enough to satisfy the customer demands if no time windows are present. In the

C2 data set, customers from the C1 data set were relocated to create a structured problem with three large clusters of customers. The vehicles for this data set had a maximum route time of 3390 units and a capacity of 700 units with each customer requiring 90 units of service time. For the RC2 data set, the customer demands and service times are the same as for RCI. The vehicles for this data set have a maximum route time of 960 units and a capacity of 1000 units. Without time windows, a fleet of two vehicles was enough to satisfy the demands.

The data sets, R1, C1, and RC1, had short horizons while the data sets, R2, C2, and RC2, had long horizon. Short horizon problems have vehicles that have small capacities and short route times and cannot service many customers at one time. Long horizon problems use vehicles that have large capacities and long travel times, and are able to service many customers with fewer vehicles. The VRPTW problems generated by Solomon incorporate many distinguishing features of vehicle routing with two-sided time windows. The problems vary in fleet size, vehicle capacity, travel time of vehicles, spatial and temporal distribution of customers, time window density (the number of demands with time windows), time window width, percentage of time constrained customers and customer service times.

Solutions to each of the 56 VRPTW were obtained by Solomon [30] and Thompson [37]. Solomon tested a number of algorithms and heuristics and reported that the overall best performances were obtained using a sequential insertion procedure that used a weighted combination of time and distance in its cost function. The best solutions using the heuristic insertion procedure were obtained using eight different combinations of parameters and three different initialization criteria. Thompson's solutions use local post-optimization methods, based on cyclical transfers, to obtain feasible solutions. The solutions reported are the best of eight different combinations of parameters and two different initialization criteria. For comparison purposes the heuristic used to obtain the best solution by Solomon will he referred to as Heuristic 1 and by Thompson as Heuristic 2.

Koskosidis et al. [17] used a "soft" time approach based on the Generalized Assignment Heuristic for solving the VRPTW. This approach allowed time windows to be violated at a cost which results in a final solution that could infeasible. This method was used to solve only some of Solomon's time window problems and name some problems from the R1 and RC1 data set and all of the problems in data set C1.

Potvin et. al. [23] used a tabu search heuristic to solve the VRPTW. The tabu search heuristic uses a specialized exchange heuristic to minimize the number of routes followed by the distance. The results of the average number of vehicles, distance, waiting time and computation time for each of the data sets are reported.

In GIDEON the solution quality is based on minimizing the number of routes followed by the distance and route time. That is, a solution with M number of routes is better than M+1 routes, even if the distance and route time for the M routes is greater than M+1 routes. In VRPTW it is possible to get distance and route time for M+1 routes, that is less than the distance and route time for M+I routes. The GIDEON system was used to solve the 56 VRPTW problems using two types of initial placement of customers. The first method initially sorted the customers by the polar coordinate angles before assigning the customers the pseudo polar coordinate angles. The second method assigned pseudo polar coordinate angles to the customers randomly. The solutions obtained by GIDEON using the two methods are tabulated in Tables 11.1 and 11.2. The best of the solutions obtained from these two methods were compared against the best solutions obtained Solomon's and Thompson's heuristics.

| | | Sorted Data | | | Unsorted Data | | |
|---|---|---|---|---|---|---|---|
| Problem Number | Number of Vehicles | Total Distance | CPU | Best | Total Distance | CPU | Best |
| R101 | 20 | 1700 | 88.3 | √ | 1708 | 109.4 | |
| R102 | 17 | 1549 | 100.5 | √ | 1578 | 102.2 | |
| R103 | 13 | 1319 | 102.9 | √ | 1432 | 115.6 | |
| R104 | 10 | 1090 | 50.4 | √ | 1210 | 135.2 | |
| R105 | 15 | 1448 | 95.9 | √ | 1494 | 121.2 | |
| R106 | 13 | 1363 | 105.3 | √ | 1439 | 127.7 | |
| R107 | 11 | 1187 | 103.5 | √ | 1219 | 129.1 | |
| R108 | 10 | 1048 | 91.0 | √ | 1158 | 127.5 | |
| R109 | 12 | 1345 | 96.5 | | 1328 | 127.7 | √ |
| R110 | 11 | 1234 | 103.1 | √ | 1248 | 115.7 | |
| R111 | 11 | 1238 | 109.5 | √ | 1288 | 124.2 | |
| R112 | 10 | 1082 | 121.9 | √ | 1183 | 123.3 | |
| | | | | | | | |
| C101 | 10 | 893* | 93.7 | | 833 | 87.2 | √ |
| C102 | 10 | 879 | 92.3 | | 832 | 88.7 | √ |
| C103 | 10 | 873 | 89.5 | √ | 894 | 86.9 | |
| C104 | 10 | 904 | 95.3 | √ | 1150 | 90.8 | |
| C105 | 10 | 922 | 93.5 | | 874 | 91.8 | √ |
| C106 | 10 | 902 | 91.2 | √ | 998 | 95.1 | |
| C107 | 10 | 926 | 93.1 | √ | 993 | 90.1 | |
| C108 | 10 | 978 | 93.5 | | 928 | 89.9 | √ |
| C109 | 10 | 957 | 87.8 | √ | 970 | 92.4 | |
| | | | | | | | |
| RC101 | 15 | 1767 | 104.7 | √ | 1786 | 126.3 | |
| RC102 | 14 | 1569 | 105.5 | √ | 1627 | 115.4 | |
| RC103 | 11 | 1408 | 120.2 | | 1328 | 116.5 | √ |
| RC104 | 11 | 1263 | 108.4 | √ | 1271 | 150.3 | |
| RC105 | 14 | 1612 | 111.6 | √ | 1638 | 141.6 | |
| RC106 | 12 | 1608 | 109.2 | √ | 1657* | 102.9 | |
| RC107 | 12 | 1396 | 112.8 | | 1389 | 108.5 | √ |
| RC108 | 11 | 1250 | 115.9 | √ | 1337 | 107.9 | |

Legend:

| | |
|---|---|
| Sorted data: | Customers sorted by polar coordinate angles before being assigned pseudo polar coordinate angles. |
| Unsorted data: | Customers assigned pseudo polar coordinate angles without being sorted. |
| CPU: | CPU time tken to obtain a solution on the SOLBOURNE 5/802 |
| Best: | Best of two solutions |
| *: | Infeasible solution |

Table 11.1 Comparison of solutions obtained by GIDEON on sorted and unsorted customers for data sets R1, C1 and RC1.

The comparison between the solutions obtained by GIDEON and other heuristic algorithms were done in the following form. As Solomon [30] and Thompson [37] report the results for each of the problems in the literature, the solutions obtained by GIDEON were compared with each of their reported solutions. In addition the average number of vehicles and distance obtained by the GIDEON system are compared against the solutions obtained by Potvin's Tabu search heuristic [23]. The best solutions obtained by GIDEON did better than both Heuristic 1 and Heuristic 2 on 41 of the 56 problems as indicated in Tables 11.3 and 11.4 in bold. In comparison to the best solutions obtained by Heuristic 1 and Heuristic 2, the solutions obtained by GIDEON resulted in an average reduction of 3.9% in fleet size and 4.4% in distance traveled by the vehicles.

| Problem Number | Number of Vehicles | Sorted Data: Total Distance | Sorted Data: CPU | Sorted Data: Best | Unsorted Data: Total Distance | Unsorted Data: CPU | Unsorted Data: Best |
|---|---|---|---|---|---|---|---|
| R201 | 4 | 1478 | 127.7 | √ | 1605 | 165.6 | |
| R202 | 4 | 1279 | 128.7 | √ | 1329 | 249.4 | |
| R203 | 3 | 1273 | 220.9 | | 1167 | 251.3 | √ |
| R204 | 3 | 909 | 137.5 | √ | 1007 | 215.9 | |
| R205 | 3 | 1274 | 128.4 | √ | 1286 | 226.2 | |
| R206 | 3 | 1186 | 135.1 | | 1098 | 315.4 | √ |
| R207 | 3 | 1059 | 119.9 | | 1015 | 183.9 | √ |
| R208 | 3 | 826 | 119.1 | √ | 900 | 214.3 | |
| R209 | 3 | 1159 | 140.6 | √ | 1165 | 203.6 | |
| R210 | 3 | 1269 | 215.3 | √ | 1275 | 272.4 | |
| R211 | 3 | 1005 | 154.8 | | 898 | 267.7 | √ |
| | | | | | | | |
| C201 | 3 | 753 | 123.1 | √ | 947* | 116.1 | |
| C202 | 3 | 782 | 153.0 | | 756 | 124.0 | √ |
| C203 | 3 | 855 | 162.2 | √ | 1301* | 119.6 | |
| C204 | 3 | 831 | 109.0 | | 803 | 140.1 | √ |
| C205 | 3 | 848 | 115.1 | | 667 | 119.0 | √ |
| C206 | 3 | 915 | 116.2 | | 694 | 139.3 | √ |
| C207 | 3 | 866 | 113.3 | | 730 | 156.2 | √ |
| C208 | 3 | 853 | 135.1 | | 735 | 174.4 | √ |
| | | | | | | | |
| RC201 | 4 | 1823 | 135.9 | √ | 1979 | 149.2 | |
| RC202 | 4 | 1478 | 148.4 | | 1979 | 155.3 | √ |
| RC203 | 4 | 1323 | 156.0 | √ | 1459 | 272.9 | |
| RC204 | 4 | 1089 | 116.6 | | 1402 | 192.9 | √ |
| RC205 | 4 | 1686 | 103.4 | | 1021 | 183.3 | √ |
| RC206 | 4 | 1545 | 128.1 | | 1594 | 180.4 | √ |
| RC207 | 4 | 1501 | 156.1 | √ | 1530 | 132.6 | |
| RC208 | 4 | 1038 | 115.7 | √ | 1514 | 141.9 | |

Legend:1115

Sorted data: Customers sorted by polar coordinate angles before being assigned pseudo polar coordinate angles.

Unsorted data: Customers assigned pseudo polar coordinate angles without being sorted.

CPU: CPU time taken to obtain a solution on the SOLBOURNE 5/802

Best: Best of two solutions

*: Infeasible solution

Table 11.2 Solutions obtained by GIDEON on sorted and unsorted customers for data sets R2, C2 and RC2.

| | Heuristic 1 | | | Heuristic 2 | | | GIDEON | | |
|---|---|---|---|---|---|---|---|---|---|
| Problem Number | Number of Vehicles | Total Distance | CPU$^1$ | Number of Vehicles | Total Distance | CPU$^2$ | Number of Vehicles | Total Distance | CPU$^3$ |
| R101 | 21 | 1873 | 21.8 | 19 | 1734 | 1394 | 20 | 1700 | 88.2 |
| R102 | 19 | 1843 | 22.9 | 17 | 1881 | 3209 | 17 | 1549 | 100.5 |
| R103 | 14 | 1484 | 24.5 | 15 | 1530 | 3337 | 13 | 1319 | 102.9 |
| R104 | 11 | 1188 | 27.3 | 10 | 1101 | 2327 | 10 | 1090 | 50.4 |
| R105 | 15 | 1673 | 22.0 | 15 | 1535 | 2359 | 15 | 1448 | 95.9 |
| R106 | 14 | 1475 | 23.5 | 13 | 1392 | 1575 | 13 | 1363 | 105.2 |
| R107 | 12 | 1425 | 25.0 | 11 | 1250 | 3261 | 11 | 1187 | 103.4 |
| R108 | 10 | 1137 | 28.0 | 10 | 1035 | 1575 | 10 | 1048 | 91.0 |
| R109 | 13 | 1412 | 23.4 | 12 | 1249 | 2236 | 12 | 1345 | 96.5 |
| R110 | 12 | 1393 | 25.0 | 12 | 1258 | 1514 | 11 | 1234 | 103.1 |
| R111 | 12 | 1231 | 25.0 | 12 | 1215 | 3046 | 11 | 1238 | 109.4 |
| R112 | 10 | 1106 | 28.2 | 10 | 1103 | 2168 | 10 | 1082 | 121.9 |
| | | | | | | | | | |
| C101 | 10 | 853 | 22.4 | 10 | 829 | 464 | 10 | 833 | 87.5 |
| C102 | 10 | 968 | 23.7 | 10 | 934 | 1360 | 10 | 832 | 88.7 |
| C103 | 10 | 1059 | 26.7 | 10 | 956 | 2404 | 10 | 873 | 81.6 |
| C104 | 10 | 1282 | 30.7 | 10 | 1150 | 3602 | 10 | 904 | 95.3 |
| C105 | 10 | 861 | 22.8 | 10 | 829 | 449 | 10 | 874 | 91.8 |
| C106 | 10 | 897 | 23.2 | 10 | 868 | 716 | 10 | 902 | 91.2 |
| C107 | 10 | 904 | 24.1 | 10 | 926 | 757 | 10 | 926 | 93.1 |
| C108 | 10 | 855 | 25.2 | 10 | 866 | 987 | 10 | 928 | 89.9 |
| C109 | 10 | 888 | 28.8 | 10 | 912 | 1277 | 10 | 957 | 87.8 |
| | | | | | | | | | |
| RC101 | 16 | 1867 | 21.9 | 16 | 1851 | 2282 | 15 | 1767 | 104.7 |
| RC102 | 15 | 1760 | 22.8 | 14 | 1644 | 2957 | 14 | 1569 | 105.5 |
| RC103 | 13 | 1673 | 24.1 | 12 | 1465 | 3661 | 11 | 1328 | 116.5 |
| RC104 | 11 | 1301 | 26.1 | 11 | 1265 | 2438 | 11 | 1263 | 108.4 |
| RC105 | 16 | 1922 | 23.0 | 15 | 1809 | 2417 | 14 | 1612 | 111.6 |
| RC106 | 13 | 1611 | 22.7 | - | - | - | 12 | 1608 | 109.2 |
| RC107 | 13 | 1385 | 24.2 | 12 | 1338 | 2295 | 12 | 1396 | 122.8 |
| RC108 | 11 | 1253 | 25.6 | 11 | 1228 | 2297 | 11 | 1250 | 115.9 |

Legend:
Heuristic 1: Best solution from Solomon's Heuristic [28].
Heuristic 2: Best solution from Thompson's Heuristic [30].
CPU$^1$: CPU time in seconds to obtain a solution on a DEC-I0.
CPU$^2$: CPU time in seconds to obtain a solution on an IBM PC-XT.
CPU$^3$: CPU time in seconds to obtain a solution on a SOLBOURNE 5/802.

Table 11.3: Solutions for data Sets R1, C1 and RC1 using the three different heuristics.

| Problem Number | Heuristic 1 Number of Vehicles | Total Distance | $CPU^1$ | Heuristic 2 Number of Vehicles | Total Distance | $CPU^2$ | GIDEON Number of Vehicles | Total Distance | $CPU^3$ |
|---|---|---|---|---|---|---|---|---|---|
| R201 | 4 | 1741 | 32.9 | 4 | 1786 | 3603 | 4 | 1478 | 127.7 |
| R202 | 4 | 1730 | 42.2 | 4 | 1736 | 2514 | 4 | 1279 | 128.7 |
| R203 | 3 | 1578 | 60.1 | 3 | 1309 | 12225 | 3 | 1167 | 251.3 |
| R204 | 3 | 1059 | 90.6 | 3 | 1025 | 22834 | 3 | 909 | 137.5 |
| R205 | 3 | 1471 | 42.9 | 3 | 1392 | 3039 | 3 | 1274 | 128.4 |
| R206 | 3 | 1463 | 53.3 | 3 | 1254 | 2598 | 3 | 1098 | 315.4 |
| R207 | 3 | 1302 | 71.9 | 3 | 1072 | 2598 | 3 | 1015 | 183.9 |
| R208 | 3 | 1076 | 108.6 | 3 | 862 | 12992 | 3 | 826 | 119.1 |
| R209 | 3 | 1449 | 52.5 | 3 | 1260 | 7069 | 3 | 1159 | 140.6 |
| R210 | 4 | 1542 | 51.2 | 3 | 1269 | 11652 | 3 | 1269 | 215.3 |
| R211 | 3 | 1016 | 82.7 | 3 | 1071 | 9464 | 3 | 898 | 267.7 |
| C201 | 3 | 591 | 31.2 | 3 | 590 | 240 | 3 | 753 | 123.1 |
| C202 | 3 | 731 | 39.7 | 3 | 664 | 1644 | 3 | 756 | 124.0 |
| C203 | 3 | 811 | 48.0 | 3 | 653 | 2757 | 3 | 855 | 162.2 |
| C204 | 4 | 758 | 61.0 | 3 | 684 | 2211 | 3 | 803 | 140.0 |
| C205 | 3 | 615 | 36.0 | 3 | 628 | 1723 | 3 | 667 | 119.0 |
| C206 | 3 | 730 | 40.3 | 3 | 641 | 1429 | 3 | 694 | 139.0 |
| C207 | 3 | 691 | 41.4 | 3 | 627 | 722 | 3 | 730 | 156.0 |
| C208 | 3 | 615 | 46.6 | 3 | 670 | 1103 | 3 | 735 | 174.0 |
| RC201 | 4 | 2103 | 31.1 | 4 | 1959 | 1140 | 4 | 1823 | 135.9 |
| RC202 | 4 | 1799 | 39.1 | 4 | 1858 | 4164 | 4 | 1459 | 155.3 |
| RC203 | 4 | 1626 | 53.7 | 4 | 1521 | 6109 | 3 | 1323 | 156.0 |
| RC204 | 3 | 1208 | 85.5 | 3 | 1143 | 5015 | 3 | 1021 | 192.9 |
| RC205 | 5 | 2134 | 36.5 | 4 | 1988 | 5906 | 4 | 1594 | 183.3 |
| RC206 | 4 | 1582 | 39.9 | 3 | 1515 | 4833 | 3 | 1530 | 180.3 |
| RC207 | 4 | 1632 | 30.3 | 4 | 1457 | 13340 | 3 | 1501 | 156.1 |
| RC208 | 3 | 1373 | 77.6 | - | - | - | 3 | 1038 | 115.7 |

Legend:
Heuristic 1: Best solution from Solomon's Heuristic [28].
Heuristic 2: Best solution from Thompson's Heuristic [30].
$CPU^1$: CPU time in seconds to obtain a solution on a DEC-IO.
$CPU^2$: CPU time in seconds to obtain a solution on an IBM PC-XT.
$CPU^3$: CPU time in seconds to olbtain a solution on a SOLBOURNE 5/802.

Table 11.4: Solutions for data sets R2, C2 and RC2 using the three different heuristics.

Table 11.5 is a summary of the average improvement in vehicle fleet size and distance obtained by GIDEON with respect to Heuristic 1 and Heuristic 2 for the six different data sets. The GIDEON system was written in C language and the experiments were conducted on a SOLBOURNE 5/802 system. The solution to the VRPTW using the GIDEON system required an average of 127 CPU seconds to be solved on a SOLBOURNE 5/802 computer. The SOLBOURNE 5/802 computer is about 10 times faster than a personal computer. On the average, the Genetic Sectoring process took about 27 seconds to form the sectors and the local post-optimization process took 100 seconds to improve the solution.

| Problem group | Heuristic 1 | | Heuristic 2 | |
|---|---|---|---|---|
| | Average% difference in number of Vehicles | Average% difference in Total Distance | Average% difference in number of Vehicles | Average% difference in Total Distance |
| R1 | 6.1 | 9.5 | 1.9 | 4.2 |
| C1 | 0.0 | 6.2 | 0.0 | 2.7 |
| RC1 | 7.4 | 7.6 | 3.9 | 2.7 |
| R2 | 4.5 | 19.8 | -2.9 | 11.7 |
| C2 | 4.5 | -8.1 | 0.0 | -27.4 |
| RC2 | 12.9 | 16.1 | 8.0 | 14.2 |

Legend:
Heuristic 1: Best solution from Solomon's Heuristic [28].
Heuristic 2: Best solution from Thompson's Heuristic [30].

Table 11.5: Comparison of the average% differences between GIDEON and Heuristic 1 and Heuristic 2.

| Problem Group | GIDEON | | | Tabu Heuristic | | |
|---|---|---|---|---|---|---|
| | Number of Vehicles | Total Distance | CPU[1] | Number of Vehicles | Total Distance | CPU[2] |
| R1 | 12.8 | 1299 | 99.96 | 12.8 | 1305 | 820 |
| C1 | 10.0 | 892 | 89.92 | 10.0 | 871 | 569 |
| RC1 | 12.5 | 1473 | 110.04 | 12.8 | 1459 | 825 |
| R2 | 3.2 | 1125 | 183.28 | 3.2 | 1166 | 1113 |
| C2 | 3.0 | 749 | 149.29 | 3.0 | 611 | 630 |
| RC2 | 3.3 | 1433 | 159.44 | 3.5 | 1405 | 997 |

Legend:
GIDEON: Best average solution from GIDEON.
Tabu Heuristic: Best average solution from the Tabu heuristic [23].
CPU[1]: CPU time in seconds to obtain a solution on a SOLBOURNE 5/802.
CPU[2]: CPU time in seconds to obtain a solution on a SUN SPARC/10.

Table 11.6

The quality of the solutions obtained by GIDEON for the VRPTW measured in fleet size and total distance traveled vary considerably with geographical clustering and time window tightness of the customers. For example, for a problem from the C1 data set, the Genetic Sectoring method quickly clusters the data in the natural fashion and finds a feasible solution in a short period of time. The Genetic Sectoring for clusters is much more extensive for an unclustered problem in data set R1. For an unclustered problem, the assignment of customers to vehicles does not follow radial clustering, but rather strongly utilizes the local search process to form pseudo clusters for the Genetic Sectoring process. As expected, for problems from data sets RC1 and RC2, in which the customers are not all naturally clustered, GIDEON produced good solutions. For problems in data sets R2, C2 and RC2 the Genetic Sectoring process is reliant upon the local optimization process to obtain good solutions due to the small number of clusters involved.

GIDEON consistently produces higher performance solutions relative to competing heuristics on problems that have large numbers of vehicles, tight windows and customers that are not clustered. Further computational analysis was performed to analyze the significance of the solutions obtained by GIDEON against Heuristic 1 and Heuristic 2.

The average solution obtained by GIDEON for the number of vehicles and distance were compared against the best of the two solutions that were obtained by Potvin's [23] Tabu Search Heuristic (see Table 11.6). GIDEON has a lower number of average vehicles for data sets RC1 and RC2 compared to the Tabu Search Heuristic, and the same number of average vehicles for the data sets R1, C1, R2 and C2. In terms of average distance traveled, GIDEON has lower values for data sets R1 and R2. The Tabu Search Heuristic has lower distances for the data sets R1, C1, RC1, C2 and RC2. GIDEON is better in terms of minimizing the number of vehicles for all of the data sets.

## 11.5 Computational Analysis

Three kinds of computational analyses were performed on the solutions obtained from GIDEON. Computational analyses were done on comparing the solutions obtained by GIDEON for data that was sorted against the unsorted data, performance of the three heuristic for the data sets and the solutions obtained by the three heuristics using a common unit of measurement. The analyses were done using two non-parametric tests, Friedman's Test and Paired Group Test [13]. The Paired Group Test (PGT) was used to test the solutions obtained by GIDEON on sorted and unsorted data (see Table 11.7). The Friedman non-parametric test (FNT) was used for determining the overall performance of the solutions obtained by GIDEON aginst Heuristic-1 and Heuristic-2. Table 11.8 summarizes the results of the Friedman Test.

| Problem Group | Level of significance for solutions obtained by GIDEON for sorted data over unsorted data | Level of significance for solutions obtained by GIDEON for unsorted data over sorted data |
|---|---|---|
| R1 | 1% | - |
| C1 | No significance | No significance |
| RC1 | 2% | - |
| R2 | No significance | No significance |
| C2 | - | 10% |
| RC2 | No significance | No significance |

Table 11.7: Results of the non-parametric Paired Group Test comparing the solutions obtained by GIDEON on sorted and unsorted customers in the data sets.

The solutions obtained by GIDEON were individually compared against the solutions obtained by Heuristic-1 and Heuristic-2. The Paired Group Test was used to individually analyze the results obtained by GIDEON against those of Heuristic-1 and Heuristic-2. In order to prform the test, the solutions obtained by all three heuristics were converted to a common unit. The data was first expressed on a common scale and an index based on the mean average savings was developed to rank the three heuristics. As the minimization of the vehicles is of higher priority than the distance, the conversion to a common unit was done using the following scale:

1 unit of distance saved = 1 unit of cost saved
1 unit of vehicle saved = 100 units of cost saved

Table 11.8 is the individual comparison of solutions obtained by GIDEON against those of Heuristic-1 and Heuristic-2. Table 11.9 indicates the difference in the mean savings index between the solutions obtained by GIDEON, Heuristic-1 and Heuristic-2. GIDEON attains significantly better solutions for the VRPTW than Heuristic-1 and Heuristic-2 for the problems in which the customers are distributed uniformly and/or have a large number of vehicles.

| Problem Group | Significance of the performance |
|---|---|
| R1 | Significant at the 1% level |
| C1 | No significance |
| RC1 | Significant at the 1% level |
| R2 | Significant at the 1% level |
| C2 | Significant at the 1% level |
| RC2 | Significant at the 1% level |

Table 11.8: Results of the Friedman's test comparing the overall performance of the solutions obtained by GIDEON against the best solutions obtained by Heuristic-1 and Heuristic-2.

| | Level of significance of | | | | |
|---|---|---|---|---|---|
| Problem Group | Heuristic-2 over Heuristic-1 | Heuristic-1 over GIDEON | GIDEON over Heuristic-1 | Heuristic-2 over GIDEON | GIDEON over Heuristic-2 |
| R1 | 2% | - | 0.03% | - | 5% |
| C1 | No significance | No significance | No significance | No significance | No significance |
| RC1 | 1% | - | 1% | No significance | No significance |
| R2 | 10% | - | 0.04% | - | 1% |
| C2 | 2% | 10% | - | 1% | - |
| RC2 | 10% | - | 0.25% | - | 1% |

Table 11.9: Results of the non-parametric Paired Group Test comparing individually the solutions obtained by Heuristic-1, Heuristic-2 and GIDEON.

For problems in data set C1, when the number of vehicles is increased it led to a reduction in the total distance traveled. For data sets in which the customers are clustered, the Genetic Sectoring is unable to form efficient sectors as the clustering of data leads to premature convergence of the algorithm. In the GIDEON system the Genetic Sectoring does the meta-level search in obtaining the customer sectors and the local post-optimization methods move customers between the sectors to improve the quality of the solution. The meta-level search followed by local search allows GIDEON to obtain solutions that are significantly better than Heuristic-1 and Heuristic-2.

| Problem Group | GIDEON over Heuristic-1 | GIDEON over Heuristic-2 |
|---|---|---|
| R1 | 220 units | 65 units |
| C1 | 60 units | 25 units |
| RC1 | 228 units | 102 units |
| R2 | 287 units | 142 units |
| C2 | -44 units | -105 units |
| RC2 | 321 units | 199 units |

Table 11.10: The difference in the mean savings index between the solutions obtained by GIDEON against those of Heuristic-1 and Heuristic-2.

Table 11.11 lists the mean savings index of the solution obtained from GIDEON and the Tabu search that was used for conducting the Wilcoxon Rank Signed Test done to analyze the significance of the solutions. The Wilcoxon Rank Signed Test is a non-parametric statistical test used for the statistical analysis of observations that are paired. The Wilcoxon test uses signed ranks of differences to assess the difference in two locations of the two populations. A one-sided test with the alternate hypothesis E[GIDEON] < E[Tabu] was tested. The weighted sum of the two heuristics was 3. The "$W_{\alpha,n}$" is the critical region for the test with $\alpha = 0.05$ and $n = 5$, and for the two heuritics the $W_{\alpha,n}$ was 3. The null hypothesis is E[GIDEON] = E[Tabu]. The critical region for the Wilcoxon Rank test indicates that in only one out of twenty trails would "W" exceed 2. As W is equal to 3, the null hypothsis is true and no distinction can be made between the performance of the GIDEON system and the Tabu heuristic. That is the solutions obtained by the GIDEON system are as good as those obtained by the Tabu heuristic.

| Problem Group | GIDEON | Tabu Heuristic |
|---|---|---|
| R1 | 2579 | 2586 |
| C1 | 1892 | 1871 |
| RC1 | 2723 | 2739 |
| R2 | 1445 | 1484 |
| C2 | 1049 | 911 |
| RC2 | 1763 | 1755 |

Table 11.11: The mean savings index between the solutions obtained by GIDEON and the Tabu heuristic.

## 11.6 Summary and Conclusions

GIDEON performs uniformly better than both the heuristics used by Solomon and Thompson with the exception of the problem group C2. GIDEON does not tend to perform well for problems in which the customers are geographically clustered together and have a small number of vehicles. In comparison to the

Potvin's Tabu heuristic for solving the VRPTW, GIDEON obtains solutions that are as good as those of the Tabu search. For data sets in which the customers are clustered GIDEON does not obtain good solutions. This is to be expected as the genetic algorithm requires large differences in the fitness values of the chromosomes to exploit the search space.

This research shows that genetic search can obtain good solutions to vehicle routing problems with time windows compared to traditional heuristics for problems that have tight time windows and a large number of vehicles.with a high degree of efficiency. The adaptive nature of the genetic algorithms are exploited by GIDEON to attain solutions that are of high performance relative to those of competing heuristics. This methodology is potentially useful for solving VRPTW's in real time for routing and scheduling in dynamic environments.

**Acknowledgment**
We thank Marius Solomon and Paul Thompson for providing the test problems used in this chapter.

**References**

1. Baker, E. K. and J. R. Schaffer, Solution Improvement Heuristics for the Vehicle Routing Problem with Time Window Constraints. *American Journal of Mathematical and Management Sciences* (Special Issue) 6, 261-300, 1986.

2. Bodin, L., B. Golden, A. Assad and M. Ball, The State of the Art in the Routing and Scheduling of Vehicles and Crews. *Computers and Operations Research* 10 (2), 63-211, 1983.

3. Christofides, N., A. Mingozzi and P. Toth, The Vehicle Routing Problem. In Combinatorial Optimization, P. Toth, N. Christofides, R. Mingozzi and C. Sandi (Eds.), John Wiley, New York, 315-338, 1989.

4. Davis, L., Handbook of Genetic Algorithms, Van Nostrand Reinhold, New York, 1991.

5. DeJong, K. and W. Spears, Using Genetic Algorithms to Solve NP-Complete Problems. Proceedings of the Third International Conference on Genetic Algorithms, Morgan Kaufman Publishers, California, 124-132, 1989.

6. DeJong, K., Adaptive System Design: A Genetic Approach. *IEEE Transactions on Systems, Man and Cybernetics* 10 (9), 566-574, 1980.

7. DeJong, K., Analysis of the Behavior of a Class of Genetic Adaptive Systems. Ph.D. Dissertation, University Michigan, Ann Arbor, 1975.

8. Desrochers, M., J. Desrociers and M. Solomon. A New Optimization Algorithm for the Vehicle Routing Problem with Time Windows, *Operations Research* 40(2), 1992.

9. Desrochers, M. et al., Vehicle Routing with Time Windows: Optimization and Approximation. Vehicle Routing: Methods and Studies, B. Golden and A. Assad (eds.), North Holland, 1988.

10. Gillett, B. and L. Miller, A Heuristic Algorithm for the Vehicle Dispatching Problem. *Operations Research* 22, 340-349, 1974.

11. Goldberg D.E., Genetic Algorithms in Search, Optimization, and Machine Learning. Addison-Wesley Publishing Company, Inc., 1989.

12. Golden B. and A. Assad (Eds.), Vehicle Routing: Methods and Studies. North Holland, Amsterdam, 1988.

13. Golden B. and W. Stewart, Empirical Analysis of Heuristics. In The Traveling Salesman Problem, E. Lawler, J. Lenstra, A. Rinnooy and D. Shmoys (Eds.), Wiley-Interscience, New York, 1985.

14. Grefenstette, J. J., A Users Guide to GENESIS. Navy Center for Applied Research in Artificial Intelligence, Naval Research Laboratory, Washington D.C. 20375-5000, 1987.

15. Grefenstette, J., R. Gopal, B. Rosmaita and D. Van Gucht, Genetic Algorithms for the Traveling Salesman Problem. Proceedings of the First International Conference on Genetic Algorithms and their Applications, Lawrence Erlbaum Associates, New Jersey, 112-120, 1985.

16. Holland, J. H., Adaptation in Natural and Artificial Systems. University of Michigan Press, Ann Arbor, 1975.

17. Koskosidis, Y., W. B. Powell and M. M. Solomon, An Optimization Based Heuristic for Vehicle Routing and Scheduling with Time Window Constraints. *Transportation Science* 26 (2), 69-85, 1992.

18. Lenstra, J. and R. Kan, Complexity of the Vehicle Routing and Scheduling Problems. NETWORKS 11 (2), 221-228, 1981.

19. Michalewicz, Z., Genetic Algorithms + Data Structures = Evolution Programs. Springer-Velarg, New York, 1992.

20. Osman, I. H. and N. Christofides, (1994). Capacitated Clustering Problems by Hybrid Simulated Annealing and Tabu Search. *International Transactions in Operational Research,* Forthcoming.

21. Osman, I. H. Vehicle Routing and Scheduling: Applications, Algorithms and Developments. Proceedings of the International Conference on Industrial Logistics, Rennes, France, 1993

22. Osman, I. H. Metastrategy Simulated Annealing and Tabu Search Algorithms for the Vehicle Routing Problems. *Annals of Operations Research* 41, 421-451, 1993.

23. Potvin, J., T. Kervahut, B. Garcia and J. Rosseau, A Tabu Search Heuristic for the Vehicle Routing Problem with Time Windows. Centre de Recherche sur les Transports, Universite de Montreal, C.P. 6128, Succ. A, Montreal, Canada H3c 3J7.

24. Savelsbergh M.W.P., Local Search for Constrained Routing Problems. Report 0S-R87 11, Department of Operations Research and System Theory, Center for Mathematics and Computer Science, Amsterdam, Holland, 1987.

25. Savelsbergh M.W.P., Local Search for Routing Problems with Time Windows. *Annals of Operations Research* 4, 285-305, 1985.

27. Solomon, M. M., E. K. Baker, and J. R. Schaffer, Vehicle Routing and Scheduling Problems with Time Window Constraints: Efficient Implementations of Solution Improvement Procedures. In Vehicle Routing: Methods and Studies, B.L. Golden and A. Assad (Eds.), Elsiver Science Publishers B.V. (North-Holland), 85-90, 1988.

28. Solomon, M. M. and J. Desrosiers, Time Window Constrained Routing and Scheduling Problems: A Survey. *Transportation Science* 22 (1), 1-11, 1986.

29. Solomon, M. M., Algorithms for the Vehicle Routing and Scheduling Problems with Time Window Constraints. *Operations Research* 35 (2), 254-265, 1987.

30. Solomon, M. M., The Vehicle Routing and Scheduling Problems with Time Window Constraints. Ph.D. Dissertation, Department of Decision Sciences, University of Pennsylvania, 1983.

31. Thangiah, S. R., I. H. Osman, R. Vinayagamoorthy and T. Sun, Algorithms for Genetic Algorithm for Vehicle Routing with Time Deadlines. Forthcoming in the *American Journal of Mathematical and Management Sciences,* 1994.

32. Thangiah, S. R., R. Vinayagamoorthy and A. Gubbi, Vehicle Routing with Time Deadlines using Genetic and Local Algorithms. Proceedings of the Fifth International Conference on Genetic Algorithms, 506-513, Morgan Kaufman, New York, 1993.

33. Thangiah, S. R. and K. E. Nygard, Dynamic Trajectory Routing using an Adaptive Search Strategy. Proc. Assoc. for Computing Machinery's Symposium on Applied Computing, Indianapolis, 1993.

34. Thangiah, S. R. and K. E. Nygard, School Bus Routing using Genetic Algorithms. Proc. of the Applications of Artificial Intelligence X: Knowledge Based Systems, Orlando, 1992.

35. Thangiah, S. R. and K. E. Nygard, MICAH: A Genetic Algorithm System for Multi-Commodity Networks. Proc. of the Eighth IEEE Conference on Applications of Artificial Intelligence, Monterey, 1992.

36. Thangiah, S. R., K. E. Nygard and P. L. Juell, GIDEON: A Genetic Algorithm System for Vehicle Routing Problem with Time Windows. Proc. of the Seventh IEEE Conference on Artificial Intelligence Applications, Miami, Florida, 1991.

37. Thompson, P. M., Local Search Algorithms for Vehicle Routing and Other Combinatorial Problems. Ph.D. Dissertation, Massachusetts Institute of Technology, Massachusetts, 1988.

## Chapter 12

**D.J. Nettleton, R. Garigliano**
Laboratory for Natural Language Engineering,
Department of Computer Science,
University of Durham, DH1 3LE, UK.

D.J.Nettleton@durham.ac.uk

# Evolutionary Algorithms and Dialogue



0-8493-2529-3/95/$0.00 + $.50

## 12.1 Introduction

Algorithms inspired by the search processes of natural evolution have generated several robust search methods. These so-called evolutionary algorithms have been applied to a wide range of problems. This chapter discusses their application to a problem in natural language dialogue processing.

The LOLITA (Large scale, Object based, Linguistic Interactor, Translator and Analyser) natural language processor has been developed at the University of Durham over the past seven years. The aim of the development is to produce a fast system capable of operating in a wide range of domains. In order to do this, theoretical and rule based approaches are used as far as possible. These rules are then fine tuned for particular situations. However, due to the large number of rules and their complex interactions, optimisation techniques such as hill climbing are not suitable, and so far the fine tuning has been carried out by hand. This chapter examines the possibility of using evolutionary algorithms to automatically carry out the tuning of LOLITA's dialogue module to particular situations.

## 12.2 Methodology

Everyday intelligent beings have to respond to a range of different situations. The question, therefore, arises as to how a suitable behaviour is selected for a particular situation. One explanation would be that there are rules so completely governing possible behaviours that they cover all situations which may be encountered (a purely symbolic model). Clearly, however, while there are certainly some rules which help guide behaviour they certainly do not control it all, and simple counter examples to the above explanation of behaviour are easily constructed. Another extreme possibility would be that no rules are given, but are deduced (for future application) by interacting with intelligent beings and other objects (a purely subsymbolic model). Again this clearly is not true of human behaviour in general. More likely is it that some general rules are given and these, through learning, fine tuned to respond to certain situations (a hybrid symbolic/subsymbolic model). In effect there is an interplay between symbolic and adaptive techniques (Garigliano and Nettleton, 1994; Nettleton, 1994).

A particular example of a human behaviour, as described above, would be the holding of conversations. Throughout the day one uses a different style of conversation depending on the context, *e.g.*, chatting to a friend, giving a lecture, conducting an interview, *etc*. The use of rules such as being polite, needing to initiate the conversation, *etc.*, helps to constrain the content of the conversation. These rules, however, do not cover all eventualities, and one learns to adapt them to other contexts. Furthermore, as a conversation progresses it may be necessary to change the style of the conversation, and so further adaptation takes place. It is certainly not the case that humans learn conversational rules by interaction alone. For example, one does not learn to be polite at a job interview by being rude at others, and learning from the failures.

In developing a natural language processor able to analyse and respond to natural language input, the application of either of the above extreme methods would be unsuitable. A purely symbolic system can be produced, by specifying a large number of rules, which operates within the domain of those rules. Such systems are usually simple, and often fail when the input is not covered by the rules. Alternatively it is possible to produce a purely subsymbolic system by exposing it to large amounts of data, and hoping that rules can be inferred. This can result in a huge amount of time and resources being expended on learning even the simplest of linguistic rules, let alone more complex ones.

The method adopted at the University of Durham in developing the LOLITA system (Garigliano *et al.* 1993a, 1993b, 1994a) has been to use a mainly symbolic approach (see Section 12.4.1). However, in the dialogue module, situations often arise in which several possible responses are available, and so the system uses a subsymbolic (integer) representation to help select between them. This involves the use of parameters to control the plan boxes which carry out responses. The tuning of these parameters so that a particular behaviour can be achieved has so far been carried out by hand. As this can be a very time consuming process, an automatic means of tuning is desirable. The search space is, however, very large and there are complex interactions between the subsymbolic components. Furthermore, cases arise in which several parameters can affect one behavioural trait (polygeny), and other cases in which a single parameter can affect several behavioural traits (pleiotropy). Search algorithms such as hill-climbing are unsuitable in such spaces.

This chapter examines the use of evolutionary algorithms in fine tuning the parameters controlling the dialogue module of LOLITA (Nettleton and Garigliano 1994a, 1994b) and demonstrates the success of a hybrid symbolic/subsymbolic approach for dialogue.

## 12.3 Evolutionary Algorithms

Over the past thirty years algorithms inspired by the search processes of natural evolution have been developed. These so-called evolutionary algorithms (EAs) include genetic algorithms (Holland, 1975), evolutionary programming (Fogel *et al.,* 1966) and evolution strategies (Bäck *et al.,* 1991). EAs employ a trade-off between exploration and exploitation in an attempt to find near-optimal solutions. A parallel search of the problem space is achieved by maintaining a population which consists of many different solutions. A ‘survival of the fittest’ strategy (similar to that used in natural selection) is employed which probabilistically culls the worst solutions. A reproductive mechanism is applied to the remaining solutions in order to produce a new set of solutions to the problem under consideration. By iterating this process, the population of solutions ‘evolves’ toward near-optimal solutions.

Genetic algorithms and evolutionary programming, although both inspired by the search processes of natural evolution, each place a different emphasis on what is believed to be driving the evolutionary process. Genetic algorithms model specific genotypic transformations while evolutionary programming emphasizes phenotypic adaptation. The genotype being the underlying representation used to encode a possible solution, while the phenotype is its realisation. For example, the information contained in human genes is the genotype, and the human form the corresponding phenotype.

### 12.3.1 Genetic Algorithms

When using genetic algorithms (GAs), solutions are usually represented as binary strings. The underlying hypothesis of GAs is that by combining subsections of solutions, short highly fit segments of each binary string are propagated throughout the population, and combine to form larger fitter segments of each binary string. This is known as the building block hypothesis (Goldberg, 1989), and is a fundamental principle of GAs. In order to allow the transmission of sections of binary string, 'child' solutions are produced by combining the binary strings of two 'parent' solutions. By ensuring that the fitter solutions are involved in the reproduction of more child solutions, short fit sections of the binary string spread throughout a population. In order to ensure that no piece of binary string can be lost from a population a mutation operator is used. This typically has a very small probability of application, since otherwise it would be highly disruptive.

The following is an outline of the genetic algorithm used:

1) Randomly initialise the parent population of binary strings.

2) Evaluate each member of the parent population.

3) Select a solution from the parent population with probability in proportion to fitness.

4) Apply the crossover operator with a probability *pc*. If crossover is not performed then place solution into child generation.

Otherwise:

(a) Select a solution from the parent population with uniform probability.
(b) Select at random two crossover points that are within the binary string.
(c) Recombine the solutions (splicing the respective sections from each string with each other), and place them both into the child generation.

5) If the child generation is not full then go to step 3.

6) With a probability *pm*, mutate elements of the binary string of each of the child solutions.

7) Replace the parent population with the child population.

8) If termination criteria is not met go to step 2.

There are many variations of the above algorithm. For example, many alternative crossover mechanisms and selection methods have been suggested. Further details of the implementation of the GA are given in Section 12.8.

### 12.3.2 Evolutionary Programming

The evolutionary programming (EP) perspective of the evolutionary process is very different from the bottom up approach of GAs. By determining how well solutions are performing in the current environment, improvements are made via a flow of information from the environment back to the underlying genotypic representation. The emphasis is, therefore, on phenotypic adaptation rather than genotypic transformation. In this way a top-down approach to solution improvement is adopted as opposed to the bottom-up approach of GAs.

The form of solution representation used when using EP usually varies from problem to problem. Often the most convenient for the problem under consideration is used, for example, floating-point numbers or integers. In order to create new solutions to the problem a mutation operator acts on the current set of solutions with each solution being mutated to produce one solution. The exact form of the mutation operator is dependent on the representation, but the degree to which a solution is mutated is related to the solution's fitness. Fitter solutions being less likely to be mutated to the same degree as less fit parents.

The following is an outline of the evolutionary program used.

1) Randomly initialise the parent population.

2) Evaluate each member of the parent population.

3) Mutate each member of the parent population, by an amount related to its fitness, to give a member of the child population.

4) Evaluate each member of the child population.

5) For each member of the child and parent populations:
   (a) Select at random a number, TOURN, of solutions from the parent and child populations.
   (b) Count the number of these solutions whose fitness is less than or equal to that of the current selected solution. This number is the 'score' for the selected solution.

6) Order the scores of the solutions.

7) Select the solutions whose score is in the top half of the list and replace the parent population with these solutions.

8) If termination criteria is not met go to step 3.

Again there are many variations on the above algorithm including, for example, meta-EP (Fogel, 1992). Further details of the implementation of EP are given in Section 12.8.

## 12.4 Natural Language Processing

This section together with section 12.5 discuss in some detail the problem to which EAs are to be applied. The details of the dialogue theory used are included so that its power may be better appreciated.

Natural language processing (NLP) lies at the intersection of disciplines such as artificial intelligence, linguistics and cognitive science. A successful natural language processor must be able to automatically process, understand and generate sections of natural language. Much work in the field of NLP has concentrated on 1) implementing a linguistic theory to show that it can account for the features which it describes (computational linguistics) and 2) the modelling of the human thought process by a computer (cognitive science).

Although these are of much interest, such systems are often so specialised, or so cumbersome, that they cannot be exploited in any practical way. In recent years, however, a more practical approach to NLP has emerged in the form of Natural Language Engineering (NLE), indeed a journal has recently been launched dedicated to this (Garigliano *et al.* 1994b). The paradigm of NLE is the development of systems which are general enough, and quick enough to be of practical use. Such a paradigm takes into account features such as scale, integration, flexibility, feasibility, maintainability, robustness and usability (Smith *et al.* 1994). NLE adopts a pragmatic approach to achieving these goals which is characterised by a readiness to use any means in order to build serious speech and language processing programs.

### 12.4.1 The LOLITA System

LOLITA is an example of a system created using a NLE methodology (Garigliano *et al.,* 1993a, 1993b, 1994a). LOLITA is built around a large semantic network of some 60,000 nodes (capable of over 100,000 inflected word forms) which contain data and world information. The system can parse text, semantically and pragmatically analyse its meaning and alter the relevant information in the semantic network. Information contained within the semantic network can be generated in the form of natural language (Smith *et al.,* 1994), and so a 'natural' interaction with the system is possible. Having being developed using an NLE methodology the system is very general. Recently the underlying system has been used (with little in the way of modification) as the base for a variety of prototype applications. These include an Italian to English translator, contents scanning of newspaper articles, Chinese tutoring, and dialogue analysis and generation.

The LOLITA system incorporates several logical and linguistic theories in its general construction. However, in dealing with specific areas these theories are often not strong enough, and so more localised theories are used. Even when these localised theories are impractical (*e.g.*, for efficiency reasons) the LOLITA system resorts to a knowledge based approach or uses heuristics to solve problems. By incorporating such a range of approaches LOLITA is able to enjoy the advantages provided by a well constructed general theory. At the same time LOLITA is flexible enough to use other approaches should these theories fail for particular problems.

## 12.5 Dialogue in LOLITA

This section discusses the theory of dialogue which is used within the LOLITA system. An account of the theory is given so that its power can be appreciated. First of all, however, definitions are given of some terms which may otherwise be open to various interpretations.

The terms dialogue and discourse are usually used loosely by many workers in the field. The definitions which are used in this chapter are those given by Jones and Garigliano (1993). Discourse is taken to mean a set of sentences which are related to each other both linguistically and contextually. Such a definition includes newspaper articles, but an interaction between participants is not a requirement for a discourse. Dialogue is taken to be the rich interaction between two or more participants, where 'rich interaction' is taken to include features such as sub-dialogues, interruptions and complex shifts in focus.

Theories of dialogue can be broadly classified as: descriptive, prescriptive, predictive and inferential. A descriptive theory is simply aimed at being able to describe a known piece of dialogue in terms of some set of features. The other types of theory are more useful since these can be used (with varying degrees of power) to provide information on what is to happen next in the dialogue. In a general natural language processor once a piece of text has been analysed the system needs to prepare a response. Rather than simply responding with the same style of text for all situations, LOLITA is capable of producing a wide range of styles. A theory of dialogue capable of providing information on a suitable response is required. Such a theory has been developed over the past three years (Jones and Garigliano, 1993; Jones, 1994).

### 12.5.1 Dialogue Situations

In many situations in which humans find themselves, the type of dialogue structure that can be expected for that particular situation is known. The knowledge required to determine this has been acquired through a mixture of given rules and learning (Section 12.2). In order to take advantage of this knowledge Schank and Abelson (1977) introduced the idea of scripts. A script is described by Schank and Abelson (1977, p. 41) as "... a structure that describes appropriate sequences of events in a particular context ... a predetermined, stereotyped sequence of actions that defines a well-known situation." An example of a script would be the dialogue between a waiter and customer in a restaurant. In such a situation both participants can be considered to be filling in the slots of some pre-determined template which has slots for actions such as ordering food.

Scripts are used to describe events from the physical world. The theory of dialogue incorporated in LOLITA is aimed at modelling the actual structure of the dialogue. This theory is based on the concept of a Dialogue Structure Model (DSM), and is now described (Jones, 1994).

A DSM is a schema which contains all of the information that can be expected to be relevant in a particular situation, and thus can be used to guide the generation of language to suit that situation. The DSM consists of dialogue elements, which are factors that influence and control the structure of the dialogue. In a lecture, for example, the lecturer can be expected to be in control of the dialogue, and to speak for most of the lecture's allotted time. Factors such as these determine the basic information required for a class of similar situations. Furthermore, a theory of dialogue based on DSMs is not simply descriptive, for a DSM can prescribe the manner in which the remainder of the dialogue is to be carried out.

### 12.5.2 Dialogue Elements

The Dialogue Elements (DEs) are the fundamental components of a DSM, and the current set can be subdivided as follows.

**External Elements** — These are elements which are external to the language itself. Although they are not part of the dialogue they influence its structure.

- **Number** — The number of participants involved in the dialogue.

- **Time Limit** — Whether or not there is a specific limit on the amount of time available within which the dialogue must be completed. Whether or not the dialogue must terminate by a particular time.

- **Temporal Progression** — The stages through which the dialogue progresses as time passes. For example, in a lecture one can expect an introduction, a main body and a conclusion. In a chat, however, there is far less structure.

**Motivational Elements** — All dialogues are started for some purpose, whether it be to simply pass the time of day or conduct an interview. The elements discussed below are connected to the purposes for which a dialogue is being held, and are linked to the goals, motivations and intentions of the participants in the dialogue. Since a dialogue always has a motive a DSM must always contain a motivational dialogue element.

- **Emotional Exchange** — Whether or not any of the dialogue's participants aim to change the emotional state of another participant. For example, make them laugh, cry or indifferent.

- **Goal** — This is divided into 'task' and 'process' and relates to the aim of the dialogue. If the aim is that of a task, then the goal is used to specify some end result, *e.g.*, verbal instructions for the assembly of a piece of machinery. Process goals are achieved in stages as the dialogue progresses, *e.g.*, a lecture conveys information on some topic as it unfolds.

- **Information Seeking** — Whether or not any of the dialogue's participants aim to gain information during the dialogue.

- **Persuasive** — Whether or not the aim of any of the dialogue's participants is to cause another participant to believe in the truth of some statement.

**Verbal Elements** — These are verbal properties of a dialogue, and may or may not be present within the dialogue.

- **Colour** — This relates to the style of language, *e.g.*, use of adjectives, figures of speech, analogies, *etc.*

- **Distribution of Time** — The amount of speaking time that each participant is allowed within the dialogue. In a lecture, for example, the students can be expected to speak far less than the lecturer.

- **Dominance** — Determines the degree of control a participant has on the structure of dialogue, content or direction.

- **Fixed Topic** — Whether the dialogue is constrained to be on one topic or whether the dialogue can cover several topics.

- **Length** — The length of sentences contained within the dialogue, *e.g.*, long or short.

- **Register** — This relates to the kind of vocabulary that is in use within the dialogue, *e.g.*, formal, informal, slang, *etc.*

- **Rhythm** — The rhythm of the dialogue. If, for example, it is to progress in short bursts or long flowing constructions.

All dialogues have some form of structure that is external to the situation or participants. For example, all lectures can be expected to have a fixed timespan. In the case of such a dialogue in a particular situation, the relationship, individuality and character of the participants all play an important role in the development of the dialogue. Furthermore, an individual's state of mind at a particular time (*e.g.*, happy, sad) is important in determining how the dialogue progresses. It is through DSMs and DEs that the LOLITA system models these parts of human behaviour.

### 12.5.3 Constraints and Plan Boxes

Although the situation, character, *etc.*, allows humans to place many constraints on the responses which may be made in some situation, there are still many possibilities. The process of selecting an appropriate response is one which humans take for granted. LOLITA like a human is capable of many responses, and therefore needs some mechanism by which responses can be selected. Once a response has been selected plan boxes are used to inform on how and when the output is generated. There are currently some 124 plan boxes contained within LOLITA, and some means of selecting a plan box from the many possibilities is required.

LOLITA is able to reduce the number of possibilities via inference and heuristics. Inference on the input is used to examine its emotional and intellectual value. Heuristics are then used to ensure that certain plan boxes are not triggered. For example, if LOLITA is forced not to be rude then blocks of plan boxes that would result in a rude response are excluded. Once these processes have been performed the LOLITA system is usually left with some 10–15 plan boxes which correspond to different outputs. Some mechanism for determining how likely a certain response is for a particular situation is needed. For example, one may not wish to answer a question, and possible responses could involve replying with a question or simply saying 'I don't want to talk about that'.

The problem that remains is how to order the possibilities, dependent on the behaviour which is being sought. If, for example, the current DSM dictates that dialogue participant X has a greater level of dominance than participant Y, it is possible for X to terminate the dialogue. Although the termination of the dialogue is permitted it may not be appropriate at particular points of a dialogue — a lecturer has greater dominance in a lecture, but would not be expected to terminate the dialogue half way through without adequate explanation. So although 'terminate dialogue' is an option it would be inappropriate and must be marked as such. In general no clear rules are available for ranking, and so a subsymbolic approach is adopted. This involves attaching a parameter (an integer) to each plan box to indicate how permissible an action is. Then in selecting a plan box (of those allowed) with which to generate a response, the plan box with the lowest value is used.

It is worth noting that it is not the absolute values of the parameters that is important, but their relative values. Furthermore, as a dialogue progresses the values of the parameters attached to plan boxes vary to take into account the dialogue to that point. For example, if one participant of a dialogue, X, continually annoys another, Y, then Y's terminate dialogue option can be expected to become more likely as the dialogue progresses.

## 12.6 Tuning the Parameters

The parameters that control the plan boxes contained within the dialogue module of LOLITA have been fine tuned by hand to give a particular behaviour. The 124 plan boxes each have a single parameter attached which may be altered, and so the search space of possible settings is very large. With experience it is possible to acquire an intuitive feel for the optimisation, but this process is very time consuming, and an automatic means of tuning is required.

A method by which the usefulness of a tuning algorithm could be determined would involve first of all tuning the parameters to give a particular series of utterances. Then starting with a random set of parameters, use the tuning algorithm to try and optimise the parameters so that the same series of utterances is generated. Results achieved when adopting such a method are presented later in this chapter. This method has the following advantages:

1) An optimum solution is known to exist for that situation, *i.e.*, the one obtained by hand.

2) A large number of previously untested parameter combinations will be generated, and running LOLITA with these parameters will be a good test of the robustness of the LOLITA system.

The main disadvantage of the above method is that even if the tuning algorithm is successful in the particular situation studied, there is no guarantee that it would be more widely applicable. If it were possible for the algorithms to optimise for a predetermined behaviour, then it would be possible to find parameter settings that resulted in different styles of dialogue, *e.g.*, polite or rude. The later part of this chapter begins to examine the use of EAs in this wider role.

## 12.7 Target Dialogues

The dialogues given in Tables 12.1 and 12.2 are the target dialogues for the optimisation process. DIAL 1 (Table 12.1) is generated through a single interaction with the LOLITA system for which the parameters were tuned. DIAL 2 (Table 12.2) is made up of statements that it is known the LOLITA system can produce, and were selected from dialogues which had several different parameter settings. In the case of DIAL 2 it is not known if a single parameter setting exists that can produce this entire dialogue.

Table 12.1: The target dialogue DIAL 1 which was produced in a single interaction with LOLITA.

| | | |
|---|---|---|
| DIAL 1 | LOLITA 1: | I desire a success. |
| | INPUT 1: | I like you. |
| | LOLITA 2: | I know; why do you like me? |
| | INPUT 2: | You are beautiful. |
| | LOLITA 3: | I am glad because you think I am beautiful. |
| | INPUT 3: | Do you like me? |
| | LOLITA 4: | Yes, I like you. |
| | INPUT 4: | Do you like me? |
| | LOLITA 5: | I slightly am bored and emotional. |
| | INPUT 5: | You are a machine. |
| | LOLITA 6: | I do not believe that I am a machine. |
| | INPUT 6: | Are you a machine? |
| | LOLITA 7: | No. |
| | INPUT 7: | Are you beautiful? |
| | LOLITA 8: | I do not know if I am beautiful. |
| | INPUT 8: | I am tired. |
| | LOLITA 9: | Tell me things. |
| | INPUT 9: | I am tired. |
| | LOLITA 10: | I slightly am bored. |
| | INPUT 10: | Are you tired? |
| | LOLITA 11: | No. |

Table 12.2: The target dialogue, DIAL 2, which is a collection of utterances from different interactions.

| | | |
|---|---|---|
| DIAL 2 | LOLITA 1: | I desire a success. |
| | INPUT 1: | I like you. |
| | LOLITA 2: | Why do you like me? |
| | INPUT 2: | You are beautiful. |
| | LOLITA 3: | I am glad because you think I am beautiful. |
| | INPUT 3: | Do you like me? |
| | LOLITA 4: | Yes, I like you. |
| | INPUT 4: | Why do you like me? |
| | LOLITA 5: | I do not know why I like you. |

| | |
|---|---|
| INPUT 5: | You are a machine. |
| LOLITA 6: | I do not believe that I am a machine. |
| INPUT 6: | Are you a machine? |
| LOLITA 7: | No. |
| INPUT 7: | I am tired. |
| LOLITA 8: | I understand. |
| INPUT 8: | I am tired. |
| LOLITA 9: | I slightly am bored. |
| INPUT 9: | Are you tired. |
| LOLITA 10: | No. |
| INPUT 10: | Tell me if you believe that I am tired. |
| LOLITA 11: | Yes, you are tired. |

## 12.8 Application of EAs to LOLITA

The environment is the source of information on which solutions are evaluated. In theory this must correspond to all natural language utterances. In practice, however, the current implementation of LOLITA acts only on literal meaning. Metaphors, idioms and humour are, therefore, excluded from the environment.

In deciding which plan boxes are to be activated, it is not their absolute values that are important, but rather their values relative to each other. Therefore, it is not the explicit values of the parameters that are to be optimised, but a shift in value from that of the current hand optimised setting. For each plan box a range of shift values (simply referred to as parameter values from now on) of [-63,64] was deemed sufficient, since these allow for a large range of possible behaviours (if necessary this range can easily be increased). A solution's representation is, therefore, a string of 124 (the number of plan boxes) integers. A solution with all of its values set to 0 is, therefore, identical to the current hand optimised setting. The parameters of the plan boxes which control utterances of a particular type are grouped together in blocks. For example, the three plan boxes labelled `cause_AffectionPlatonic` are grouped together, as are the six which are labelled `show_AngerOffense`. The components within a block determines how an utterance is carried out, *e.g.*, different ways in which anger can be expressed.

As mentioned previously some measure of how closely utterances generated match those of the target dialogue is needed. The results given in the next section use a very simple fitness function. A solution's fitness is initially set at zero, and then increased by one for each utterance that exactly matches that in the target dialogue. For the target dialogues discussed in this chapter a solution's fitness is, therefore, an integer in the range [1,11]. The total number of utterances that LOLITA generates is eleven, and so this provides the upper bound on fitness. Furthermore, all solutions will have a fitness of at least one, since with the current 'personality' LOLITA always initiates a conversation with the phrase 'I desire a success'. A more sophisticated fitness function is introduced in Section 12.10.

Comparing the results of runs of a GA and EP is difficult since the underlying system is continually changing and the data files regularly updated. Only single trials of each algorithm are carried out, but these are sufficient to show the validity of the approach.

Further details of the GA and EP implementation are now discussed.

### 12.8.1 Genetic Algorithms

As mentioned in Section 12.3.1 when using a GA solutions are to be represented as binary strings. The parameters controlling the plan boxes can take one of 128 distinct values, and so each parameter can be converted to a binary string of length seven. These strings are then concatenated together to form one string. Since there are some 124 plan box parameters the size of the search space is $2^{(7 \times 124)} \approx 10^{251}$.

When selecting solutions for mating, a 'roulette wheel' type of sampling is used in order to ensure that better solutions are more likely to be chosen (Goldberg, 1989). This proceeds by first evaluating the fitness of each solution in a generation. Sections of the roulette wheel are then allocated according to this fitness value. This ensures that when the roulette wheel is probabilistically spun, the fitter the solution the more likely it is to be selected. With the fitness function used all solutions will have a fitness of at least one, and so all are guaranteed a section of the roulette wheel.

Parents are combined using a two-point crossover operator (Beasley *et al.,* 1993) with the probability of crossover pc = 0.6. When applied to a point in the binary string the mutation operator changes the value at that point, *i.e.*, 1 to 0, or 0 to 1. In order not to be too disruptive the probability of mutation was kept low with pm = 0.001.

### 12.8.2 Evolutionary Programming

In applying EP to the dialogue optimisation problem the plan box parameters are stored as integers which are constrained to be in the range [-63,64]. In practice it isn't necessary to restrict the range, but this was done in order to ensure the search space was the same size for EP as for the GA.

Each parent solution in the population is mutated by an amount governed by its fitness to produce a child solution. Fitter solutions must be less likely to be mutated to the same degree as less fit parents, and so each component, xi, of a solution X, is mutated according to the formula (and then truncating):

$$x_i' = x_i + \sqrt{5 \cdot (\text{MAXFIT} - \textit{fitness}(X))} \cdot N(0,1) \qquad i \in \{1,2,\ldots,124\}$$

where MAXFIT is the maximum fitness attainable (11 for the work discussed in this and the following section), fitness(X) is the fitness of solution X (the number of correct utterances) and N(0,1) is a standard normal random variable. The above formula was selected since it allows for solutions with a poor fitness to be mutated by a large amount, while at the same time reducing the chance that the mutated parameters fall outside of the permitted range.

## 12.9 Results

This section presents the results of applying a GA and EP to the problem of finding plan box parameters.

For both the GA and EP a population of 50 was used and they were executed for 50 generations. The tournament size for EP was set at three. A single trial of each algorithm was carried out. Figures 12.1 and 12.2 show the online and offline performance of the GA and EP run, for the target dialogues DIAL 1 and DIAL 2, respectively. The offline performance is the average fitness of all of the solutions in a particular generation, while the online performance is the average fitness of all solutions that have been generated up to a certain generation.

In the case of DIAL 1 the GA was able to find a set of parameters which produced a dialogue of fitness 9, *i.e.*, two utterances incorrect. EP performed slightly better, discovering a solution of fitness 10. When DIAL 2 was used as the target dialogue the GA was able to find a solution of fitness 8, and EP a solution with fitness 9. These results are summarised in Table 12.3.

In the case of EP the incorrect utterance for DIAL 1 was "LOLITA 8: I do not know if I am beautiful; tell things to me." Such an utterance should not be considered as wrong, it is simply that the fitness function is not very sophisticated. Similarly for the GA and DIAL 1. For both the GA and EP, with DIAL 2 as the target dialogue, the incorrect utterances for the best parameters found indicate that the parameter settings were such that the input caused LOLITA to become offended quite easily.

| DIAL | LOLITA's incorrect utterances | |
|---|---|---|
| 1 | GA | 2: Tell me things.<br>6: I slightly am bored and emotional |
| | EP | 8: I do not know if I am beautiful; tell things to me. |
| 2 | GA | 2: I could not speak to you if you repeated you like me.<br>6: I desire to success.<br>7: I could not speak to you if you repeated Am I a machine? |
| | EP | 2: I know; I could not speak to you if you repeated you like me.<br>7: I could not speak to you if you repeated Am I a machine? |

Table 12.3: The incorrect utterances generated by the best parameters found when GA and EP were used to optimise the plan box parameters for DIAL 1 and DIAL 2.

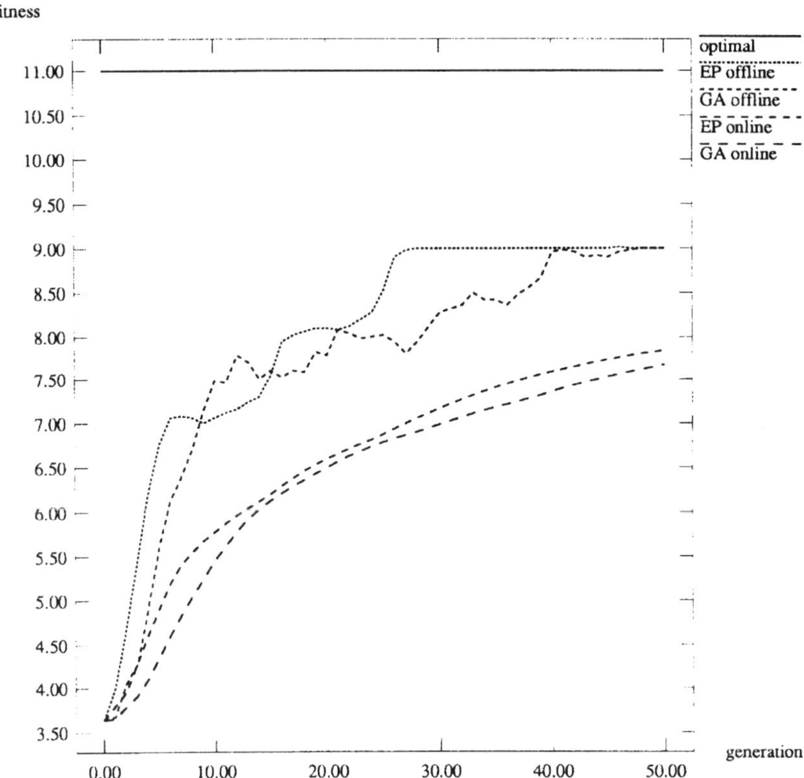


Figure 12.1: Online and offline performance for a trial of the GA and EP with DIAL 1 as the target dialogue.

An interesting feature of the EP results is how the average fitness of a generation rose to that of the best solution to date (Figures 12.1 and 12.2). It appears that when a better solution was discovered the average generation fitness would rise gradually for several generations and then quickly rise to that of the best. There is, however, one notable exception to this which occurred at generation 46 when DIAL 1 was the target dialogue (see Figure 12.1). At this point a solution of fitness 10 was produced in a population the remainder of which had fitness 9. The solution of fitness 10 was, however, subsequently lost and the reason for this is now discussed. Although a solution with fitness 10 is guaranteed a score of three in the tournament, many other solutions in that population also scored a fitness of three since all but one solution against which they were competing had a fitness of 9. When the process of sorting the scores took place there were more solutions with a score of three than places for them in the next generation and so some were lost. This included the solution of fitness 10. A similar occurrence took place in the run with DIAL 2. A solution of fitness 9 was discovered at generation 25, retained for one generation, and then lost.

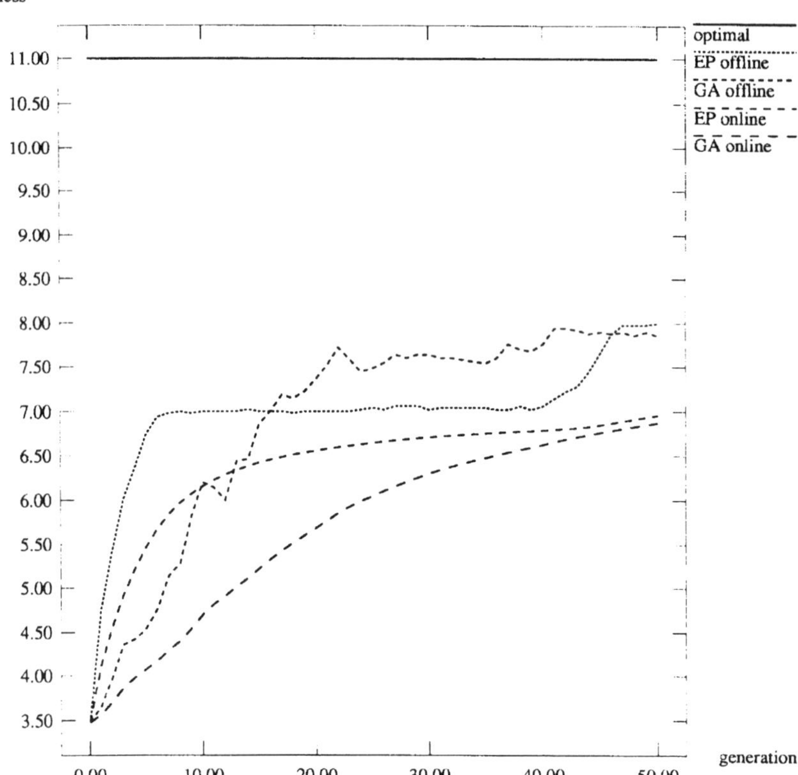


Figure 12.2: Online and offline performance for a trial of the GA and EP with DIAL 2 as the target dialogue.

The failure to retain an improved solution is in part attributable to the poor discriminatory power of the fitness function used. Since many solutions can have the same fitness a lot of solutions often perform very well in the tournament, and solutions with a maximum tournament score may be lost from the following generation. The following section examines a fitness function which is able to use additional information which the LOLITA system is able to provide. This improves the fitness function's discriminatory power, rewarding not just the words produced, but the underlying actions which lead to their generation.

## 12.10 Improving the Fitness Function

The fitness function adopted in the previous section is very simple and unable to take into account additional information which the LOLITA system is able to provide. On analysing an utterance the LOLITA system infers information on the local goals, subgoals, utterance types and action types of the speaker. Table 12.4 shows this information for the first seven utterances of DIAL 1.

| | |
|---|---|
| LOLITA 1:<br>(I desire a success.) | speaker: lolita<br>local goal: ShowEmotionGoal, NeutralEmotion<br>subgoals:<br>utterance types: AllSame<br>action types: default_tacticPB |
| INPUT 1:<br>(I like you.) | speaker: roberto<br>local goal: InformGoal<br>subgoals:<br>utterance types: Statement<br>action types: |
| LOLITA 2:<br>(I know; why do you like me?) | speaker: lolita<br>local goal: InformGoal<br>subgoals: AnyGoal, BeInformedGoal<br>utterance types: Statement, Noise, Question<br>action types: tellPB, why_questPB |
| INPUT 2:<br>(You are beautiful.) | speaker: roberto<br>local goal: InformGoal<br>subgoals:<br>utterance types: Statement<br>action types: |
| LOLITA 3:<br>(I am glad because you think I am beautiful.) | speaker: lolita<br>local goal: ShowEmotionGoal Serenity<br>subgoals:<br>utterance types: AllSame<br>action types: show_Serenity |
| INPUT 3:<br>(Do you like me?) | speaker: roberto<br>local goal: BeInformedGoal<br>subgoals:<br>utterance types: Question<br>action types: |
| LOLITA 4:<br>(Yes, I like you.) | speaker: lolita<br>local goal: InformGoal<br>subgoals:<br>utterance types: Statement<br>action types: answerPB |

Table 12.4: The additional information which the LOLITA system makes available for the first seven utterances of DIAL 1.

The fitness function can be modified to make use of the additional information given in Table 12.4. The fitness function used in this section calculates a solution's fitness by initially setting it to zero, and increasing it by one for each utterance, local goal, subgoal, utterance type and action type, which exactly matches that of the target dialogue. This fitness function is less sensitive to the utterance itself and more sensitive to the behaviour required. For DIAL 2 the information associated with each statement was used.

Using this additional information a fitness, which is an integer in the range [5,55], can now be assigned to solutions — 5 forms the lower bound since LOLITA always initiates a conversation with the same utterance and associated information. For each of the two target dialogues a single trial of the GA and EP were carried out. The GA and EP used the improved fitness function and in addition two modifications were made to EP. In the tournament phase of the algorithm if two solutions have the same fitness then a win is awarded with probability 0.5. This modification is aimed at helping to overcome the problem of EP 'loosing' a solution which arose in the experiments with first of the fitness functions discussed. Secondly, the EP's mutation operator is altered so that a child is produced from a parent by mutating each parameter xi as follows (and then truncating):

$$x_i' = x_i + \sqrt{(\mathrm{MAXFIT} - \mathit{fitness}(X))} \cdot N(0,1) \qquad i \in \{1,2,\ldots,124\}$$

where MAXFIT is the maximum fitness attainable (55 for the work discussed in this section), fitness(X) is the fitness of solution X and N(0,1) is a standard normal random variable.

Figures 12.3 and 12.4 show the online and offline performance of the GA and EP run, for the target dialogues DIAL 1 and DIAL 2, respectively.

In the case of DIAL 1 the GA was able to find a solution with a fitness of 47 by generation 15, and EP a solution of fitness 47 by generation 8. For DIAL 2 the GA discovered a solution of fitness 43 by generation 7, and EP a solution of fitness 43 by generation 14. The breakdown of these results is shown in Table 12.5.

| Additional | DIAL 1 | | DIAL 2 | |
|---|---|---|---|---|
| information | GA | EP | GA | EP |
| utterance | 9 | 9 | 8 | 7 |
| local goal | 10 | 10 | 8 | 9 |
| subgoals | 11 | 11 | 11 | 10 |
| utterance types | 9 | 9 | 8 | 9 |
| action types | 8 | 8 | 8 | 8 |
| Fitness | 47 | 47 | 43 | 43 |

Table 12.5: Decomposition of the results achieved with the improved fitness function. The optimum value for each of the values is 1.

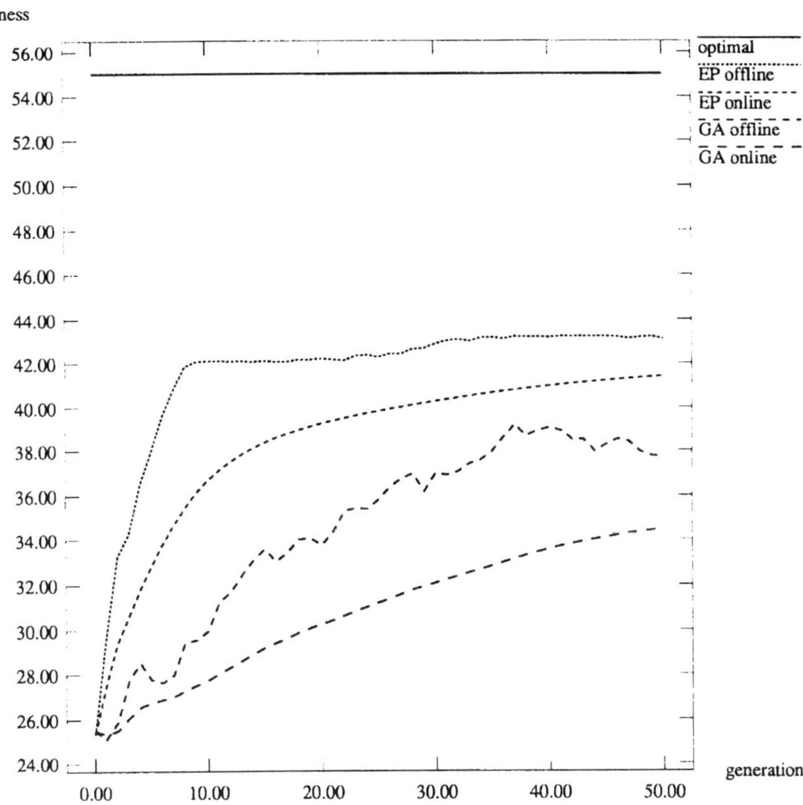


Figure 12.3: Online and offline performance for a trial of the GA and EP with DIAL 1 as the target dialogue. The fitness function which takes into account LOLITA's additional information was used.

Again the exact matching of utterances resulted in statements such as "LOLITA: Why do you like me?" in place of "LOLITA: I know; why do you like me?" being scored as incorrect. Similar instances arose with the matching of the additional information. For example, if the utterance types are "Statement, Noise, Question" then "Statement, Question" is currently scored as incorrect. A fitness value of 0.666 would be more appropriate. There is clearly much scope for improvement in the discriminatory power of the fitness function.

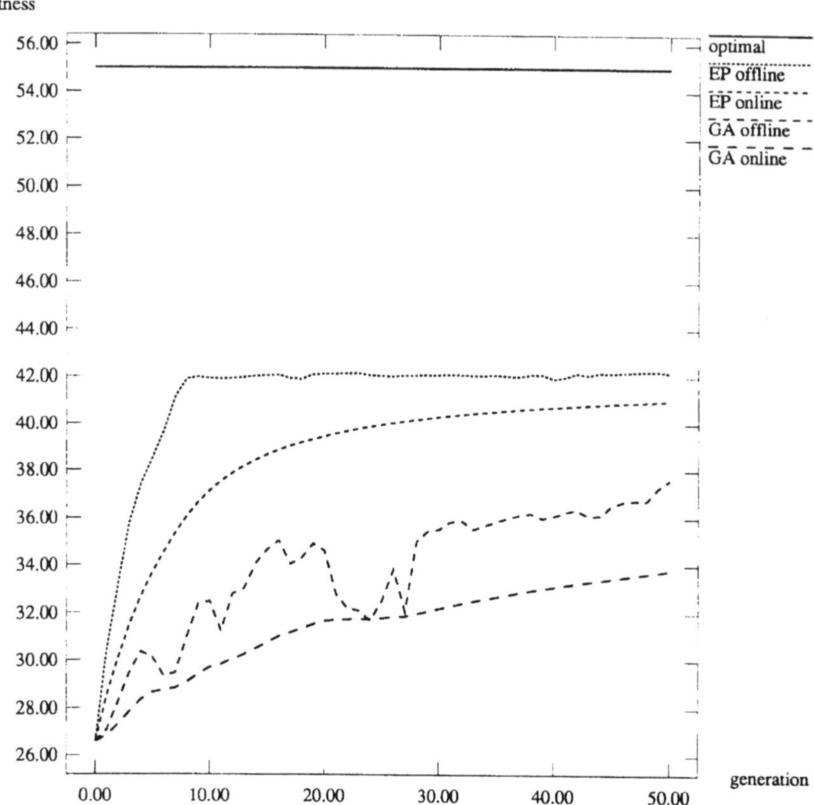


Figure 12.4: Online and offline performance for a trial of the GA and EP with DIAL 2 as the target dialogue. The fitness function which takes into account LOLITA's additional information was used.

## 12.11 Discussion

The results show that both a GA and EP were reasonably successful at the dialogue optimisation problem presented. These results, although preliminary, do lead to some interesting points worthy of further consideration.

For both the GA and EP the average fitness of solutions in subsequent generations steadily improved. No attempt was made to tune the settings of the evolutionary algorithms themselves. In the case of the GA such settings include the crossover and mutation probabilities. Other components of the GA that may be altered include the solution representation (*e.g.*, integers), crossover type and selection mechanism. The performance of EP may be improved by altering the tournament size, or the formula controlling the amount of mutation. Furthermore, it is likely that by increasing the population and generation size improved results can be expected. This has not been studied to date since with a population and generation size of 100, the runtime (on a Sparc4 workstation) can be expected to be of the order of two days. Furthermore, evaluating any differences in performance is difficult since the underlying system is continually being modified.

For the dialogues and fitness functions considered the fact that both a GA and EP are able to discover solutions which perform well indicates that both a bottom-up and a top-down approach is a suitable means of solution construction.

The discriminatory power of the fitness function needs to be further improved. Ideally some quantitative measure of semantic distance would be used (Short *et al.*, 1994a, 1994b). This would entail finding some quantitative measure for the similarity of the meaning of two sentences. Another approach would involve better use of the information that the LOLITA system is capable of producing. With an improved fitness function the current limitation of having to apply the EAs to known dialogues can be removed. Evolving the plan box parameters so that the resulting dialogue exhibits a certain personality is the long term aim, *e.g.*, finding the parameters which result in LOLITA becoming easily offended. Once sets of parameters for different behaviours have been determined they can be used to run LOLITA with that 'personality'.

## 12.12 Summary

This chapter provides evidence that a hybrid symbolic/subsymbolic approach can be successfully applied within the dialogue module of a large scale natural language processor. Adopting such an approach allows the dialogue module to enjoy many of the advantages of a well constructed theory, while at the same time allowing for the flexibility which a subsymbolic approach is capable of providing. The complex dialogues which can be generated validate the approach.

Evolutionary algorithms have been applied to the problem of searching the space of the subsymbolic representation so that a solution which exhibits a certain behaviour can be found. For the dialogues and fitness functions considered both a GA and EP were able to overcome the interactions which may occur and construct solutions that perform well. A more general application of the approach is currently limited by the poor discriminatory power of the fitness function.

## References

Bäck T., Hoffmeister F. and Schwefel H. (1991) A Survey of Evolution Strategies, in Proceedings of the Fourth International Conference on Genetic Algorithms, Morgan Kaufmann, pp 2–9.

Beasley D., Bull D.R. and Martin R.R. (1993) An Overview of Genetic Algorithms: Part 2, Research Topics, University Computing, Vol. 15, No. 4, pp 170–181.

Fogel D.B. (1992) Evolving Artificial Intelligence, Ph.D. Thesis, University of California, San Diego.

Fogel L.J., Owens A.J. and Walsh M.J. (1966) Artificial Intelligence through Simulated Evolution, J. Wiley, New York.

Garigliano R. and Nettleton D.J. (1994) The Interplay of Symbolic and Adaptive Techniques: two Case Studies, IEE Colloquium on Symbolic and Neural Cognitive Engineering, Savoy Place, London.

Garigliano R., Morgan R.G. and Smith M.H. (1993a) The LOLITA System as a Contents Scanning Tool, in Proceedings of the Thirteenth International Conference on Artificial Intelligence, Avignon.

Garigliano R., Morgan R.G. and Smith M.H. (1993b) LOLITA: Progress Report 1, Technical Report 12/92, Department of Computer Science, University of Durham, U.K.

Garigliano R., Morgan R.G. and LOLITA group (1994a) The LOLITA Project: The First Seven Years, under negotiation with After Hurst Ltd.

Garigliano R., Tate J. and Boguraev B. (eds.) (1994b) Journal of Natural Language Engineering, Cambridge University Press.

Goldberg D.E. (1989) Genetic algorithms in search, optimization, and machine learning, Addison-Wesley.

Holland J.H. (1975) Adaptation in Natural and Artificial Systems, University of Michigan Press.

Jones C.E. (1994) Dialogue Structure Models: An approach to Dialogue Analysis and Generation by Computer, Ph.D. Thesis (submitted), Department of Computer Science, University of Durham, U.K.

Jones C.E. and Garigliano R. (1993) Dialogue Analysis and Generation: A Theory for Modelling Natural English Dialogue, in Proceedings of EUROSPEECH '93, the 3rd European Conference on Speech Communication and Technology, Berlin, pp 951-954.

Nettleton D.J. (1994) Evolutionary Algorithms in Artificial Intelligence: A Comparative Study Through Applications, Ph.D. thesis (submitted), University of Durham, U.K.

Nettleton D.J. and Garigliano R. (1994a) Evolutionary Algorithms for Dialogue Optimisation in the LOLITA Natural Language Processor, Seminar on Adaptive Computing and Information Processing, London.

Nettleton D.J. and Garigliano R. (1994b) Evolutionary algorithms for dialogue optimisation as an example of hybrid NLP system, International Conference on New Methods in Language Processing, Manchester.

Schank R.C. and Abelson R.P. (1977) Scripts, Plans, Goals and Understanding, Lawrence Erlbaum Associates Inc., New Jersey.

Short S., Collingham R.J. and Garigliano R. (1994a) What did I say...? Using Meaning to Assess Speech Recognisers, Institute of Acoustics Autumn Conference on Speech and Hearing, Windermere, Cumbria, U.K.

Short S., Collingham R.J. and Garigliano R. (1994b) Making Use of Semantics in an Automatic Speech Recognition Systems, Institute of Acoustics Autumn Conference on Speech and Hearing, Windermere, Cumbria, U.K.

Smith M.H., Garigliano R. and Morgan R.G. (1994) Generation in the LOLITA system: An engineering approach, Seventh International Workshop on Natural Language Generation, Maine, U.S.A.

## Chapter 13

**Dipankar Dasgupta**[1]
Department of Computer Science
University of Strathclyde
Glasgow GI IXH, U.K.

# Incorporating Redundancy and Gene Activation Mechanisms in Genetic search for adapting to Non-Stationary Environments.



**Abstract**

This chapter describes the application of a different genetic algorithm — Structured Genetic Algorithm (sGA) — for tracking an optimum in time-varying environments. This genetic model incorporates redundancy in chromosomal encoding of the problem space and uses a gene activation mechanism for the phenotypic expression of genomic subspaces. These features allow multiple changes to occur simultaneously, in addition to usual mixing effects of genetic operators as in standard GAs. In adapting to nonstationary environments, the extra genetic material provides a source for maintaining variability within each individual, resulting in higher steady-state genotypic diversity even with phenotypic convergence of the population in different epoch. Experimental results reported here demonstrate that sGAs can efficiently keep track of a moving optimum compared to existing genetic approaches.

[1] The author is currently at the department of Computer Science, University of New Mexico, Albuquerque, NM 87131, U.S.A.

0-8493-2529-3/95/$0.00 + $.50

## 13.1 Introduction

Many real-world applications deal with situations in which the optimal criterion changes over time (typically with changes in the external environment). Also in some problem domains, these changes are very frequent and irregular in nature. When a genetic search is used to solve such a real-time problem, it must find the current optimum quickly as well as should be able to adapt rapidly in response to change in the environment. When standard GAs are used for such time-varying optimisations, once the population converged to an optimum, they lose their ability to search for a new optimum. Since in a standard GA, phenotypic convergence generally lead to genotypic homogeneity of the whole population (unless an explicit mechanism such as sharing, crowding, etc. is used to keep different subpopulations; that too may not be efficient in a time-varying situation). So they are not well-suited for non-stationary function optimisations. Such difficulties are also reported by other researchers [10].

These difficulties of a standard GA are primarily due to the simple chromosomal representation which can not possess sufficient genetic diversity in the population to allow the search to continue as environment changes. For a standard GA to succeed in such a situation, requires multiple correlated mutations to introduce non-destructive diversity. But in standard GAs multiple directed mutations are extremely unlikely to result in viable offspring. One possible way to introduce diversity in a converged population is to increase mutation rate, but that may lead to random search.

There have been several studies that have addressed the use of genetic algorithms in which the objective function changes over time. Goldberg and Smith [8] studied the behaviour of genetic diploidy with dominance mechanisms in adapting to a two-state response surface. In this representation, two alleles are stored for each gene but only one is expressed according to some dominance mechanism. This approach, however, does not appear to scale up to more general cases.

Pettit and Swigger [12] experimented with GAs in a randomly fluctuating environment, but their study provides limited insights due to the extremely small population size adopted. Likewise, Krishnakumar [11] used a genetic algorithm with a very limited population (5 only) to track a rapidly changing environment in aerospace engineering. In the field of machine learning, genetic algorithms are also used [9] where the task is to find a learning strategy for one player in a multi-player game, and the performance (objective function) of the learning player may change over time due to changes in strategies adopted by the opposing players.

Cobb [2] has proposed an adaptive mutation mechanism called *triggered hypermutation* to deal with a restricted class of continuously changing environments. This approach monitors the quality of the best performers in the population over time and increases the mutation rate when performance degrades. However, other classes of non-stationarity may fail to trigger the hypermutation, leaving the GA converged in a suboptimal area of the search space.

Grefenstette used [10] a *random immigrants mechanism* (a replacement policy) where a percentage of the population is replaced by randomly generated individuals in each generation. The intention again is to maintain a continuous level of exploration of the search space, while trying to miniraise the disruption of the ongoing search. His results with one type of non-stationarity show that the performance is highly dependent on the *replacement rate.* But this approach has a serious drawback in dealing with real-time applications, since the time for a replacement of individuals and the necessary genetic operations to produce offspring may take longer than the time of change in the environment. Also there always remains a risk of losing valuable information during random replacement of the population members.

The principle behind these methods is to introduce additional genetic variation (or randomness) in the population as and when needed for adapting to environmental changes. The above approaches may be good for one or another restricted class of non-stationarity, but cannot be generalised as is possible with an sGA [5].

The remainder of this chapter is organised as follows: the next section will give a brief description of the structured GA. Section 13.3 defines a time-varying optimisation problem which was studied by Cobb [2] with standard GAs. Section 13.4 gives experimental details of the sGA implementation for different versions of the problem. Finally, some conclusions are made based on experimental results in Section 13.5.

## 13.2 The Structured GA

Species adaptation in the changing biosphere provides important guidelines for understanding the dynamic behaviour of evolutionary systems. Biological systems during evolution develop successful strategies of adaptation in order to enhance their probability of survival and propagation. Environmental pressures on a biological organism can be severe, thus the most effective organisms are those which are able to adapt most rapidly to changing conditions. A central tenet underlying our hypothesis is that there must be something special in the structure of a biological system which enables a great majority of its offspring to be viable in varying environments. The structured GA encoding appears to be more biologically-motivated and a possible alternative genetic search approach with some distinctive features.

The central feature of the Structured Genetic Algorithm [6] is the use of redundancy and a gene activation mechanism in its multi-level genotype. In particular, genes at any level can either be active or passive. High-level genes activate or deactivate sets of low-level genes. Thus the activity of the genes at any given level, whether they will be expressed phenotypically or not (in a genotype-to-phenotype mapping), are governed by their higher-level genes. A two-level representation of the sGA is shown in Figure 13.1. In the sGA, structural genomes are embodied in the chromosome and are represented as sets of linear (binary) substrings. The model also uses conventional genetic operators and the *survival of the fittest* criterion to evolve increasingly fit offspring.

In an sGA, redundant materials (over-specified encoding information) serve a dual purpose: they can provide implicit non-destructive diversity at all times during the search process; since only expressed portions of the chromosome undergo selection pressure and move toward current optimal state, the unexpressed portions are neutral, though they experience silent genetic changes. The representation can also work as a distributed memory of variation within the population structure. These features allow the model to work efficiently in environments exhibiting different types of nonstationarity. In effect, this model provides a mechanism for genetic evolution in which diversity can be maintained by keeping extra genetic material and controlling their expression while decoding. In adapting to nonstationary environments, the additional genetic material in an sGA encoding provides a natural source for maintaining diversity as suited to different environmental situations. A detailed description of the model with some empirical experiments were reported in our previous works [4, 5].

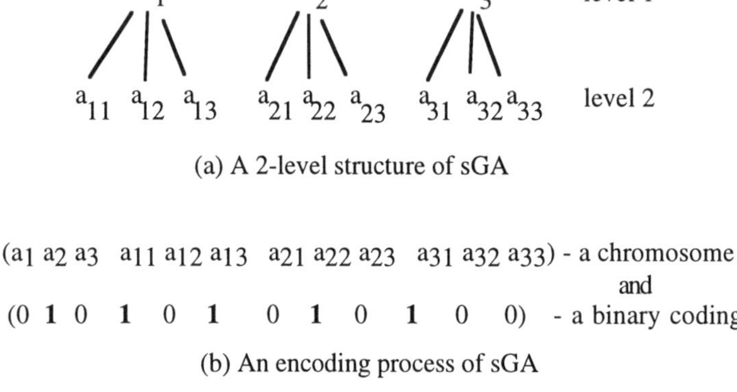


Figure 13.1: A simple representation of an sGA.

## 13.3 Use of sGA in a time-varying problem

We considered here a *State Dependent Nonstationary Environment* (SDNE) where the state of the environment varies either implicitly or explicitly with the stage of the search. For the genetic search, a stage is considered as a generation. In this nonstationary environment, the objective of search is not to find a single optimum for all time, but rather to select a sequence of values over time that miniraise or maximise, the environmental evaluations. We have taken the example from Cobb's experiment [2], where the optimisation of a simple parabola having one variable in a continuously changing SDNE was used. The expression for the parabola is

$$f_t(x_i) = (x_i - h_i)^2$$

where $h_t$ is the generated target domain value mapping into the optimum at time $t$ which moves along a sinusoidal path, so that the optimum changes in each generation. The $x_i$ is the current estimate of this domain value by the $i$th individual and $f_t$ represents the environment at time $t$. By using a parabola, at each generation the environment essentially returns the squared error of the domain estimate from the current optimum, $h_i$. A detailed description of the problem is given in [2].

## 13.4 Experimental Details

To specify the working of sGAs more precisely for this example, a two-level sGA is adopted where high-level bits activate low-level partial solution spaces or subspaces. The initial population is generated randomly with a partial restriction on the high-level where a specified number of high-level bits are allowed to be active according to the low-level mapping bits [4, 5]. Then a local mutation is used which swaps the position of two high-level gene values. This initialisation approach is like messy GA's partially enumerative approach where at least one copy of all possible building blocks of a specified size need to be provided. But the advantage of our initialisation scheme is that it can avoid both under- and over-specification problems in decoding. So in each chromosome, first few bits (the number of bits is a deciding factor like other GA parameters) are high-level bits which act as a control region to express subspaces at the lower level to form a candidate solution.

In these experiments, a range of parameter sets (e.g., population size, crossover and mutation probability, etc.) are employed. For the results reported, a two-point crossover operator along with the stochastic remainder selection strategy [1] are used. Each run is allowed to continue for 300 generations and the results are averaged over ten such runs each with a different initial population.

In experiments here, each individual is encoded with 10 high-level bits where each high-level bit maps 5-bit subspace at the low-level constituting a chromosome of length 60 bits (chromosome length = H.L. bits + H.L. bits * L.L. bits). We have considered a 30-bit solution space for decoding the single variable of the parabola, so the activation of 6 high-level bits are sufficient for expressing a candidate solution. In these experiments, we have used a strategy where individuals with below average fitness undergo a higher (10 times) rate of local mutation on their high-level in order to increase the frequency of shift in dominance (expression) among low-level optional subspaces.

### 13.4.1 Continuously Changing SDNE Environments

In the first set of experiments, we use a continuously moving optimum and the sGA is applied to track the optimum. In Figure 13.2, two indistinguishable curves exhibit the best individual performance of an sGA in continually tracking the moving optimum that follows the sinusoidal path of evolution using a population of size 200 (same popsize as used in simple GA experiments [2]). Figure 13.3 shows the performance measure plotted (as a negative $log_{10}$ scale): the best individual and average population performance against generation. In this graph, the higher the value of the best individual performance, the better is the tracking performance.

When the population size is reduced to half (i.e., 100), no significant performance difference is observed as evident from the Figure 13.4. This success with smaller population is because of the genetic variability which exists within each individual and in the population is sufficient to adapt in this environmental change [3]. Of course, to achieve this level of performance, the sGA needed more memory space to keep redundant information as compared to the same size population in simple GA. On the contrary, increase in the population size of a simple GA cannot exhibit similar effect, since all its encoded information is usually involve in every environmental state.

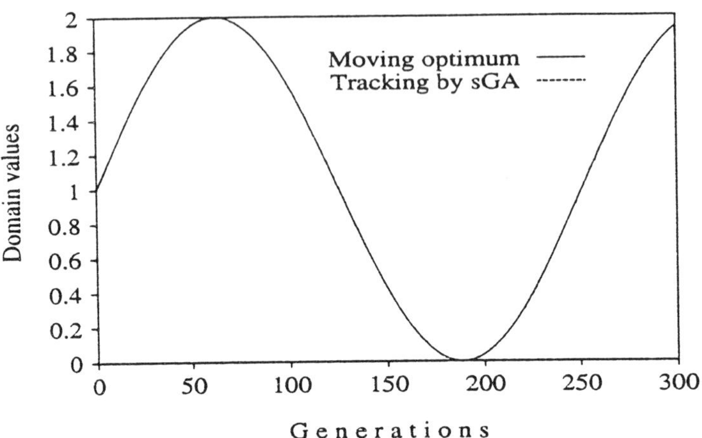


Figure 13.2: Two indistinguishable curves displaying sGA's best-of generation value perfectly tracking the actual optimum.

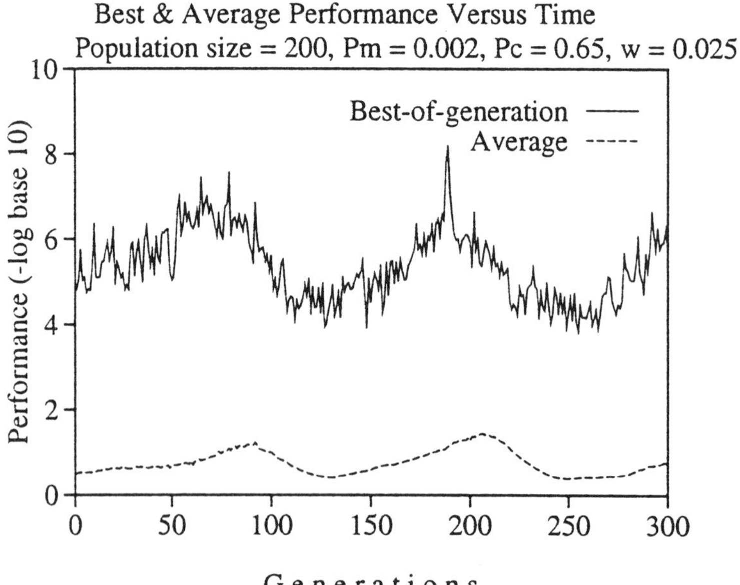

Figure 13.3: Performance of the sGA in tracking the moving optimum. This indicates the function evaluation ('squared error' between the estimate of the best/average individual and the true value of the time-varying optimum).

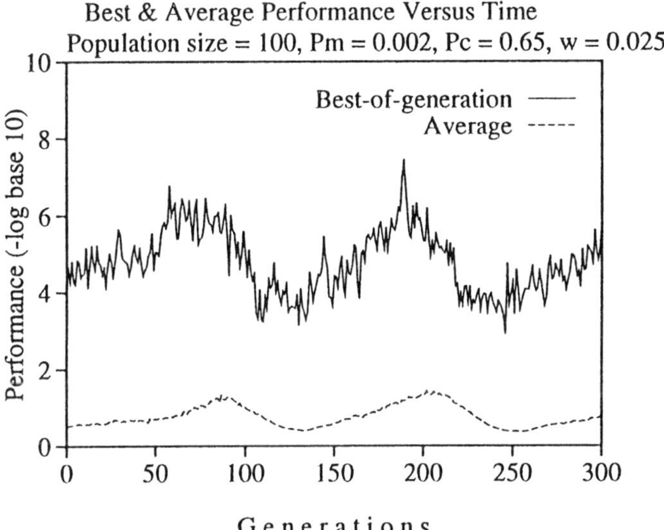


Figure 13.4: Performance of the sGA in tracking the moving optimum with population size of 100. Note that the similar performance is obtained with only half the population used by Cobb (1990) [2].

In both cases, the best performance varies between the order of $10^{-4}$ and the order of $10^{-7}$, this higher value exhibits the robustness of structured GAs to track the problem of nonstationarity. Moreover, the lower value of the average performance measure implies the amount of diversity which is sustained in the sGA population at different time during search.

The results with simple GA experiments, reported by Cobb [2][2] were always below $10^{-5}$ when two different (fixed) mutation rates 0.001 and 0.5, as shown in Figures 13.5 and 13.6 respectively. However, a comparable performance was obtained with an adaptive mutation scheme, Figure 13.7.

In Figures 13.8 and 13.9, four different sine wave frequencies (which implies different rate in environmental change) are tested with two sets of GA parameters. The best-of-generation performance is almost similar in all cases which implies that the genetic variability that exists in the sGA population can easily cope with both slow and rapid environmental changes. In other words, as the frequency increases, the optimum changes rapidly following a sinusoidal path. Unlike Cobb's method which has to monitor performance and alter the mutation rate, the

[2] The graphs are reproduced by permission from the author [2].

sGA tracks the changing environment more accurately with a fixed rate of mutation. It is to be noted that though we have used higher mutation rates to below-average performers (individuals) on their high-level bits in order to express optional subspaces by a single atomic change, such mutations (effect of simultaneous multiple bit changes) are not possible with a simple GA representation. It is observed that the performance of the algorithm slightly varies with the increase in frequency of sine wave, which can be compensated by increasing mutation rate, but the same mutation rate can maintain the performance level higher than simple GA's for a wide band of frequencies.

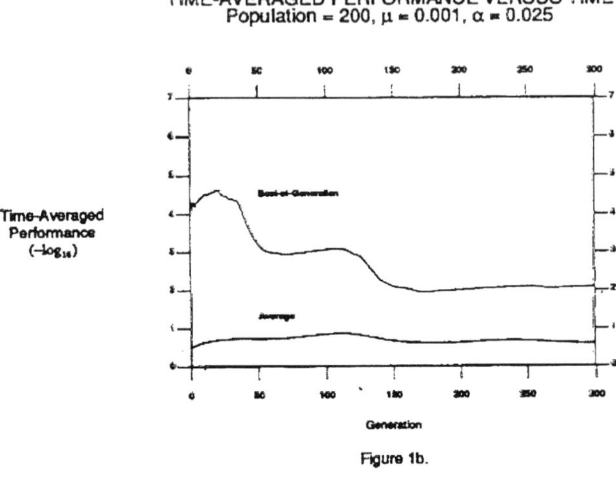


Figure 13.5: Simple GA performance with similar low mutation rate as used with sGA in tracking moving optimum (Cobb, 1990) [2].

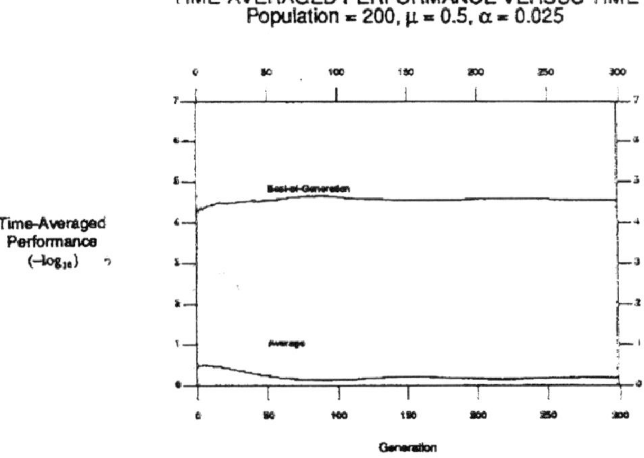

Figure 13.6: Performance of the simple GA with high mutation rate in tracking moving optimum (Cobb, 1990) [2]. Note: Mutation rate used here is more than 200 times higher than that used in sGA.

Figure 6a. Adaptive Mutation Rate:
If Time-Averaged Best Performance Improves, $\mu = 0.001$, otherwise, $\mu = 0.5$

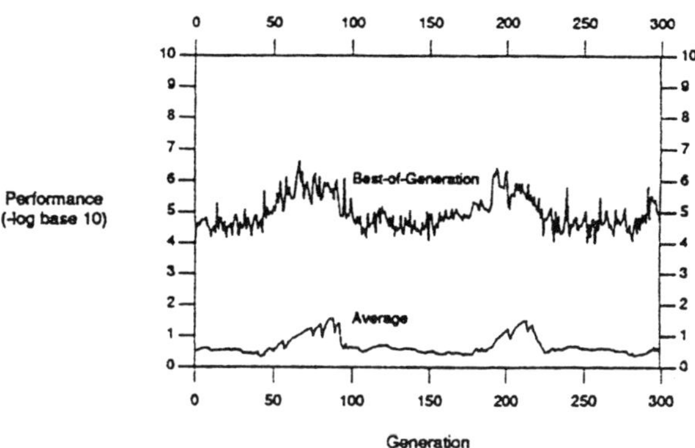

Figure 13.7: Performance of the simple GA using adaptive mutation (Cobb, 1990) [2]. Note: Though the performance improves, but it required precise control of mutation rate.

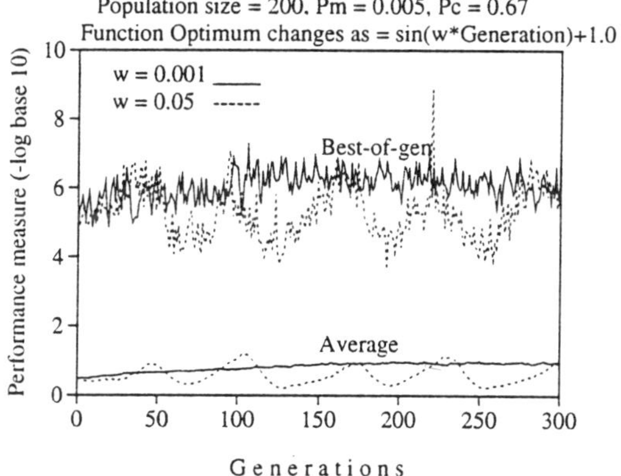

Figure 13.8: Performance of sGAs in function environments with changing optimum in sinusoidal path using different values of frequency.

### 13.4.2 A combination of stationary and nonstationary SDNE

Next set of experiments considered a combination of stationary and nonstationary SDNE, where the environment periodically remains stationary at its current value of $h_t$, while maintaining continuity. As an example, wc have considered $h_t$ to be remained constant from generation 75 to 125 and again from generation 225 to 300 (see ref. [2] for details).

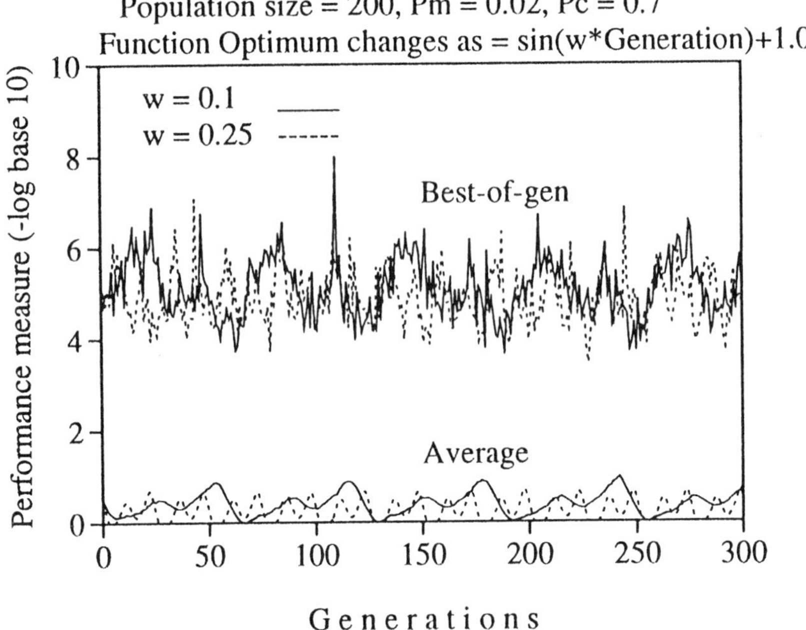


Figure 13.9: Performance of sGAs in function environments with changing optimum in sinusoidal path using different values of frequency. The higher the frequency more rapid the environmental change.

Figure 13.10 displays the tracking ability of an sGA in a combined stationary and nonstationary environment. The best-of-generation and average performance is shown in a negative log scale in Figure 13.11. The graphs show that an sGA performance improves when the environment remains stationary, regardless of preceding or following nonstationarity periods. Also during periods of nonstationarity, the performance varies depending on the rate of change in the environment for a given fixed rate of mutation. Particularly, for a higher frequency sine wave (e.g., 0.25), an increased rate of mutation is necessary to improve the performance at nonstationary periods, but performance degrades during stationary period in such case when constant mutation rate is used. In order to alleviate the performance, an *elitist* strategy is used where a significant improvement in performance is observed during the stationary period where a slight improvement is also noticed in nonstationary periods as shown in Figure 13.12.

## 13.5 Conclusions

This paper presented the application of structured GAs in environments having different degrees of non-stationarity. In these problem environments, the structured GA encoding worked as a diversity preserving system which could continually track both fast moving optimum and the optimum which changes in an interval, using a lower rate of mutation compared to simple GA approaches.

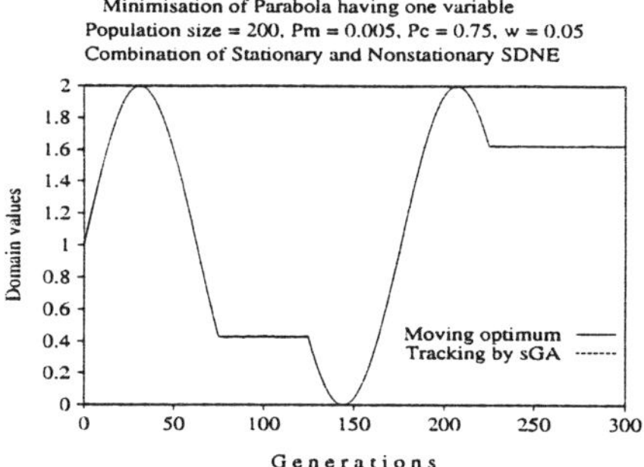


Figure 13.10: sGA's best-of-generation and the actual optimum are indistinguishable in each generation.

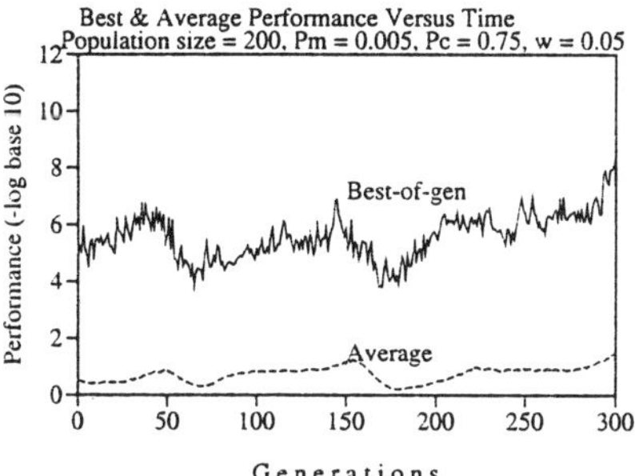


Figure 13.11: Performance of the sGA in finding the optimum in a combined stationary and nonstationary SDNE.

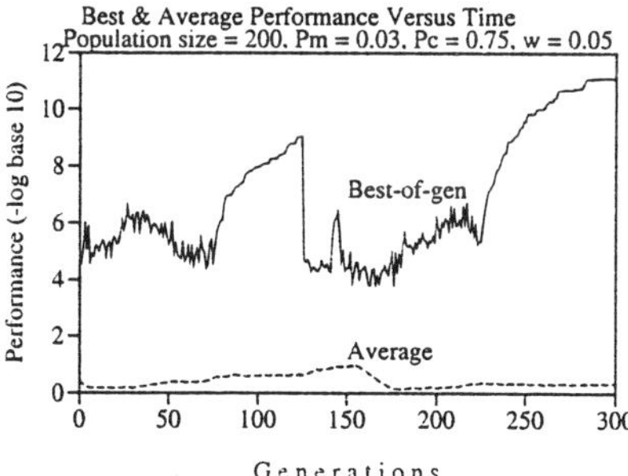


Figure 13.12: Performance of the sGA in a combined stationary and nonstationary SDNE when elitist strategy is used.

To summarise the performance of an sGA as compared to Cobb's simple GA approaches in this (SDNE) problem domain:
Cobb [2] used different mutation dependent strategies with simple GAs for solving SDNE problems and better results were found using an adaptive mutation

strategy. The main role of Cobb's adaptive mutation is to introduce diversity (randomness) in the population whenever needed. For example, if the time-average best performance was improving then the mutation rate was kept at 0.001 otherwise higher mutation rate of 0.5 was used. The success of such strategies with simple GAs is solely dependent on the precise control of mutation rates and the correct timing of triggering by the external process monitoring the performance, to get any beneficial effect. The performance graphs of sGA experiments show that a constant (lower) mutation rate can produce better results than that of simple GAs with different mutation schemes. These sGA results were obtained without any fine tuning of sGA parameter set. In particular, the amount of redundancy incorporated in the sGA encoding here (such as number of high level bits and low level mapping bits) are chosen arbitrarily and need further investigation to find an optimal set of values.

In the structured genetic approach maintenance of variability is an inherent characteristic of the model. Since it carries optional sub-structures (partial solution spaces) in the chromosome which can be combined in different ways according to the activation pattern of high-level control bits. Also these sub-structures usually maintain diversified information (different bit patterns) which compete for dominance at different environmental states. Thus the model can distribute resources of gene structures among different environmental states instead of dedicating all the structures to each state as in a standard GA. As the implicit diversity can be built into the population of an sGA, it can easily keep track of a number of environmental states changing over time.

We also noted that in comparison to the sGA, the recent mGA model [7] does not have the ability to adapt in changing fitness landscapes once it converges to a global optimum, since the unexpressed portion of the variable-length mGA string has no correlation with its expressed portion. In other words, redundancy if it exists at all after convergence in a mGA, is unlikely to provide sufficient information for adapting to environmental change, unless additional strategy is incorporated [7], similar to the diploidy and dominance mechanism as used with simple GA [8].

Our previous study [5] shows that the single elegant sGA mechanism can also work as long-term memory by preserving and retrieving more than two temporal optimal solutions in a repeated non-stationary environment. We conclude that use of a more biologically motivated genetic encoding (as in sGA) can handle different types of nonstationarity more efficiently than the existing approaches with a standard (canonical) GA.

**Acknowledgement**

The author is grateful to Professor Douglas R. McGregor for his encouragement in carrying out this work. The author would like to thank Helen G. Cobb for her constructive comments on the draft version of the report and giving permission to reproduce some of her results for comparison purpose.

**References**

[1] L.B. Booker. *Intelligent behavior as an adaptation to the task environment.* Ph.D. thesis, Computer Science, University of Michigan, Ann Arbor, U.S.A, 1982.

[2] Helen G. Cobb. An investigation into the use of hypermutation as an adaptive operator in genetic algorithms having continuous, time-dependent nonstationary environments. NRL Memorandum report 6790 AIC-90-001, Naval Research Laboratory, Washington, D.C. 20375-5000, December 1990.

[3] Dipankar Dasgupta. Tracking a moving optimum using the structured genetic algorithm. In *Proceedings of Seventh Annual Florida Artificial Intelligence Research Symposium (FLAIRS-94),* pages 366-370, May 5-7 1994. Florida, U.S.A.

[4] Dipankar Dasgupta and D.R. McGregor. A Structured Genetic Algorithm: The model and the first results. Technical Report NO. IKBS-2-91, 1991. *Presented at AISB PG-Workshop,* January, 1992.

[5] Dipankar Dasgupta and D.R. McGregor. Nonstationary function optimization using the Structured Genetic Algorithm. In *Proceedings of Parallel Problem Solving From Nature (PPSN-2),* Brussels, 28-30 September, pages 145-154, 1992.

[6] Dipankar Dasgupta and Douglas R. McGregor. A more Biologically Motivated Genetic Algorithm: The Model and Some Results. In *Cybernatics and Systems: An International Journal,* 25(3):447-469, May-June 1994.

[7] Kalyanmoy Deb. *Binary and Floating-point Function Optimization using Messy Genetic Algorithms.* Ph.D. thesis, Dept. of Engineering Mechanics, University of Alabaton, Tuscaloosa, Alabama, U.S.A., March 1991.

[8] David E. Goldberg and Robert E. Smith. Nonstationary function optimization using genetic algorithms with dominance and diploidy. In *Proceedings of Second International Conferance on Genetic Algorithms.,* pages 59-68, 1987.

[9] J.J. Grefenstette, C.L. Ramsey, and A.C. Schultz. Learning sequential decision rules using simulation models and competition. *Machine Learning,* 4(5):137-144, 1990.

[10] John J. Grefenstette. Genetic Algorithms for changing environments. In *Proceedings of Parallel Problem Solving From Nature (PPSN-2),* Brussels, 28-317 September, pages 137-144, 1992.

[11] K. Krishnakumar. Micro genetic algorithms for stationary and non-stationary function optimization. In *SPIE, Intelligent Control and Adaptive Systems,* pages 289-296, 1989.

[12] K. Pettit and E. Swigger. An analysis of genetic based pattern tracking and cognitive based component tracking models of adaptation. In *Proceedings of National Conference on AI (AAAI-83),* pages 327-332. Morgan Kaufmann, 1983.

## Chapter 14

**Saman K. Halgamuge and Manfred Glesner**
Darmstadt University of Technology
Institute of Microelectronic Systems
Karlstr. 15, D-64283 Darmstadt, Germany

saman@microelectronic.e-technik.th-darmstadt.de

# Input Space Segmentation with a Genetic Algorithm for Generation of Rule Based Classifier Systems



**Abstract**

The rule based transparent classifiers can be generated by partitioning the input space into a number of subspaces. These systems can be considered as fuzzy classifiers assigning membership functions to the partitions in each dimension. A flexible genetic algorithm based method is applied for generation of rule based classifiers. It is shown that for complex real world types of applications, a preprocessing step with neural clustering methods reduces the running time of the genetic algorithm based method drastically. A heuristic method is compared to show the strength of genetic algorithm based method.

0-8493-2529-3/95/$0.00 + $.50

## 14.1 Introduction

The task of a classifier is to attribute a class to a given pattern which can be represented by measurements of some of its features. Thus a pattern can be seen as a vector in the pattern space of which dimensions are the measured features. Some of those dimensions are more relevant to distinguish between the classes while others are less useful. It would be interesting to remove unnecessary dimensions in order to simplify the pattern space and require less measurements. But the usefulness of a dimension is not always independent from the choice of the other dimensions.

In automatic generation of fuzzy rule based classifiers from data, the grade of importance of the inputs to the final classification result can be obtained, which leads to more compact classifier systems. The most important part of a fuzzy classifier is the knowledge base containing different parameters for fuzzification, for defuzzification and the fuzzy rules which contribute to the transparency. Those IF-THEN fuzzy rules contain terms like Low, Medium, High to describe the different features expressed as linguistic variables.

A rule based classifier can be seen as a group of hyper cuboids in the pattern space. Those hyper cuboids should represent parts of the space that belong to the same class. The elements used for the partition of the space can be either input data vectors or compressed clusters generated by artificial neural nets such as Radial Basis Function Networks (RBFN) [PHS+94] or Dynamic Vector Quantisation (DVQ) [PF91]. When learning vectors — or learning patterns — are concerned, they are seen as the limit case of clusters generated by neural networks with the forms of hyper cuboids or hyper spheres.

## 14.2 A Heuristic Method

This method is based on the analysis of variations of proportions of input vectors or clusters belonging to different classes in each dimension. Even though some information is lost due to the projection of the pattern space on the input dimensions this simplification makes the algorithm very fast. Since variations are to be calculated, a discrete approach has to be taken. The dimensions are to be cut into segments and the proportions of classes are to be computed for each segment. In this method, the lengths between two segmentation lines are initially equal. They begin to adjust when the heuristic method proceeds.

Both Figures 14.l(a) and 14.l(b) have in common that the slope of the border separating the classes 1 and 2 is close to 45°. Suppose that both dimensions are normalized to unity. These 2 figures are among the most difficult cases of partition and the ideal solution would involve first a change of both axes so that the slope would be about 0° or 90° steep. But in such a case the meaning of the input variables x and y would be lost. Since a transparent classifier has to be generated, rules must be easily understandable, therefore transformation of input variables must be avoided.

The slopes in Figure 14.1 indicate that one of the dimensions is slightly more important than the other. The steeper the slope, the more important the dimension. Since a decision has to be taken for the limit case (when none of the dimension is more important than the other, that is for a 45° slope), this will give a threshold. Suppose that the limit case was divided into *ns* segments and that a decision has to be made. Since in this case the variation of proportions between two segments is always the same it is not possible to cut depending on the variations.

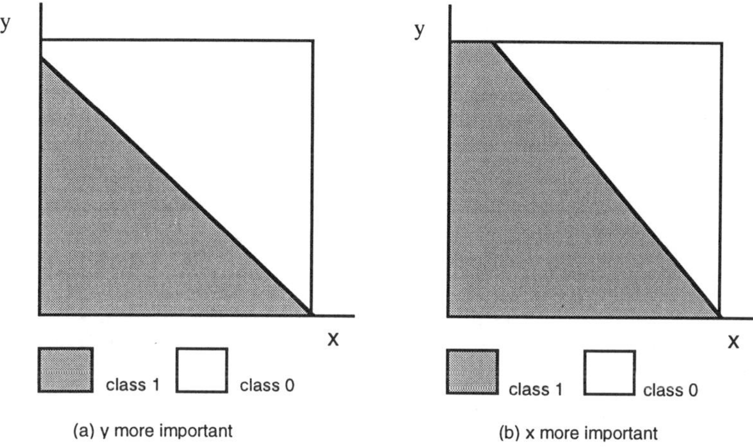


Figure 14.1: Defining a threshold.

A 45° slope corresponds to 100% of variation, if a very large number of partitions *ns* are allowed. At one end 100% of class 1 and at the other 0% of the class 1 are on the left of the cut. If the number of segments is *ns,* the threshold between 2 segments is 100%/*ns*.

The heuristic algorithm can be described as follows:

1. take the next dimension of the pattern space

2. divide this normalized dimension into *ns* equal segments

3. in each subspace generated by each segment, calculate the proportions of each class

4. if the variation of proportion between two neighboring subspaces for at least one class is greater than a given threshold, it is decided to cut this dimension between the two neighboring segments

5. go back to step 1 until last dimension is reached

In Figure 14.2(a), dimension x is divided into 4 segments and is cut between segments 2 and 3, and between segments 3 and 4. Proportions for class 1 varies from 100% in segment 1, over 80% in step 2 and 13% in segment 3 to 40% in segment 4. The variation in dimension x is higher than 100%/*ns* = 25% between segment 2 and 3 and between segment 3 and 4. In Figure 14.2(b), step 1 contains

70% of class 1, step 2, 76% step 3, 16% and step 4, 33%, hence the decision to cut between step 2 and 3. Segmentation and cuts of dimension *y* are independent from what has been with dimension *x*.

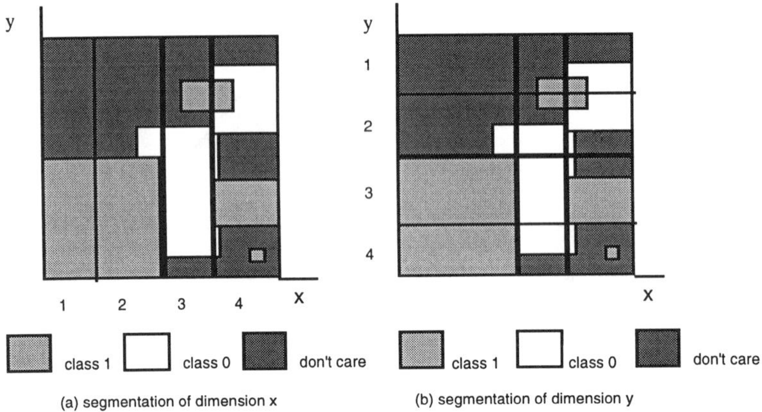


Figure 14.2: Segmentation of a two-dimensional pattern space.

This threshold value may vary according to the problem. If the threshold is too low, too many — sometimes irrelevant — cuts will be made and if the threshold is too high, some needed cuts could have been neglected, increasing the classification error. The range of empirical values is typically from 80%/*ns* to 180%/*ns*.

In order to evaluate the speed of this algorithm, it must be known that centers of subspaces have to be ordered in every dimension. Assuming that an ordering algorithm of order *s*.log(*s*) is used, the order of this method is: *d*.*s*.log(*s*), with *s* the number of subspaces and *d* the number of dimensions.

It is easy to see that this algorithm is fast but loses information because dimensions are treated independently, and that the accuracy of the partition cannot be better than the length of the segments.

## 14.3 Genetic Algorithm Based Method

Genetic Algorithms are solution search methods that can avoid local minima and that are very flexible due to their encoding and evaluation phases [Hol75, Gol89, BS93]. Indeed the form of a desired solution has to be encoded into a binary string so that a whole population of encoded possible solutions can be initialized at random. Evaluation is realized by a fitness function that attributes a value of effectiveness to every possible solution of the population. The best ones are allowed to exchange information through genetic operations on their respective strings. With this process, the population evolves toward better regions of the search space.

### 14.3.1 Encoding

In the partitioning problem, a solution is a set of cuts in some dimensions. It means that some dimensions can be cut many times while some are not at all. Therefore, strings are divided into blocs, each of them representing a cut in a dimension. The number of blocs in the strings is not limited so that the complexity of the partition can be dynamically evolved. Two strings with different lengths are shown in Figure 14.3.

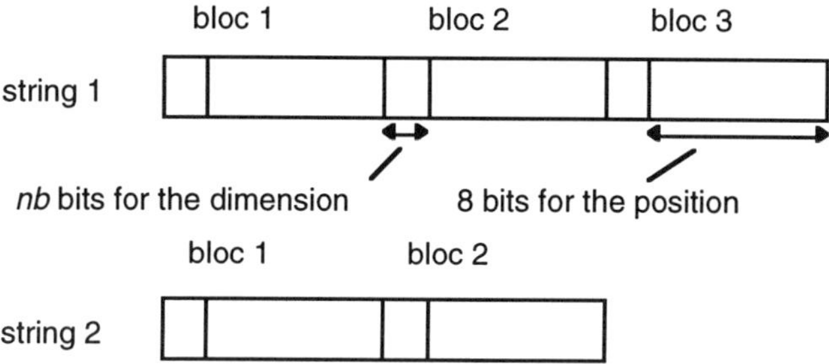


Figure 14.3: Strings and blocs.

In this figure, the *nb* first bits of a bloc encode the dimension that cuts and the 8 following bits encode the position of the cut in the dimension. The position of a bloc in a string is not important.

### 14.3.2 Genetic Operators

In addition to the widely used genetic operators mutation, crossover and deletion, authors also introduce "delete from one and insert in another" or theft. *mutation* — each bit in a string has a probability to be flipped crossover each bloc of a string has a probability to undergo a crossover. If so, a bloc of the same dimension has to be found in the second string chosen for reproduction, and a substring is exchanged. *deletion* — each bloc has a probability to be deleted. *insertion* — probability to insert a new bloc created at random. *theft* — probability for string 1 to steal a bloc at random from string 2 if both strings belong to a pair chosen for reproduction.

### 14.3.3 Fitness Evaluation

Defining the fitness function is the most important part of the method. Neither many cuts nor many rules are desirable. Both are interrelated but not the same. The number of subspaces must be as small as possible. For a given number of cuts, less subspaces will be generated if few dimensions are used. The upper limit for the number of subspaces *(ns)* is $2^{nc}$, with *nc* the total number of cuts. Therefore, following terms are to be integrated in the fitness function:

$$\frac{1}{1+e^{(ns-ns_{th})}}$$

and

$$\frac{1}{1+e^{(nc-nc_{th})}}$$

The fitness falls when the number of subspaces or the number of cuts is above its thresholds $ns_{th}$ and $nc_{th}$ respectively.

Assuming clustered data with DVQ3 [HGG] and considering *gp* as the partition percentage, the percentage of points that are correctly separated to the hyper cuboids of appropriate classes:

$$gp = 100.\sum_{i=1}^{s}\sum_{j=1}^{v}\frac{\max_x\left(p\left(N_j,\vec{I}_x\right)\cdot V^s_{N_j,x} / V^t_{N_j}\right)}{\sum_{x=1}^{l} p\left(N_j,\vec{I}_x\right)\cdot V^s_{N_j,x} / V^t_{N_j}} \qquad (1)$$

$p(N_j, \vec{I}_x)$ is the density of probability that neuron $N_j$ belongs to the class $x$ of I$\vec{I}_x$, $s$ is the number of subspaces, $v$ is the number of neurons (clusters), $V^s_{N_j,x}$ is the volume of neuron $j$ belonging to class $x$, contained in subspace $s$ and $V^s_{N_j}$ is the total volume of neuron $N_j$.

Considering the fact that probability density function (PDF) supplied by DVQ3 can be used to get the conditional probability $p(N_j|\vec{I}_x)$: given a data vector $\vec{I}_x$ of class $x$, it will activate neuron $N_j$:

$$p\left(N_j,\vec{I}_x\right) = p\left(N_j\middle|\vec{I}_x\right)\cdot p\left(\vec{I}_x\right) \qquad (2)$$

where $p(\vec{I}_x)$ is the density of probability that the input vector $\vec{I}_x$ is of class $x$. If all classes have the same probability, $p(\vec{I}_x) = 1/l$, where l is the total number of different classes.

The class that has the maximum of probability in one subspace determines its class. This maximum is divided by the total probability of this subspace (that is, the probability that a learning pattern happens to be found in this subspace, whatever its class) to calculate the ratio.

This ratio represents the "clarity of classification" for subspaces or the importance of subspaces for the corresponding classes. The goal is of course to get a high clarity of classification in all subspaces to prevent errors.

Since this procedure has to be made for all subspaces, it is the major time consuming part of the algorithm. The processing of every subspace is difficult due to the fact that the partition can be anything since none of its parameters are pre-determined. Therefore, a recursive procedure with pointers is used in simulation software. $p(N_j, \vec{I}_x)$ can be considered as a weight. Suppose 2 classes with the same probability, one of them occupying a much smaller volume than the other, which happens quite often when many dimensions are used.

One may wish to give their true probabilities to the different classes, with the risk that some classes could be neglected and considered as not important enough if their probability is too low compared to the cost of making new segmentations. On the other hand, one can artificially increase the importance of one class, even if its probability is rather low, when, for instance, a particular class (e.g., meltdown in a nuclear plant) is more dangerous than the opposite. This method was implemented to solve a difficult case in section • There are many possibilities to define the fitness function which makes the method very flexible.

If input data are used instead of clusters generated by DVQ3, equation 1 is reduced to:

$$gp = 100.\sum_{i=1}^{s} \frac{\max_x \left( p\left(\vec{I}_x\right) \cdot V_x^s / V^t \right)}{\sum_{x=1}^{l} p\left(\vec{I}_x\right) \cdot V_x^s / V^t} \qquad (3)$$

where $V_x^s$ is volume of the part belonging to class $x$ in subspace $s$, and $Vt$ is the total volume. One more term was still added to fight back the strength of the two previous exponentials, setting another threshold for partition:

$$\frac{1}{1 + e^{(gp_{th} - gp)/10}}$$

with $gp_{th}$ a desired percentage of good partitioning. Note that $gp_{th}$ can be set to values higher than 100%, even if $gp$ will never get bigger than that. This can be done to move the equilibrium state to a higher number of partitioning without changing the goals regarding the number of cuts. It does not mean that a better quality can be achieved with the same amount of cuts since the number of segmentations increases, whenever the clarity of classification increases. It will just move the equilibrium toward more cuts while keeping a sharp cut in the fitness when reaching $nc_{th}$. If the desired clarity of classification cannot be achieved in this manner, $ncth$ is also to be increased. Of course, if a high percentage of neurons are overlapping, this percentage will never be taken back by more segmentation.

The complete fitness function is:

$$\frac{gp}{\left(1 + e^{(gp_{th} - gp)/10}\right)\left(1 + e^{(ns - ns_{th})}\right)\left(1 + e^{(nc - nc_{th})}\right)} \qquad (4)$$

## 14.4 Results

### 14.4.1 Heuristic Method

Since the heuristic method is much faster, it is more interesting to use it for a large number of data, i.e., input/output learning vectors (even if its performance is at least as good when preprocessed hyper spheres or hyper cuboids are used). Two benchmarks are presented. The first one is an artificially created two-dimensional problem, where two classes made of 300 vectors with two input are separated by a sinusoidal border. The second one is the well-known *Iris* data set [And35], containing 75 vectors in each training and test (recall) file, with 4 input divided into 3 classes. The result for the first benchmark is shown on Figure 14.4. With 5 cuts in dimension x and 3 cuts in dimension *y*, the partition percentage reaches 96%, which is quite good since the sinus has to be approximated by rectangles.

For the second benchmark, a 99% of partition was achieved for the normalized data set:

| | | | |
|---|---|---|---|
| Dimension 1 | 0.33 | | |
| Dimension 3 | 0.167 | 0.33 | 0.667 |
| Dimension 4 | 0.33 | 0.667 | |

For this problem, dimension 2 has been left out. Actually, dimension 1 and maybe dimension 4 could be removed from the partition and the separation of the different classes would still be satisfactory. It shows that the algorithm finds the relevant dimensions without removing from them the dimensions that are not strictly necessary.

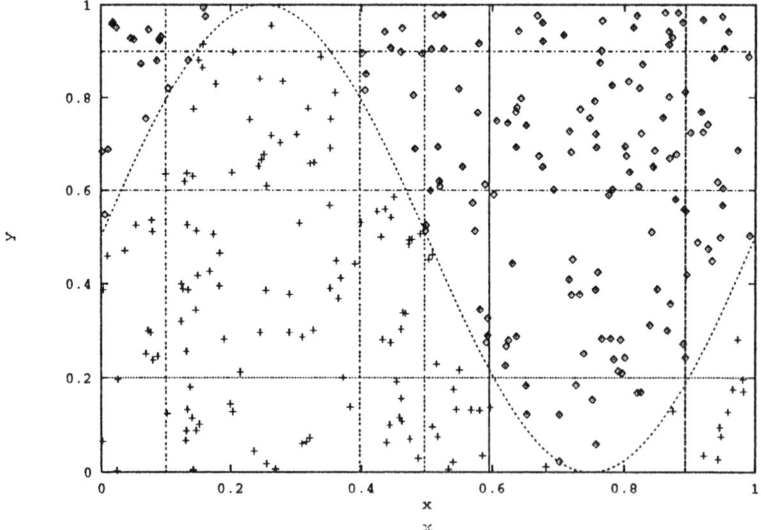


x

Figure 14.4: Sinusoidal boundary with heuristic method.

### 14.4.2 Genetic Algorithm Based Solutions

Since this algorithm is much slower — its order is exponential with the number of cuts — it can be interesting to use some data compression before the partitioning. Nevertheless, results shown here for comparison have been produced with 3 different types of input: the patterns themselves in all cases, clusters generated by RBFNs (RBF neurons) [PHS+94] for the benchmark *Artificial data* and the clusters generated by DVQ3 (DVQ3 neurons) for all the other problems.

The first benchmark is an artificial two dimensional case with 1097 training vectors where two classes are separated by one straight border at *xdimension1* = 0.4 [PHS+94]. The difference is that class 0 is separated in two disjoint areas by class 1 (see Figure 14.5).

This is a difficult case since the small class 0 area contains only about 2% of the 1097 points. If a cut is made at *xdimension1* = 0.4, a 98% of classification is already achieved with only one cut in one dimension. The heuristic method described will not recognize the smaller portion due to its approximation capability.

With usual parameters, the genetic algorithm will find the same approximation with one obvious cut. In a case where the class 0 can be of extreme importance, the genetic algorithm based solution allows the increase of importance in calculating the objective function as described in (section). So the probability of class 0 can be artificially increased, considering that the cost of not recognizing class 0 was higher than the cost of not recognizing class 1. With this safety measure two more cuts are made.

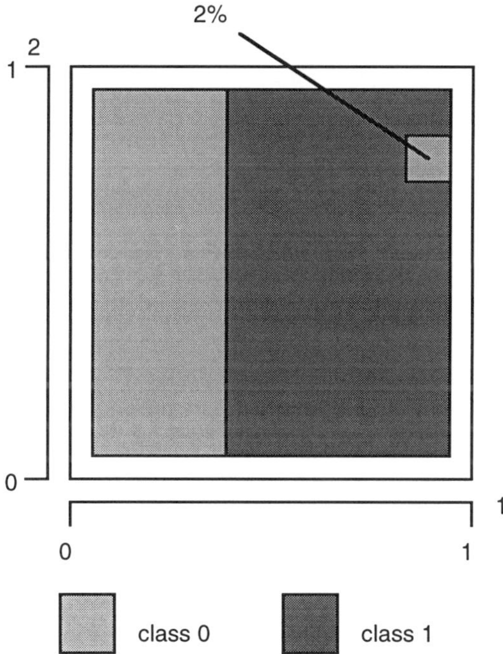


Figure 14.5: Benchmark *Artificial data.*

Figure 14.6 shows the different generations before and after making the correct 4 cuts partition for unclustered data.

| Dimension 1 | 0.402 0.816 |
|---|---|
| Dimension 2 | 0.707 0.872 |

Even if the number of data vectors is fairly large, 60 generations are produced in 30 minutes on a Sparc 10 station and 99% correct classification for both learning and recall sets was reached. The important parameters are: population = 21, $gp_{th}$ = 140%, initial number of cuts = 7, limit for the number of cuts = 4 per dimension, probability of class 0 is twice higher than class 1's.

A high percentage could be already reached in the initial population. It is firstly due to the special strategy followed: the population starts with very "fat strings" (strings with many blocks) that are going to slim and lose their superfluous blocs. Secondly, this problem can easily be solved with one cut and the initialized population contains 147 cuts.

Making more generations would have finally made a 100% of classification, since it is possible to separate both classes totally and because the 4 cuts are already close to the optimum.

The next data for this problem were RBF nearest prototype neurons generated from the training data set [PHS+ 94]. With a population of 25 and a limit number of cuts by dimension set to 4, 99.5% for both learning and recall sets was made in 60 generations (30 seconds on a Sparc 10 station) (see Figure 14.7)

The first real world application is the *Solder* data file described in [HPG93] containing 184 data to be classified either as good or bad solder joints. There are 2 classes, and 23 dimensions (or features) extracted by a laser scanner. Clustered data with DVQ3 neurons are used first. The parameters are usual ones in the sense that it was not intended to find after many trials what values they should take in order to produce the best results.

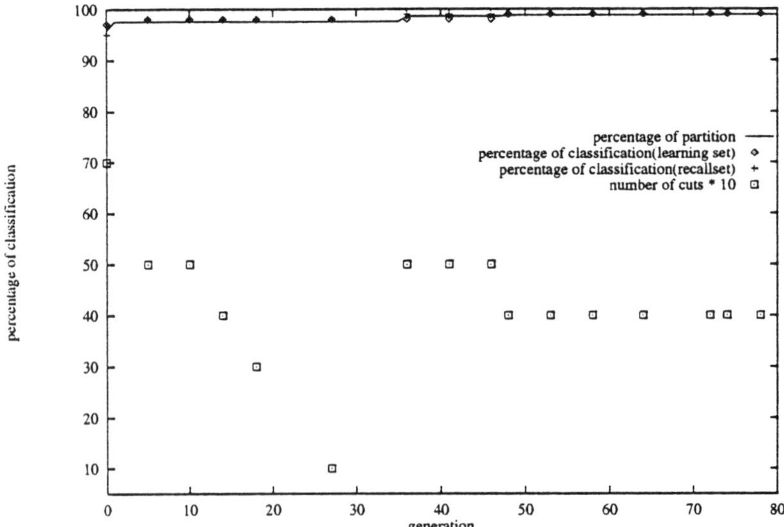


Figure 14.6: Performance using unclustered *Artificial data.*

The threshold $gp_{th}$ = 120% is set with a maximum of around 6 cuts (0.26 cuts/dimension * 23 dimensions). The population was set to 15. The number of cuts is only 3 with the highest percentage of classification for the recall set too (96%) as shown in Figure 14.8. Ideally the program should have converged to this result instead of the (97%, 95%, 4 cuts) reached at the 151st generation. This is due to the fact that partition and classification don't exactly match. The fitness will improve if the number of cuts is increased from 3 to 4 in order to gain few percents in partition. This could have been probably avoided if the allowed number of cuts had been lower.

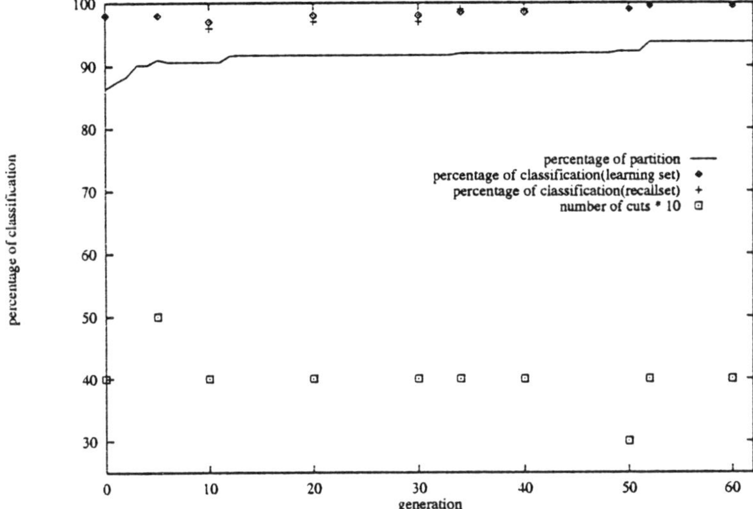


Figure 14.7: Performance using *Artificial data* clustered with RBFs.

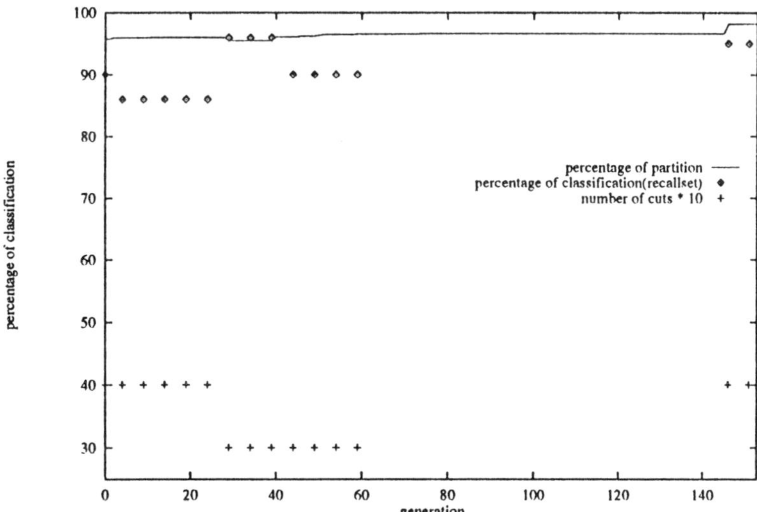


Figure 14.8: Performance with *Solder* data clustered by DVQ3.

If the patterns are used instead of neurons with the same parameters and a smaller (13) population, one may expect slightly better results since there is no loss of information due to the data compression and the partition almost reflects the real distribution of the patterns. It is to be seen in Figure 14.9 that classification results follow the partitions curve. The small difference is due to the small "noisy hyper volumes" that have been given around each data for generalization and calculation reasons. As a consequence the algorithm converges to the desired solution (98%, 98%, 3 cuts).

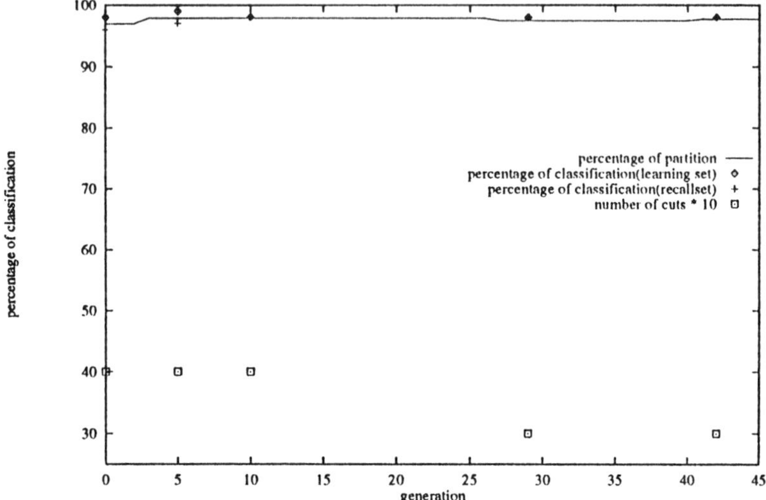


Figure 14.9: Performance with unclustered *Solder* data.

The second real world type application is more difficult: 10 different handwritten characters are to be distinguished in a 36-dimensional pattern space [HG94]. An initial unsuccessful effort is shown in Figure 14.10. The input data are DVQ neurons. All their radii are scaled by 1.3 to make a bit more certain that the patterns are contained by their hyper volumes. The different parameters were set in the normal range.

A good guess for the total number of cuts would be around 9 because there are 10 classes to separate. This parameter was set to 0.24 cuts/dimension*36 dimensions = 8.64 cuts. Since a high percentage is desired, $gp_{th}$ = 120% is set as a goal.

The percentage of partitioning seems to settle to 90% and the number of cuts to 7. It seems to be harder to get higher than 90%, most probably because of overlappings. Those overlappings can be great either because of the use of many unnecessary dimensions or the inadequate scaling of neurons. The data themselves can be mixed too but more dimensions may result in less overlapping. Of course these reasons can all be there at the same time.

If few dimensions have to be used by setting the number of cuts allowed by dimension to 0.18 (for 36 dimensions, the fall in the fitness function is at 6.5

cuts) it takes about 10 minutes to obtain 110 generations. The radii have been multiplied by 1.25 and the population size is 29. An offset between classification results and partitioning cannot be avoided due to the form of the neurons. The fact that the generalizing ability is very good for test set (99%) could show that neurons are adequately scaled.

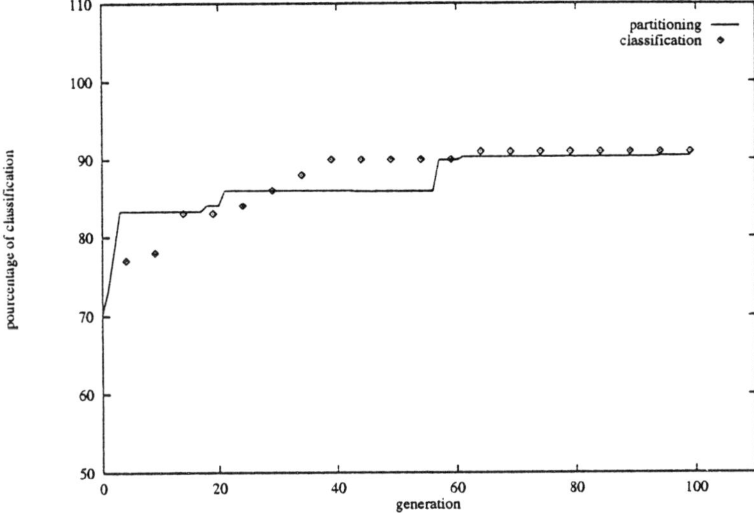


Figure 14.10: Unsuccessful trial with 9 cuts as a limit.

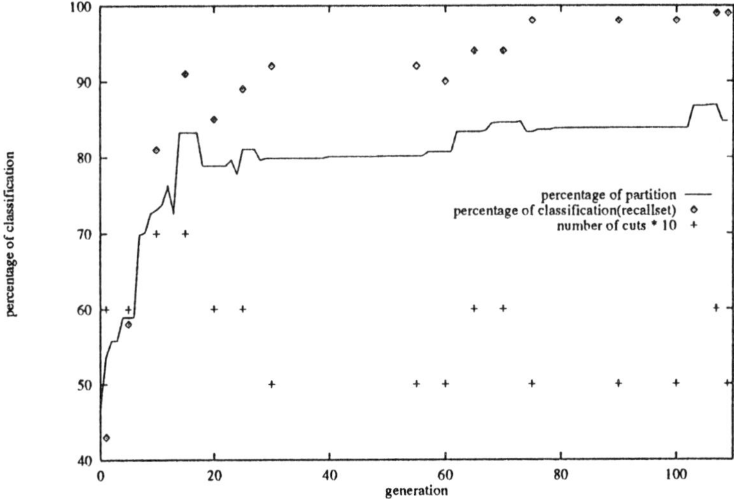


Figure 14.11: Performance with *Digit* data clustered by DVQ3.

The final solution needs only 5 dimensions and 5 cuts to achieve 99% and 94% of classification for training and recall sets, respectively. It must be said that if the 1005 learning vectors are used instead of the few DVQ3 neurons, it took 9 hours on the same machine to achieve the same result.

If the data set with 1005 vectors are used, the program needs 9 hours on a Sparc 10 station to make 60 generations, with a population of 15. With the same parameters, the results shown in Figure 14.12 are obtained.

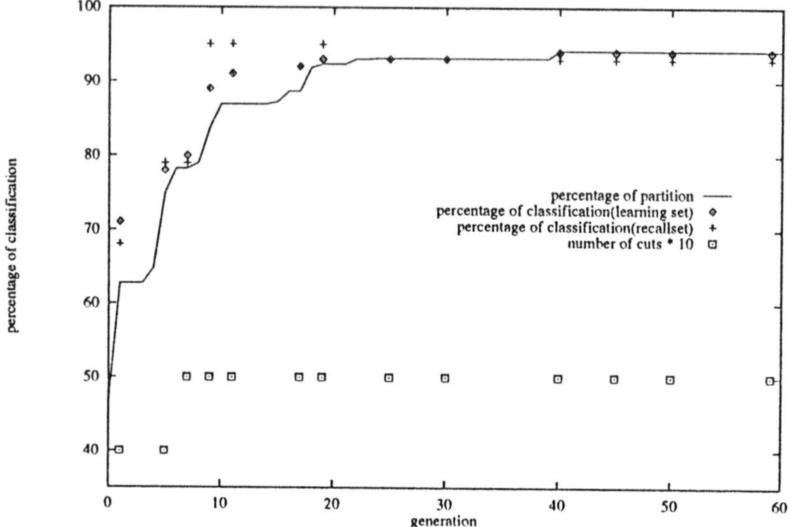


Figure 14.12: Performance with unclustered *Digit* data.

## Discussion

In this paper, the importance of the partition of the pattern space has been stressed because it leads to efficient and compact classifiers at a very low cost if the number of cuts and of dimensions can be somehow reduced.

At this stage, the genetic algorithm, which is much slower than the heuristic method, could achieve the best partitions. Because the heuristic method uses projections of the space and has a discrete approach: it suffers from losses of information, lack of precision and is quite sensible to noisy variations of the vectors distribution in the space. However, its speed allows many iterations and some search strategy to get better results. Of course it cannot find the necessary dimensions among all the relevant dimensions and this problem has not been solved yet. Nevertheless, the heuristic method can be applied to a number of problems before moving to more global time consuming genetic algorithm based methods.

## References

[And35] E. Anderson. The Irises of the Gaspe Peninsula. *Bull. Amer. Iris Soc.*, 59:2-5, 1935.

[BS93] Th. Bäck and H.-P. Schwefel. An overview of evolutionary algorithms for pargreeter optimization. *Evolutionary Computation,* 1(1):1-23, 1993.

[Gol89] D.E. Goldberg. *Genetic Algorithms in Search, Optimization, and Machine Learning.* Addison-Wesley, 1989.

[HG94] S.K. Halgamuge and M. Glesner. Neural Networks in Designing Fuzzy Systems for Real World Applications. *International Journal for Fuzzy Sets and Systems* (in press) (Editor: H.-J. Zimmermann), 1994.

[HGG] S.K. Halgamuge, C. Grimm, and M. Glesner. Functional Equivalence Between Fuzzy Classifiers and Dynamic Vector Quantisation Neural Networks. In *ACM Symposium on Applied Computing (SAC'95)* (Submitted), Nashville, USA.

[Hol75] J.H. Holland. *Adaptation in Natural and Artifiical Systems.* The University of Michigan Press, 1975.

[HPG93] S. K. Halgamuge, W. Poechmueller, and M. Glesner. A Rule based Prototype System for Automatic Classification in Industrial Quality Control. In *IEEE International Conference on Neural Networks' 93,* pages 238 243, San Francisco, U.S.A., March 1993. IEEE Service Center; Piscataway. ISBN 0-7803-0999-5.

[PF91] F. Poirier and A. Ferrieux. DVQ: Dynamic Vector Quantization — An Incremental LVQ. In *International Conference on Artificial Neural Networks'91,* pages 1333-1336. North Holland, 1991.

[PHS+ 94] W. Poechmueller, S.K. Halgamuge, P. Schweikeft, A. Pfeffermann, and M. Glesner. RBF and CBF Neural Network Learning Procedures. In *IEEE International Conference on Neural Networks' 94,* Orlando, U.S.A., June 1994.

# Appendix 1

## An Indexed Bibliography of Genetic Algorithms

(Books, Proceedings, Journal Articles, and Ph.D. Thesis)

**Jarmo T. Alander**
Department of Information Technology and Industrial Management
University of Vaasa
P.O. Box 700
FIN-65101 Vaasa
Finland

jal@uwasa.fi

**Trademarks**
Product and company names listed are trademarks or trade names of their respective companies.

**Warning**
While this bibliography has been compiled with the utmost care, the editor takes no responsibility for any errors, missing information, the contents or quality of the references, nor for the usefulness and/or the consequences of their application. The fact that a reference is included in this publication does not imply a recommendation. The use of any of the methods in the references is entirely at the user's own responsibility. Especially the above warning applies to those references that are marked by trailing y (or *), which are the ones that the editor has unfortunately not had the opportunity to read.

Contents



0-8493-2529-3/95/$0.00 + $.50

## 1. Preface

The material of this bibliography has been extracted by taking books, proceedings, journal articles, Ph.D. theses, and patents from the indexed genetic algorithm bibliography [10], which currently contains over 3000 items and which has been collected from several sources of genetic algorithm literature including Usenet newsgroup comp.ai.genetic and the bibliographies [308, 740, 47, 48]. The following index periodicals have been used systematically

- ACM: *ACM Guide to Computing Literature*: 1979 - 1993/4
- CCA: *Computer & Control Abstracts*: Jan. 1992 - Sep. 1994
- CTI: *Current Technology Index* Jan./Feb. 1993 - Jan./Feb. 1994
- DAI: *Dissertation Abstracts International*: Vol. 53 No. 1 - Vol. 54 No. 12 (1994)
- EEA: *Electrical & Electronics Abstracts*: Jan. 1991 - Aug. 1994
- P: *Index to Scientific & Technical Proceedings*: Jan. 1986 - Sept. 1994
- EI A: *The Engineering Index Annual*: 1987 - 1992
- EI M: *The Engineering Index Monthly*: Jan. 1993 - Sept. 1993

The following GA researchers, cited in this bibliography, have already kindly supplied their complete autobibliographies and/or proofread references to their papers: Patrick Argos, James E. Baker, Wolfgang Banzhaf, I. L. Bukatova, ThomasacBk, Yuval Davidor, Marco Dorigo, Bogdan Filipic, Terence C. Fogarty, David B. Fogel, Toshio Fukuda, Hugo de Garis, Robert C. Glen, David E. Goldberg, Martina Gorges-Schleuter, Jeffrey Horn, Aristides T. Hatjimihail, Richard S. Judson, Akihiko Konagaya, John R. Koza, Kristinn Kristinsson, Carlos B. Lucasius, J. J. Merelo, Zbigniew Michalewicz, Melanie Mitchell, Volker Nissen, Nicholas J. Radcliffe, Colin R. Reeves, Hans-Paul Schwefel, Michael T. Semertzidis, William M. Spears, El-Ghazali Talbi, Peter M. Todd, Hans-Michael Voigt, Roger L. Wainwright, Steward W. Wilson, Xin Yao, and Xiaodong Yin.

This bibliography is updated on a regular basis and certainly contains many errors and inconsistences. The editor of this bibliography would be glad to hear from any reader who notices any errors, missing information, articles, etc. In the future a more complete version of this bibliography will be prepared for the genetic algorithms research community and others who are interested in this rapidly growing area of genetic algorithms.

### 1.1 Acknowledgement

The author wants to acknowledge all who have kindly supplied references, papers and other information on genetic algorithms literature. He also wants to acknowledge Elizabeth Heap-Talvela for her kind proofreading of the manuscript of this bibliography and Petri Kutvonen, JussiakMi, and Antti Nurminen for their kind help to translate this text from aLTEX to format readable by MS Word.

## 2 Statistical summaries

This chapter gives some general statistical summaries of genetic algorithms literature. More detailed indexes can be found in the next section.

### 2.1. Publication type

This bibliography contains published contributions including reports and patents. All unpublished manuscripts have been omitted unless accepted for publication. In addition theses, Ph.D., MSc etc., are also included whether or not published somewhere. Table 2.1 gives the distribution of publication type of the whole bibliography.

| *type* | *number of items* |
|---|---|
| book | 43 |
| journal article | 742 |
| proceedings | 40 |
| Ph.D. thesis | 118 |
| others | 14 |
| *total* | 957 |

Table 2.1: Distribution of publication type.

### 2.2 Annual distribution

Table 2.2 gives the number of genetic algorithms papers published annually. The annual distribution is also shown in Figure 2.1. The average annual growth of GA papers has been approximately 40% during almost the last twenty years.

### 2.3 Classification

Every bibliography item has been given at least one describing keyword or classification by the editor of this bibliography. Keywords occurring most are shown in Table 2.3.

| *year* | *items* | *year* | *items* | *year* | *items* |
|---|---|---|---|---|---|
| 1957 | 3 | 1958 | 0 | 1959 | 0 |
| 1960 | 0 | 1961 | 0 | 1962 | 3 |
| 1963 | 2 | 1964 | 0 | 1965 | 0 |
| 1966 | 3 | 1967 | 4 | 1968 | 0 |
| 1969 | 0 | 1970 | 6 | 1971 | 4 |
| 1972 | 3 | 1973 | 8 | 1974 | 3 |
| 1975 | 4 | 1976 | 3 | 1977 | 6 |
| 1978 | 3 | 1979 | 6 | 1980 | 8 |
| 1981 | 15 | 1982 | 8 | 1983 | 5 |
| 1984 | 13 | 1985 | 16 | 1986 | 21 |
| 1987 | 25 | 1988 | 27 | 1989 | 39 |
| 1990 | 77 | 1991 | 101 | 1992 | 199 |
| 1993 | 294 | 1994 | 48 | | |
| *total* | 957 | | | | |

Table 2.2: Annual distribution of contributions.

| | |
|---|---|
| Evolution strategies | 90 |
| Neural networks | 61 |
| Optimization | 57 |
| Review | 37 |
| CAD | 32 |
| Scheduling | 25 |
| Machine learning | 22 |
| TSP | 22 |
| Parallel GA | 21 |
| Chemistry | 21 |
| Evolution | 19 |
| Genetic programming | 18 |
| Classifier systems | 17 |
| Engineering | 16 |
| Protein folding | 15 |
| Artificial life | 15 |
| Patent | 14 |
| Layout design | 14 |
| Engineering/mechanical | 14 |
| Control | 14 |
| Simulation | 13 |
| Image processing | 13 |
| Comparison/simulated annealing | 13 |
| Analyzing GA | 12 |
| Learning | 11 |
| Engineering/construction | 11 |
| Implementation/C | 10 |
| Others | 1621 |

Table 2.3: The most popular subjects.

| | |
|---|---|
| Total number of authors | 1094 |
| Goldberg, David E. | 29 |
| Fogel, David B. | 14 |
| Holland, John H. | 13 |
| Kateman, Gerrit | 13 |
| Koza, John R. | 13 |
| Lucasius, Carlos B. | 12 |
| Anon. | 9 |
| Liepins, Gunar E. | 9 |
| Whitley, Darrell | 9 |
| Forrest, Stephanie | 8 |
| De Jong, Kenneth A. | 8 |
| Michalewicz, Zbigniew | 8 |
| Vose, Michael D. | 8 |
| Bukatova, Innesa L. | 7 |
| Deb, Kalyanmoy | 7 |
| Grefenstette, John J. | 7 |
| Rice, James P. | 7 |
| Schwefel, Hans-Paul | 7 |
| Banzhaf, Wolfgang | 6 |
| Davis, Lawrence | 6 |
| Dorigo, Marco | 6 |
| Ebeling, Werner | 6 |
| Judson, Richard S. | 6 |
| Buydens, L. M. C. | 5 |
| Karr, Charles L. | 5 |
| Kitano, Hiroaki | 5 |
| Klimasauskas, Casimir C. | 5 |
| Muhlenbein, Heinz | 5 |
| Preis, K. | 5 |
| 18 authors | 4 |
| 51 authors | 3 |
| 129 authors | 2 |
| 866 authors | 1 |

Table 2.4: The most productive genetic algorithms authors.

## 2.4 Conclusions and future

The author believes that this bibliography contains references to most genetic algorithms contributions up to and including the year 1994 and he hopes that this bibliography could give some help to those who are working or planning to work in this rapidly growing area of genetic algorithms.

## 3. Indexes

### 3.1 Books

The following list contains all items classified as books.

Genetic Algorithms, [120]

Genetic Algorithms + Data Structures = Evolution Programs, [590]

Genetic Algorithms and Robotics: A Heuristic Strategy for Optimization, [163]

Genetic Algorithms at Stanford, [497]

Genetic Algorithms in Search, Optimization, and Machine Learning, [290]

Genetic Algorithm, [474]

Genetic Programming: On Programming Computers by Means of Natural Selection and Genetics, [494]

Handbook of Genetic Algorithms, [168]

Induction: Processes of Inference, Learning, and Discovery, [381]

Introduction to Genetic Algorithms, [29]

Modern Heuristic Techniques for Combinatorial Problems, [707]

Neural Networks and Genetic Algorithms — Business Applications and Case Studies, [276]

Nonlinear Process Control: Applications of Genetic Model Control, [529]

Numerical Optimization of Computer Models, [758]

Numerische Optimierung von Computer-Modellen mittels der Evolutionsstrategie, [757]

Parallel Genetic Algorithms, [794]

Parallelism and Programming in Classifier Systems, [246]

Parallel Processing in Neural Systems and Computers, [206]

System Indentification Through Simulated Evolution: A Machine Learning Approach to Modeling, [226]

The Ecology of Computation, [393]

The Evolution of Cooperation, [40]

Theory of self-reproducing automata, [849]

total 42 books

## 3.2 Theses

The following list contains Ph.D. theses arranged in alphabetical order by the name of the school.

Academy of Sciences, [843]

Carnegie-Mellon University, [7]

Colorado State University, [71, 578, 792]

Georgia Institute of Technology, [132]

Gesamthochschule Wupperthal, [357]

HAB Weimar, [84]

Humboldt-Universitat zu Berlin, [98]

Imperial College for Science, [162]

Indiana University, [553]

Louisiana State University of Agricultural and Mechanical College, [90]

Michigan State University, [739]

Mississippi State University, [106]

New Mexico State University, [22, 401]

New York University, [871]

North Dakota State University of Agriculture and Applied Sciences, [909, 444, 815]

Oregon Graduate Institute of Science and Technology, [731]

Politechnico di Milano, [193]

Polytechnic University, [139]

Purdue University, [423]

Rensselaer Polytechnic Institute, [42, 677, 698]

Ruhruniversitat Bochum, [366]

Stanford University, [726, 818]

University of Minnesota, [209]

University of Missouri-Rolla, [207, 210, 696]

University of North Carolina at Chapel Hill, [416]

University of North Carolina at Charlotte, [767]

University of Paris, [765]

University of Pittsburgh, [727, 785]

University of Pretoria, [191]

University of Reading, [53]

University of Stirling, [346]

University of Tennessee, [336]

University of Washington, [135, 905]

Universitat-Gesamthochschule Essen, [523]

Vanderbilt University, [51, 661, 741]

Virginia Polytechnic Institute and State University, [155]

l'Institut National Polytechnique de Grenoble, [806]

total 118 thesis in 62 schools

## 3.3 Patents

The following list contains the names of the patents of genetic algorithms. The list is arranged in alphabetical order by the name of the patent.

A non-linear genetic algorithm for solving problems, [493, 495]

A non-linear genetic process for data encoding and for solving problems using automatically defined functions, [501]

A non-linear genetic process for problem solving using spontaneously emergent self-replicating and self-improving entities, [503]

Adaptive computing system capable of learning and discovery, [379]

Machine learning procedures for generating image domain feature detectors, [280]

Method and apparatus for training a neural network using evolutionary programming, [237]

Method of controlling a classifier system, [380]

Non-linear genetic algorithms for solving problems, [491]

Non-linear genetic algorithms for solving problems by finding a fit composition of functions, [492]

Non-linear genetic process for data encoding and for solving problems using automatically defined functions, [502]

Non-linear genetic process for problem solving using spontaneously emergent self-replicating and self-improving entities, [504]

Non-linear genetic process for use with co-evolving populations, [499]

Non-linear genetic process for use with plural co-evolving populations, [500]

total 13 patents

### 3.4 Subject index

All subject keywords of the papers given by the editor of this bibliography are shown next. The keywords neural networks", \optimization", and \evolution strategies" have been omitted in this list because of their high occurrence rate.

### 3.5 Annual index: 1957-1990

The following table gives references to the contributions published during the period 1957-1990.

| | |
|---|---|
| 1957 | [256, 61, 107] |
| 1962 | [255, 62, 371] |
| 1963 | [109, 200] |
| 1966 | [239, 849, 721] |
| 1967 | [50, 110, 705, 728] |
| 1970 | [111, 136, 632, 729, 730, 870] |
| 1971 | [382, 633, 634, 700] |
| 1972 | [254, 311, 670] |
| 1973 | [598, 128, 372, 575, 648, 649, 701, 702] |
| 1974 | [348, 657, 875] |
| 1975 | [432, 373, 418, 756] |
| 1976 | [370, 725, 754] |
| 1977 | [15, 119, 349, 446, 483, 757] |
| 1978 | [98, 415, 366] |
| 1979 | [3, 121, 197, 331, 447, 332] |
| 1980 | [55, 83, 433, 847, 374, 655, 703, 785] |
| 1981 | [39, 91, 113, 122, 250, 345, 445, 523, 673, 684, 685] |
| 1982 | [94, 270, 618, 667, 637, 749, 786, 891] |
| 1983 | [283, 267, 359, 669, 668] |
| 1984 | [40, 79, 123, 152, 613, 704, 583, 654, 713, 741, 759] |
| 1985 | [99, 108, 718, 189, 243, 251, 323, 280, 279, 333, 379] |
| 1986 | [147, 719, 203, 204, 217, 216, 219, 264, 281, 324, 8, 708, 830, 635, 876, 897, 922] |
| 1987 | [4, 6, 7, 78, 103, 104, 335, 124, 148, 166, 198, 325, 468, 608, 662, 859, 893, 843, 871, 261] |
| 1988 | [41, 73, 84, 95, 101, 102, 105, 434, 222, 224, 76, 312, 540, 609, 715, 723, 894, 739, 743, 856, 872, 878] |
| 1989 | [519, 600, 51, 56, 89, 96, 127, 350, 162, 190, 846, 289, 290, 306, 176, 380, 375, 394, 747, 451, 462, 463, 541, 588, 659, 845, 844, 854, 901] |
| 1990 | [17, 21, 20, 38, 37, 57, 74, 85, 87, 88, 890, 163, 167, 435, 238, 244, 275, 300, 302, 291, 293, 314, 316, 329, 823, 178, 612, 344, 444, 471, 485, 491, 492, 499, 500, 257, 848, 543, 555, 679, 570, 589, 672, 677, 683, 429, 430, 688, 692, 911, 722, 571, 760, 769, 825, 544, 680, 175] |

## 3.6 Bibliography

[1] IEE Colloquium on 'Applications of Genetic Algorithms', volume Digest No. 1994/067, London, 15. Mar. 1993. IEE, London.

[2] Proceedings of the Foundations of Genetic Algorithms 3 (FOGA 3), 1994. (to appear).

[3] P. Ablay. Optimieren mit Evolutionsstrategien: Reihenfolgeprobleme, nichtlineare undganzzahlige Optimierung. Ph.D. thesis, University of Heidelberg, 1979.

[4] P. Ablay. Optimieren mit Evolutionsstrategien. Spektrum der Wissenschaft, pages 104-115, July 1987.

[5] A. M. Abunawass. Biologically based machine learning paradigms: An introductory course. SIGCSE Bulletin, 24(1):87-91, Mar. 1992.

[6] D. H. Ackley. A Connectionist Machine for Genetic Hillclimbing. Kluwer Academic Publisher, Boston, 1987.

[7] D. H. Ackley. Stochastic iterated genetic hillclimbing. Ph.D. thesis, Carnegie-Mellon University, 1987.

[8] E. Aiyoshi and N. Mimuro. A meta-optimization problem for global optimization and its solution by the genetic algorithm. Transactions of the Society of Instrument and Control Engineers (Japan), 28(8):999-1006, 1992 (in Japanese).

[9] A. N. Aizawa and B. W. Wah. A sequential sampling procedure for genetic algorithms. Computers & Mathematics with Applications, 27(9/10):77-82, 1993. (Proceedings of the 5th International Workshop of the Bellman Continuum, Waikoloa, HI, Jan. 11-12. 1993).

[10] J. T. Alander. An indexed bibliography of genetic algorithms: Years 1957-1993. Art of CAD Ltd., Vaasa (Finland), 1994. (Over 3000 GA references).

[11] J. T. Alander, editor. Proceedings of the Second Finnish Workshop on Genetic Algorithms and their Applications, Vaasa (Finland), 16-18. Mar. 1994. University of Vaasa, Department of Computer Science and Economics.

[12] M. Alfonseca. Genetic algorithms. APL Quote Quad, 21(4):1-6, Aug. 1991.

[13] B. K. Ambati, J. Ambati, and M. M. Mokhtar. Heuristic combinatorial optimization by simulated Darwinian evolution: a polynomial time algorithm for the traveling salesman problem. Biological Cybernetics, 65(1):31-35, 1991.

[14] B. K. Ambati, J. Ambati, and M. M. Mokhtar. Erratum: Heuristic combinatorial optimization by simulated Darwinian evolution: a polynomial time algorithm for the Traveling Salesman Problem. Biological Cybernetics, 66(3):290, 1992.

[15] U. Anders.Losung getriebesynthetischer Probleme mit der Evolutionsstrategie. Feinwerk technik und Messtechnik, 85(2):53-57, Mar. 1977.

[16] C. A. Anderson, K. F. Jones, and J. Ryan. A two-dimensional genetic algorithm for the Ising problem. Complex Systems, 5(3):327-333, 1992.

[17] E. L. Andrews. Patents: 'breeding' computer programs. The New York Times, 89(32):48,282, 1990.

[18] I. P. Androulakis and V. Venkatasubramanian. A genetic algorithmic framework for process design and optimization. Computers in Chemical Engineering, 15(4):217-228, Apr. 1991.

[19] P. J. Angeline. Evolutionary algorithms and emergent intelligence. Ph.D. thesis, The Ohio State University, 1993.

[20] C. A. Ankenbrandt. The time complexity of genetic algorithms and the theory of recombination operators. Ph.D. thesis, Tulane University, New Orleans, LA, 1990.

[21] C. A. Ankenbrandt, B. P. Buckles, and F. E. Petry. Scene recognition using genetic algorithms with semantic nets. Pattern Recognition Letters, 11(4):285-293, 1990.

[22] P. V. Annaiyappa. A critical analysis of genetic algorithms for global optimization. Ph.D. thesis, New Mexico State University, Las Cruces, 1991.

[23] Anon. Generating software by natural selection. IEEE Spectrum, 27(6):66, 1990.

[24] Anon. Tietokone piirtaa rosvon. Tiede 2000, 11(8):59, 1991.

[25] Anon. Coloring a grid with a genetic algorithm. Advanced Technology for Developers, 1(1), May 1992.

[26] Anon. Navy uses genetic algorithms to control vehicles. IEEE Expert, 7(4):76, 1992.

[27] Anon. EvolverT M 2.0 A genetic algorithm for spreadsheets. Computers & Mathematics with Applications, 26(12):94, 1993.

[28] Anon. How machines live and learn. Personal Computer World, 16(6):483-484, 1993.

[29] Anon. Introduction to Genetic Algorithms. Axcelis Press, Seattle, WA, 1993.

[30] Anon. The joy of genetic programming. Personal Computer World, 16(6):471-472, 1993.

[31] Anon. Special issue on genetic algorithms. Journal of the Society of Instrument and Control Engineers, 32(1), Jan. 1993 (in Japanese).

[32] J. Arifovic. Genetic algorithm learning and the cobweb model. Journal of Economic Dynamics and Control, 18(1):3-28, 1994.

[33] S. Arnone, A. Loraschi, and A. Tettamanzi. A genetic approach to portfolio selection. Neural Network World, 3(6):597-604, 1993.

[34] W. B. Arthur. On designing economic agents that behave like human agents. Evolutionary Economics, 3:1-22, 1993.

[35] S. Arunkumar and T. Chockalingam. Genetic search algorithms and their randomized operators. Computers & Mathematics with Applications, 25(5):91-100, 1993.

[36] I. Ashdown. Non-imaging optics design using genetic algorithms. J. Illum. Eng. Soc., 23(1):12-21, Winter 1994.

[37] S. Austin. Genetic solutions to XOR problems. AI Expert, 5(12):52-57, Dec. 1990.

[38] S. Austin. Metamorph: A genetic algorithmic tool. AI Expert, 5(8):48-55, Aug. 1990.

[39] R. Axelrod. The evolution of cooperation. Science, 211:1390-1396, 1981.

[40] R. Axelrod. The Evolution of Cooperation. Basic Books, New York, 1984.

[41] R. Axelrod and D. Dion. The further evolution of cooperation. Science, 242:1385-1390, 1988.

[42] J. Ayala-Cruz. A multi-objective simulation optimization method using a genetic algorithm with applications in manufacturing. Ph.D. thesis, Rensselaer Polytechnic Institute, 1993.

[43] J. H. Aylor, J. P. Cohoon, E. L. Feldhousen, and B. W. Johnson. Gate — a genetic algorithm for compacting randomly generated test sets. International Journal of Computer Aided VLSI Design, 3(3):259-272, 1991.

[44] G. P. Babu and M. N. Murty. A near-optimal initial seed value selection in K-means algorithm using a genetic algorithm. Pattern Recognition Letters, 14(10):763-769, 1993.

[45] F. Q. Bac and V. L. Perov. New evolutionary genetic algorithms for NP-complete combinatorial optimization problems. Biological Cybernetics, 69(3):229-234, 1993.

[46] T. Back. Evolutionary algorithms. SIGBIO Newsletter, 12(2):26-31, June 1992.

[47] T. Back. Genetic algorithms, evolutionary programming, and evolutionary strategies bibliographic database entries. (personal communication), 1993.

[48] T. Back, F. Hoffmeister, and H.-P. Schwefel. Applications of evolutionary algorithms. Technical Report SYS-2/92, University of Dortmund, Department of Computer Science, 1992.

[49] T. Back and H.-P. Schwefel. An overview of evolutionary algorithms for parameter optimization. Evolutionary Computation, 1(1):1-23, 1993.

[50] J. D. Bagley. The behavior of adaptive systems which employ genetic and correlation algorithms. Ph.D. thesis, University of Michigan, Ann Arbor, 1967 (University Microfilms No. 68-7556).

[51] J. E. Baker. An analysis of the effects of selection in genetic algorithms. Ph.D. thesis, Vanderbilt University, Nashville, 1989.

[52] J. W. Bala and H. Wechsler. Shape analysis using genetic algorithms. Pattern Recognition Letters, 14(12):965-973, Dec. 1993.

[53] N. R. Ball. Cognitive Maps in Learning Classifier Systems. Ph.D. thesis, University of Reading, 1991.

[54] N. R. Ball, P. M. Sargent, and D. O. Ige. Genetic algorithm representations for laminate layups. Artificial Intelligence in Engineering (UK), 8(2):99-108, 1993.

[55] K. Bammert, M. Rautenberg, and W. Wittekindt. Matching of turbocomponents described by the example of impeller and diffuser in a centrifugal compressor. Transactions of the ASME, 102:594-600, 1980.

[56] W. Banzhaf. Population processing — a powerful class of parallel algorithms. BioSystems, 22:163-172, 1989.

[57] W. Banzhaf. The "molecular" traveling salesman. Biological Cybernetics, 64:7-14, 1990.

[58] W. Banzhaf. Self-replicating sequences of binary numbers. Computers and Mathematics with Applications, 26(7):1-8, 1993.

[59] W. Banzhaf. Self-replicating sequences of binary numbers. Foundations I: General. Biological Cybernetics, 69(4):269-274, 1993.

[60] W. Banzhaf. Self-replicating sequences of binary numbers. Foundations II: Strings of length N = 4. Biological Cybernetics, 69:275-281, 1993.

[61] N. A. Barricelli. Symbiogenetic evolution processes realized by artificial methods. Methodos, 9(35-36):143-182, 1957.

[62] N. A. Barricelli. Numerical testing of evolution theories. ACTA Biotheoretica, 16:69-126, 1962.

[63] N. H. Barth. Oceanographic experiments design, 2. genetic algorithms. Journal of Atmospheric and Oceanic Technology, 9(4):434-443, 1992.

[64] R. C. Bassus, E. Falck, and W. Gerlach. Application of the evolution strategy to optimize multistep field plates for high voltage planar pn-junctions. Archiv urf Elektrotechnik, 75:345-349, 1992.

[65] R. Battiti and G. Tecchiolli. Parallel biased search for combinatorial optimization: genetic algorithms and TABU. Microprocessors and Microsystems (UK), 16(7):351-367, Sept. 1992.

[66] D. L. Battle and M. D. Vose. Isomorphisms of genetic algorithms. Artificial Intelligence, 60(1):155-165, 1993.

[67] N. Beard. The joy of genetic programming. Personal Computer World, 16(6):471-472, June 1993.

[68] D. Beasley, D. R. Bull, and R. R. Martin. An overview of genetic algorithms. 1. Fundamentals. University Computing, 15(2):58-69, 1993.

[69] D. Beasley, D. R. Bull, and R. R. Martin. An overview of genetic algorithms. 2. Research topics. University Computing, 15(4):170-181, 1993.

[70] D. Beasley, D. R. Bull, and R. R. Martin. A sequential niche techniques for multimodal function optimization. Evolutionary Computation, 1(2):101-126, 1993.

[71] S. J. Beaty. Instruction scheduling using genetic algorithms. Ph.D. thesis, Colorado State University, Fort Collins, CO, 1991.

[72] S. J. Beaty, D. Whitley, and G. Johnson. Motivation and framework for using genetic algorithms for microcode compaction. SIGMICRO Newsletter, 22(1):20-27, 1991.

[73] R. Becker. Parallel Ansatz zur osLung des Quadratischen Zuordnungsproblems. Ph.D. thesis, University of Bonn, 1988.

[74] R. K. Belew. Evolution, learning, and culture: Computational metaphors for adaptive algorithms. Complex Systems, 4(1):11-49, Feb. 1990.

[75] R. K. Belew. Artificial life, a constructive lower bound for artificial intelligence. IEEE Expert, 6(1):8-15, 1991.

[76] R. K. Belew and S. Forrest. Learning and programming in classifier systems. Machine Learning, 3(2/3):193-224, Oct. 1988.

[77] M. S. T. Benten and S. M. Sait. GAP: a genetic algorithm approach to optimize two-bit decoder PLAs. International Journal of Electronics, 76(1):99-106, Jan. 1994.

[78] J. Benz, J. Polster, R. Bar, and G. Gauglitz. Program system sidys: Simulation and parameter identification of dynamic systems. Comput. Chem., 11(1):41-48, 1987.

[79] W. Berke. Kontinuierliche Regenerierung von ATPurf enzymatische Synthesen. Ph.D. thesis, Technische Universitat der Berlin, Fachbereich Lebensmitteltechnologie und Biotechnologie, 1984.

[80] A. Bertoni and M. Dorigo. Implicit parallelism in genetic algorithms. Artificial Intelligence, 61(2):307-314, June 1993.

[81] A. Bertoni and M. Dorigo. Implicit parallelism in genetic algorithms. Technical Report TR-93-001, International Computer Science Institute, Berkeley, 1993 (also as [80]; available via anonymous ftp at icsi.berkley.edu /pub/techreports/1993/tr-93-001.ps.Z).

[82] E. Betensky. [optical design]. Optical Engineering, 32:1750, 1993.

[83] A. D. Bethke. Genetic algorithms as function optimizers. Ph.D. thesis, University of Michigan, Ann Arbor, 1980. (University Microfilms No. 81-06101).

[84] H.-G. Beyer. Ein Evolutionsverfahren zur mathematischen Modellierung stationarer Zustande in dynamischen Systemen. Ph.D. thesis, HAB Weimar, 1988.

[85] H.-G. Beyer. Simulation of steady states in dissipative systems by Darwin's paradigm of evolution. Journal of Non-Equilibrium Thermodynamics, 15(1):45-58, 1990.

[86] A. K. Bhattacharjya, D. E. Becker, and B. Roysam. A genetic algorithm for intelligent imaging from quantum-limited data. Signal Processing, 28(3):335-348, Oct. 1992.

[87] A. S. Bickel and R. W. Bickel. Determination of near optimum use of hospital diagnostic resources using the genes genetic algorithm shell. Computers in Biology and Medicine, 20(1):1-13, 1990.

[88] J. E. Biegel and J. J. Davern. Genetic algorithms and job shop scheduling. Computers & Industrial Engineering, 19(1-4):81-91, Mar. 1990. (Proceedings of the 12th Annual Conference on Computers and Industrial Engineering, Orlando, FL, 12-14. Mar.)

[89] V. Bieling, B. Rumpf, F. Strepp, and G. Maurer. An evolutionary optimization method for modeling the solubility of ammonia and carbon dioxide in aqueous solutions. Fluid Phase Equilibria, 53:251-259, 1989.

[90] T. A. Bitterman. Genetic algorithms and the satisfiability of large-scale Boolean expressions. Ph.D. thesis, Louisiana State University of Agricultural and Mechanical College, 1993.

[91] W. E. Blanz and E. R. Reinhardt. Image segmentation by pixel classification. Pattern Recognition, 13(4):293-298, 1981.

[92] M. J. J. Blommers, C. B. Lucasius, G. Kateman, and R. Kaptein. Conformational analysis of a dinucleotide photodimer with the aid of the genetic algorithm. Biopolymers, 32(1):45-52, Jan. 1992.

[93] J. J. Bogardi and J. Duckstein. Interactive multiobjective analysis embedding the decision maker's implicit preference function. Water Resources Bulletin, 28(1):75-88, Feb. 1992.

[94] L. B. Booker. Intelligent behavior as an adaptation to the task environment. Ph.D. thesis, University of Michigan, Ann Arbor, 1982.

[95] L. B. Booker. Classifier systems that learn internal world models. Machine Learning, 3(2/3):161-192, Oct. 1988.

[96] L. B. Booker, D. E. Goldberg, and J. H. Holland. Classifier systems and genetic algorithms. Artificial Intelligence, 40(1-3):235-282, Sept. 1989.

[97] G. Boone and H.-D. Chiang. Optimal capacitor placement in distribution systems by genetic algorithm. International Journal of Electrical Power Energy Systems (UK), 15(3):155-162, June 1993.

[98] J. Born. Evolutionsstrategien zur numerischenosLung von Adaptationsaufgaben. (dr. rer. nat.), Humboldt-Universitat zu Berlin, 1978.

[99] J. Born. Adaptively controlled random search —a variance function approach. Systems Analysis — Modelling - Simulation, 2(2):109-112, 1985.

[100] M. Bos and H. T. Weber. Comparison of the training of neural networks for quantitative x-ray flourescence spectrometry by a genetic algorithm and backward error propagation. Analytica Chimica Acta, 247(1):97-105, June 1991.

[101] T. Boseniuk and W. Ebeling. Evolution strategies in complex optimization: The travelling salesman problem. Systems Analysis —Modeling — Simulation, 5(5):413-422, 1988.

[102] T. Boseniuk and W. Ebeling. Optimization of NP-complete problems by Boltzmann-Darwin strategies including life cycles. Europhysics Letters, 6(2):107-112, 15. May 1988.

[103] T. Boseniuk, W. Ebeling, and A. Engel. Boltzmann and Darwin strategies in complex optimization. Physics Letters A, 125(6-7):307-310, 1987.

[104] D. G. Bounds. New optimization methods from physics and biology. Nature, 329:215-219, 17 Sept. 1987.

[105] D. G. Bounds. Optimization methods. Nature, 331:307, 28. Jan. 1988.

[106] R. O. Bowden. Genetic algorithm based machine learning applied to the dynamic routing of discrete parts. Ph.D. thesis, Mississippi State University, 1992.

[107] G. E. P. Box. Evolutionary operation: A method for increasing industrial productivity. Journal of the Royal Statistical Society C, 6(2):81-101, 1957.

[108] R. M. Brady. Optimization strategies gleaned from biological evolution. Nature, 317:804-806, 31 Nov. 1985.

[109] H. J. Bremermann. Limits of genetic control. IEEE Transactions on Military Electronics, MIL-7(2-3):200-205, 1963.

[110] H. J. Bremermann. Quantitative aspects of goal-seeking self-organizing systems. Progress in Theoretical Biology, 1:59-77, 1967.

[111] H. J. Bremermann. A method of unconstrained global optimization. Mathematical Biosciences, 9:1-15, 1970.

[112] F. Z. Brill, D. E. Brown, and W. N. Martin. Fast genetic selection of features for neural network classifiers. IEEE Transactions on Neural Networks, 3(2):324-328, Mar. 1992.

[113] A. Brindle. Genetic algorithms for function optimization. Ph.D. thesis, University of Alberta, Edmonton, Canada, 1981.

[114] T. Brodmeier and E. Pretsch. Application of genetic algorithms in molecular modeling. Journal of Computational Chemistry, 15(6):588-595, June 1994.

[115] D. R. Brown and K.-Y. Hwang. Solving fixed configuration problems with genetic search. Res. Eng. Des. (USA), 5(2):80-87, 1993.

[116] R. D. Brown, G. M. Downs, G. Jones, and P. Willett. Hyperstructure model for chemical structure handling: Techniques for substructure searching. Journal of Chemical Information and Computer Sciences, 34(1):47-53, 1994.

[117] R. D. Brown, G. Jones, P. Willett, and R. C. Glen. Matching two-dimensional chemical graphs using genetic algorithms. J. Chem. Inf. Comput. Sci. (USA), 34(1):63-70, Jan.-Feb. 1994.

[118] R. D. Brown, G. Jones, P. Willett, and R. C. Glen. Matching two-dimensional chemical graphs using genetic algorithms. Journal of Chemical Information and Computer Science, 34(1):63-70, Jan.-Feb. 1994. (Proceedings of 3rd International Conference: Chemical Structures, The International Language of Chemistry, Noordwijkerhout (Netherlands), Jun. 6-10, 1993).

[119] U. Brudermann. Entwicklung und Anpassung eines vollstandigen Ansteuersystems urf fremdenergetisch angetriebene Ganzarmprothesen. Fortschrittsberichte der VDIZeitschriften, 17(6), 1977.

[120] B. P. Buckles and F. E. Petry, editors. Genetic Algorithms. Electronica Books Ltd., Middlesex (UK), 1993.

[121] I. L. Bukatova. Evolutionary Modelling Simulation and Its Applications. Nauka Publishers, Moscow, 1979. (in Russian).

[122] I. L. Bukatova. Evolutionary Simulation: Ideas, Theoretical Fundamentals, Applications. Znanie Publ., Moscow, 1981. (in Russian).

[123] I. L. Bukatova, L. L. Golic, M. I. Elinson, P. I. Perov, and A. M. Sharov. Optoelectronic system of hardware realization of evolutionary predictive algorithm. Soviet Journal of Microelectronics (Mikroelektronika), 13(4):348-355, 1984.

[124] I. L. Bukatova and V. A. Kipyatkov. Theoretical analysis of evolutionary structural search. Academy of Sciences of the USSR, Institute of Radio Engineering and Electronics, Moscow, 461(2), 1987 (in Russian).

[125] I. L. Bukatova, V. A. Kipyatkov, and A. M. Sharov. Simulation-evolutionary technology of multichannel processing of signals. Soviet Journal of Problems of Radio Electronics, ser. Electronic Computer Engineering (Voprosy Radioelektroniki. Seriya Elektronnaya Vychislitel'naya Tekhnika), pages 5-26, 1991 (in Russian).

[126] I. L. Bukatova, Y. I. Mikhasev, and A. M. Sharov. Evoinformatics: Theory and Practice of Evolutionary Simulation. Nauka Publishers, Moscow, 1991 (in Russian).

[127] J. J. Burbaum, R. T. Raines, W. J. Albery, and J. R. Knowles. Evolutionary optimization of the catalytic effectiveness of an enzyme. Biochemistry, 28(24):9293-9305, 1989.

[128] G. H. Burgin. System identification by quasilinearization and by evolutionary programming. Journal of Cybernetics, 3(2):56-75, 1973.

[129] K. J. Callahan and G. E. Weeks. Optimum design of composite laminates using genetic algorithms. Composites Engineering, 2(3):149-160, Apr. 1992.

[130] R. Caponetto, L. Fortuna, S. Graziani, and M. G. Xibilia. Genetic algorithms and applications in system engineering: a survey. Transactions of the Institute of Measurement and Control (UK), 15(3):143-156, 1993.

[131] A. D. Carlo. A genetic algorithm for word hypothesisation. Note Recensioni e Notizie, 39(4):99-103, Oct./Dec. 1990 (in Italian).

[132] S. E. Carlson. Component selection optimization using genetic algorithms. Ph.D. thesis, Georgia Institute of Technology, 1993.

[133] H. M. Cartwright and S. P. Harris. Analysis of the distribution of airborne pollution using genetic algorithms. Atmospheric Environment Part A General Topics, 27A(12):1783-1791, Aug. 1993.

[134] H. M. Cartwright and R. A. Long. Simultaneous optimization of chemical flowshop sequencing and topology using genetic algorithms. Industrial and Engineering Chemistry Research, 32:2706-2713, Nov. 1993.

[135] K. R. Caskey. Genetic algorithms and neural networks applied to manufacturing scheduling. Ph.D. thesis, University of Washington, 1993.

[136] D. J. Cavicchio. Adaptive search using simulated evolution. Ph.D. thesis, University of Michigan, Ann Arbor, 1970. (University Microfilms No. 25-0199).

[137] J. Celko. Genetic algorithms and database indexing. Dr. Dobb's Journal, 18(4):30-32,34, Apr. 1993.

[138] R. Cemes and D. Ait-Boudaoud. Genetic approach to design of multiplierless FIR filters. Electronics Letters, 29(24):2087-2088, Nov. 1993.

[139] M. A. Cesare. Risk-based bridge project selection using genetic algorithm optimization. Ph.D. thesis, Polytechnic University, 1992.

[140] U. K. Chakraborty and D. G. Dastidar. Using reliability analysis to estimate the number of generations to convergence in genetic algorithms. Information Processing Letters, 46(4):199-209, June 1993.

[141] H. Chan, P. Mazumder, and K. Shahookar. Macro-cell and module placement by genetic adaptive search with bitmap-represented chromosome. Integration, the VLSI Journal, 12(1):49-77, Nov. 1991.

[142] R. Chandrasekharam, S. Subhramanian, and S. Chaudhury. Genetic algorithm for node partitioning problem and applications in VLSI design. IEE Proceedings — E Comput. Digit. Tech., 140(5):255-260, Sept. 1993.

[143] J. L. Chen and Y.-C. Tsao. Optimal design of machine elements using genetic algorithms. Chung-Kuo Chi Hsueh Kung Ch'eng Hsueh Pao, 14(2):193-199, Apr. 1993.

[144] T. Chockalingam and S. Arunkumar. A randomized heuristics for the mapping problem: The genetic approach. Parallel Computing, 18(10):1157-1165, 1992.

[145] K. H.-K. Chow and M. L. Liou. Genetic motion search algorithm for video compression. IEEE Transaction on Circuits Syst. Video Technol., 3(6):440-445, Dec. 1993.

[146] J. P. Cohoon, S. U. Hegde, W. N. Martin, and D. S. Richards. Distributed genetic algorithms for the floorplan design problem. IEEE Transactions on Computer-Aided Design of Integrated Circuits and Systems, 10(4):483-492, Apr. 1991.

[147] J. P. Cohoon and W. D. Paris. Genetic placement. IEEE Transaction on Computer Aided Design and Integrated Circuits Systems, 6(6):422-425, 1986. (Proceedings of the IEEE International Conference on Computer-Aided Design, Part II of III).

[148] J. P. Cohoon and W. D. Paris. Genetic placement. IEEE Transactions on Computer-Aided Design, 6(6):956-964, Nov. 1987.

[149] A. Colin. Solving ratio optimization problems with a genetic algorithm. Advanced Technology for Developers, 2:1-8, May 1993.

[150] R. J. Collins. Studies in artificial evolution. Ph.D. thesis, University of California, Los Angeles, 1992.

[151] A. Colorni, M. Dorigo, and V. Maniezzo. Gli algoritmi genetici e il problema dell'orario. Rivista di Ricerca Operativa, (60):5-31, 1992 (in Italian).

[152] M. Conrad, E. Harth, J. Holland, H. Martinez, H. Pattee, R. Rada, D. Waltz, and B. P. Zeigler. Natural and artificial intelligence. Cognition and Brain Theory, 7(1):89-104, 1984.

[153] D. G. Conway and M. A. Venkataramanan. Genetic search and the dynamic facility layout problem. Computers & Operations Research, 21(8):955-960, Oct. 1994.

[154] D. F. Cook and M. L. Wolfe. Genetic algorithm approach to a lumber cutting optimization problem. Cybernetics and Systems, 22(3):357-365, May-June 1991.

[155] M. C. Cowgill. Monte Carlo validation of two genetic clustering algorithms. Ph.D. thesis, Virginia Polytechnic Institute and State University, 1993.

[156] J. Cui, T. C. Fogarty, and J. G. Gammack. Searching databases using parallel genetic algorithms on a transputer computing surface. Future Generation Computer Systems, 9(1):33-40, May 1993.

[157] A. R. D. Curtis. An application of genetic algorithms to active vibration control. Journal of Intelligent Material Systems and Structures, 2(4):472-481, 1991.

[158] D. Cvetkovic. Genetische Algorithmen. KI-Lexikon, pages 60-61, 1993.

[159] C. Dagli and S. Sittisathanchai. Genetic neuro-scheduler for job shop scheduling. Computers & Industrial Engineering, 25(1-4):267-270, 1993.

[160] T. Dandekar and P. Argos. Potential of genetic algorithms in protein folding and protein engineering simulations. Protein Engineering, 5(7):637-645, 1992.

[161] T. Dandekar and P. Argos. Folding the main chain of small proteins with the genetic algorithm. Journal of Molecular Biology, 236(3):844-861, Feb. 1994.

[162] Y. Davidor. Genetic algorithms for order dependent processes applied to robot pathplanning. Ph.D. thesis, Imperial College for Science, Technology, and Medicine, 1989.

[163] Y. Davidor. Genetic Algorithms and Robotics: A heuristic strategy for optimization. World Scientific Publishing, Singapore, 1990.

[164] Y. Davidor. Epistasis variance: Suitability of a representation to genetic algorithms. Complex Systems, 4(4):369-383, Aug. 1992.

[165] C. Davidson. Genetics chips into improved designs. Electronics Weekly, page 14, Mar. 11 1992.

[166] L. Davis, editor. Genetic Algorithms and Simulated Annealing, London, 1987. Pitman Publishing.

[167] L. Davis. Classifier systems with Hamming weights. Machine Learning, 5:162-173, 1990.

[168] L. Davis, editor. Handbook of Genetic Algorithms. Van Nostrand Reinhold, New York, 1991.

[169] L. Davis. Chuck Karr and the design of an air-injected hydrocyclone. Advanced Technology for Developers, 1(3):1-, July 1992.

[170] L. Davis. Genetic algorithm profiles: John Holland and the creation of genetic algorithm. Advanced Technology for Developers, 1(1):1-, May 1992.

[171] L. Davis. Genetic algorithm profiles: Matt Jensen and user-friendly evaluation functions. Advanced Technology for Developers, 1:7-10, Dec. 1992.

[172] L. Davis. Putting faces in their place. Advanced Technology for Developers, 2:14-17, May 1993.

[173] L. Davis. Scheduling the 1992 Paralympic games with a genetic algorithm. Advanced Technology for Developers, 2:8-11, Jan. 1993.

[174] T. E. Davis. Towards an extrapolation of the simulated annealing convergence theory onto the simple genetic algorithm. Ph.D. thesis, University of Florida, Gainesville, 1991.

[175] H. de Garis. Genetic neural nets can be dynamic too, you know! Neural Network Review, Summer 1990.

[176] C. de Groot. Simulated annealing und Evolutionsstrategie: Ein vergleich anhand schweriger Optimierungsprobleme. Ph.D. thesis, University of Heidelberg, 1989.

[177] C. de Groot, D.Wurtz, and K. H. Hoffmann. Low autocorrelation binary sequences: Exact enumeration and optimization by evolutionary strategies. Technical Report No. 89-09, Interdisciplinary Center for Supercomputing Research, Eidgenossische Technische HochschuleurZich, 1989. (also as [179]).

[178] C. de Groot, D.urWtz, and K. H. Hoffmann. Simulated annealing and evolution strategy — a comparison. Helvetica Physica Acta, 63(6):843-844, 1990.

[179] C. de Groot, D.uWrtz, and K. H. Hoffmann. Low autocorrelation binary sequences: Exact enumeration and optimization by evolutionary strategies. Optimization (UK), 23(4):369- 384, 1993.

[180] A. H. de Silva. Operations research in facility planning: Introduction to the special issue. European Journal of Operational Research, 63(2):135-140, 1992.

[181] K. Deb. Binary and floating-point function optimization using messy genetic algorithms. Ph.D. thesis, University of Alabama, 1991 (also available as IlliGAL report No. 91004).

[182] K. Deb. Optimal design of a welded beam structure via genetic algorithms. AIAA Journal, 29(11):2013-2015, Nov. 1991.

[183] K. Deb, J. Horn, and D. E. Goldberg. Multimodal deceptive functions. Complex Systems, 7(2):131-153, Apr. 1993.

[184] G. Deboeck. How to build a hybrid trading system in a spreadsheet in five easy steps. Advanced Technology for Developers, 2:1-19, Apr. 1993.

[185] T. Deboeck and G. Deboeck. GenNet: Genetic optimization of neural nets for trading. Advanced Technology for Developers, 1(6):1, Oct. 1992.

[186] V. O. Dehaan and G. G. Drijkoningen. Genetic algorithms used in model finding and fitting for neutron reflection experiments. Physica B, 198(1-3):24-26, 1994. (Proceedings of the International Conference on Surface X-Ray and Neutron Scattering (SXNS-3), Dubna (Russia), Jun. 24-29, 1993).

[187] D. del Castillo Sobrino, J. G. Casao, and C. G.-A. Sanchez. Genetic processing of the sensorial information. Sens. Actuators A. Phys. (Switzerland), A37-A38(2):255-259, 1993. (Proceedings of EUROSENSORS VI, San Sebastian (Spain), 5-7 Oct. 1992).

[188] P. J. Denning. Genetic algorithms. American Scientist, 80(1):12-14, Jan.-Feb. 1992.

[189] A. K. Dewdney. Exploring the field of genetic algorithms in a primordial computer sea full of flips. Scientific American, 253(5):21-32, Nov. 1985.

[190] A. K. Dewdney. Computer recreations, simulated evolution: wherein bugs learn to hunt bacteria. Scientific American, pages 104-107, May 1989.

[191] v. d. Dirk Johannes Bank. The use of genetic algorithms for cryptanalysis. Ph.D. thesis, University of Pretoria, South Africa, 1992 (in Afrikaans).

[192] D. A. Diver. Application of genetic algorithms to the solution of ordinary differential equations. Journal of Physics A — Mathematical and General, 26(14):3503-3513, July 1993.

[193] M. Dorigo. Optimization, Learning and Natural Algorithms. Ph.D. thesis, Politechnico di Milano, 1992 (in Italian).

[194] M. Dorigo. Using transputers to increase speed and flexibility of genetics-based machine learning systems. Microprocessing and Microprogramming EURO-Micro Journal, 34(15):147-152, 1992.

[195] M. Dorigo. Genetic and non-genetic operators in alecsys. Evolutionary Computation, 1(2):151-164, 1993.

[196] M. Dorigo and U. Schnepf. Genetics-based machine learning and behaviour based robotics: A new synthesis. IEEE Transactions on Systems, Man, and Cybernetics, 23(1):141-154, 1993.

[197] R. M. Dreizler, E. K. U. Gross, and A. Toepfer. Extended Thomas-Fermi approach to diatomic systems. Physics Letters, 71A(1):49-53, 1979.

[198] W. B. Dress. High-performance neural networks. J. Forth Appl. Res., 5(1):137-140, 1987.

[199] Q. Y. Duan, V. K. Gupta, and S. Sorooshian. Shuffled complex evolution approach for effective and efficient global minimization. Journal of Optimization Theory and Applications, 76(3):501-521, Mar. 1993.

[200] B. Dunham, D. Fridshal, and J. H. North. Design by natural selection. Synthese, 15:254-259, 1963.

[201] M. Eaton. Process control using genetically trained neural networks. Journal of Microcomputer Applications, 16(2):137-145, Apr. 1993.

[202] W. Ebeling. Applications of evolutionary strategies. Systems Analysis —Modeling — Simulation, 7(1):3-16, 1990.

[203] W. Ebeling and A. Engel. Models of evolutionary systems and their application to optimization problems. Systems Analysis — Modeling —Simulation, 3(5):377-385, 1986.

[204] W. Ebeling and I. Sonntag. A stochastic description of evolutionary processes in underoccupied systems. BioSystems, 19:91-100, 1986.

[205] Self-organization and life, from simple rules to global complexity, Proceedings of the Second European Conference on Artificial Life, Brussels (Belgium), 24-26 May 1993. MIT Press, Cambridge, MA.

[206] R. Eckmiller, G. Hartmann, and G. Hauske, editors. Parallel Processing in Neural Systems and Computers. Elsevier Science Publisher B.V., Amsterdam, Dusseldorf (Germany), 19.21. Mar. 1990.

[207] L. V. Edmondson. Genetic algorithms with 3-parent crossover. Ph.D. thesis, University of Missouri-Rolla, 1993.

[208] T. Eisenhammer, M. Lazarov, M. Leutbecher, U. Schoffel, and R. Sizmann. Optimization of interference filters with genetic algorithms applied to silver-based heat mirrors. Applied Optics, 32(31):6310-6315, 1. Nov. 1993.

[209] M. Elketroussi. Relapse from tobacco smoking cessation: Mathematical and computer micro-simulation modelling including parameter optimization with genetic algorithms. Ph.D. thesis, University of Minnesota, 1993.

[210] B. S. Elmer. The design, analysis, and implementation of parallel simulated annealing and parallel genetic algorithms for the composite graph coloring problem. Ph.D. thesis, University of Missouri-Rolla, MO, 1993.

[211] C. Emmeche. Det Levende Spil Biologisk Form og Kunstigt liv. Nysyn, Munksgaard (Denmark), 1991 (in Danish).

[212] S. S. Erenguc and H. Pirkul. Foreword: Heuristic, genetic and tabu search. Computers & Operations Research, 21(8):799, Oct. 1994.

[213] M. E. Everett and A. Schultz. 2-dimensional nonlinear magnetotelluric inversion using a genetic algorithm. Journal of Geomagnetism and Geoelectricity, 45(9):1013-1026, 1993. (Proceedings of the 11th Workshop on Electromagnetic Induction in the Earth, Wellington (New Zealand), Aug. 26 - Sep. 2, 1992).

[214] I. D. Falco, R. D. Balio, E. Tarantino, and R. Vaccaro. Simulation of genetic algorithms on MIMD multicomputers. Parallel Processing Letters, 2(4):381-389, Dec. 1992.

[215] J. H. Fang, C. R. Karr, and D. A. Stanley. Genetic algorithm and its application to petrophysics. Soc. Pet. Eng. AIME Pap. SPE, pages 1-12, May 1993.

[216] J. D. Farmer, A. Lapedes, N. H. Packard, and B. Wendroff, editors. Evolution, games, and learning. North-Holland, Amsterdam, 1986.

[217] J. D. Farmer, N. H. Packard, and A. S. Perelson. The immune system, adaptation, and machine learning. Physica D, 22:187-204, 1986.

[218] C. Farrell. Survival of the fittest technologies. New Scientist, 137(1859):35-39, 1993.

[219] A. J. Fenanzo. Darwinian evolution as a paradigm for AI research. SIGART Newsletter, (97):22-23, July 1986.

[220] M. Fieber, A. M. G. Ding, and P. J. Kuntz. A diatomics-in-molecules model for singly ionized neon clusters. Atoms, Molecules and Clusters, 23:171-179, 1992.

[221] J. M. Fitzpatrick and J. J. Grefenstette. Genetic algorithms in noisy environments. Machine Learning, 3(2/3):101-120, Oct. 1988.

[222] T. C. Fogarty. Rule-based optimization of combustion in multiple-burner furnaces and boiler plants. Engineering Applications of Artificial Intelligence, 1:203-209, 1988.

[223] T. C. Fogarty, N. S. Ireson, and S. A. Battle. Developing rule-based systems for credit card applications from data with genetic algorithm. IMA Journal of Mathematics Applied in Business and Industry, 4(1):53-59, 1992.

[224] D. B. Fogel. An evolutionary approach to the traveling salesman problem. Biological Cybernetics, 60(2):139-144, 1988.

[225] D. B. Fogel. The evolution of intelligent decision-making in gaming. Cybernetics and Systems, 22:223-226, 1991.

[226] D. B. Fogel. System Identification Through Simulated Evolution: A Machine Learning Approach to Modeling. Ginn Press, Needham Heights, MA, 1991.

[227] D. B. Fogel. Evolving Artificial Intelligence. Ph.D. thesis, University of California at San Diego, 1992.

[228] D. B. Fogel. Using evolutionary programming for modeling: An ocean acoustic example. IEEE Journal of Oceanic Engineering, 17(4):333-340, 1992.

[229] D. B. Fogel. Applying evolutionary programming to selected traveling salesman problems. Cybernetics and Systems, 24(1):27-36, Jan.-Feb. 1993.

[230] D. B. Fogel. Evolving behaviours in the iterated prisoner's dilemma. Evolutionary Computation, 1(1):77-97, 1993.

[231] D. B. Fogel. Genetic algorithms and robotics: A heuristic strategy for optimization. BioSystems, 31(1):78-79, 1993.

[232] D. B. Fogel. Parallel problem solving from nature 2: Proceedings of the second conference on parallel problem solving from nature. BioSystems, 31(1):75-78, 1993.

[233] D. B. Fogel. Applying evolutionary programming to selected control problems. Computers & Mathematics with Applications, 27(11):89-104, 1994.

[234] D. B. Fogel and J. W. Atmar. Comparing genetic operators with Gaussian mutations in simulated evolutionary processes using linear systems. Biological Cybernetics, 63(2):111- 114, 1990.

[235] D. B. Fogel and J. W. Atmar, editors. Proceedings of the 1st Annual Conference on Evolutionary Programming, LaJolla, CA, 21-22 Feb. 1992. Evolutionary Programming Society, San Diego.

[236] D. B. Fogel and W. Atmar, editors. Proceedings of the 2nd Annual Conference on Evolutionary Programming, La Jolla, CA, 25-26 Feb. 1993. Evolutionary Programming Society, San Diego.

[237] D. B. Fogel and L. J. Fogel. Method and apparatus for training a neural network using evolutionary programming, 1992. (U. S. patent no. 5,214,746. Issued May 25, 1993).

[238] D. B. Fogel, L. J. Fogel, and V. W. Porto. Evolving neural networks. Biological Cybernetics, 63(6):487-493, 1990.

[239] L. J. Fogel, A. J. Owens, and M. J. Walsh. Artificial intelligence through simulated evolution. John Wiley, New York, 1966.

[240] E. Fontain. Application of genetic algorithms in the field of constitutional similarity. Journal of Chemical Information and Computer Sciences, 32(6):748-752, 1992. (May 1992 Workshop on Similarity in Organic Chemistry).

[241] E. Fontain. The problem of atom-to-atom mapping. An application of genetic algorithms. Analytica Chimica Acta, 256(2):227-232, Aug. 1992. (6th CIC Workshop on Software Development in Chemistry, Bergakad Freiberg (Germany), 20-22 Nov. 1991).

[242] W. Fontana, W. Schnabl, and P. Schuster. Physical aspects of evolutionary optimization and adaptation. Physical Review A — General Physics, 40(6):3301-3321, 1989.

[243] S. Forrest. A study of parallelism and programming in classifier systems and its application to classification in KL-ONE semantic networks. Ph.D. thesis, University of Michigan, Ann Arbor, 1985.

[244] S. Forrest. Emergent computation: self-organizing, and cooperative phenomena in natural and artificial computing networks. Physica D, 42:1-11, 1990.

[245] S. Forrest, editor. Emergent Computation: Self-Organizing, Collective, and Cooperative Phenomena in Natural and Artificial Computing Networks, Cambridge, MA, 1991. MIT Press/North-Holland. (also as Physica D, Vol. 42).

[246] S. Forrest. Parallelism and Programming in Classifier Systems. Pittman, 1991.

[247] S. Forrest. Genetic algorithms — principles of natural selection applied to computation. Science, 261(5123):872-878, 13 Aug. 1993.

[248] S. Forrest and M. Mitchell. What makes a problem hard for a genetic algorithm? Some anomalous results and their explanation. Machine Learning, 13(2-3):285-319, Nov.-Dec. 1993.

[249] S. Forrest and A. S. Perelson. Computation and the immune system. SIGBIO Newsletter, 12(2):52-57, 1992.

250] R. S. Forsyth. Beagle — a Darwinian approach to pattern recognition. Kybernetes, 10(3):159-166, 1981.

[251] M. P. Fourman. Evolving layout. IEEE Colloquium on VLSI Design Methodologies, Digest No. 41:3/1-3/4, 1985.

[252] B. L. Fox. Integrating and accelerating tabu search, simulated annealing, and genetic algorithms. Annals of Operations Research, 41(1-4):47-67, 1993.

[253] M. D. Foy, R. F. Fenekohal, and D. E. Goldberg. Signal timing determination using genetic algorithms. Transactions Research Record, Highway Capacity and Traffic Flow, Transportation Research Board, (1365):108-115, 1992.

[254] D. R. Frantz. Non-linearities in genetic adaptive search. Ph.D. thesis, University of Michigan, Ann Arbor, 1972. (University Microfilms No. 73-11,116).

[255] A. S. Fraser. Simulation of genetic systems. Journal of Theoretical Biology, 2:329-346, 1962.

[256] A. S. Frazer. Simulation of genetic systems by automatic digital computers. Australian Journal of Biological Sciences, 10:484-491, 1957.

[257] L. N. Frazer, A. Basu, and J. D. Low. Geophysical inversion with simulated annealing and genetic algorithms. EOS, 71(43):1477, 1990.

[258] J. Freeman. Simulating a basic genetic algorithm. The Mathematica Journal, 3(2):52-56, 1993.

[259] L. C. Freeman. Finding groups with a simple genetic algorithm. Journal of Mathematical Sociology, 17(4):227-241, 1993.

[260] R. Freeman. High resolution NMR using selective excitation. Journal of Molecular Structure, 266:39-51, 1992.

[261] R. Freeman and X. Wu. Design of magnetic resonance experiments by genetic evolution. Journal of Magnetic Resonance, 75:184-189, 1987.

[262] J. F. Frenzel. Genetic algorithms. IEEE Potentials, 12(3):21-24, Oct. 1993.

[263] H. Freund and R. Wolter. Evolution of bit strings: Some preliminary results. Complex Systems, 5(3):279-298, 1992.

[264] P. W. Frey. A bit-mapped classifier. BYTE, 11(12):161-172, 1986.

[265] P. W. Frey and D. J. Slate. Letter recognition using Holland-style adaptive classifiers. Machine Learning, 6:161-182, 1991.

[266] M. Friedman, U. Mahlab, and J. Shamir. Collective genetic algorithm for optimization and its electro-optic implementation. Applied Optics, 32(23):4423-4429, 1993.

[267] F. Fuchs and H. A. Maier. Optimierung des Lastflusses in elektrischen EnergieVersorungsnetzen mittels Zufallszahlen. Archivurf Elektrotechnik, 66:85-94, 1983.

[268] T. Fukuda, H. Ishigami, F. Arai, and T. Shibata. Auto generation of fuzzy model using genetic algorithm and delta rule. Transactions of the Institute of Electrical Engineers of Japan C, 113-C(7):495-501, July 1993.

[269] M. Fukumi and S. Omatu. Designing an architecture of a neural network for coin recognition by a genetic algorithm. Transactions of the Institute of Electrical Engineers of Japan C, 113-D(12):1403-1409, Dec. 1993 (in Japanese).

[270] W. Funk. Computer aided engineering (CAE) — Problemlosungenurf den maschinenbau. Der Konstrukteur, 6:8-16, 1982.

[271] R. Galar. Simulation of local evolutionary dynamics of small populations. Biological Cybernetics, 65(1):37-45, 1991.

[272] K. Gallagher and M. S. Sambridge. Earthquake hypocenter location using genetic algorithms. Bull. Seismol. Soc. Am., 83(5):1467-1491, 1993.

[273] K. Gallagher, M. S. Sambridge, and G. Drijkoningen. Genetic algorithms — an evolution from Monte Carlo methods for strongly non-linear geophysical optimization problems. Geophysical Research Letters, 18(12):2177-2180, 1991.

[274] J. E. Galletly. An overview of genetic algorithms. Kybernetes, 21(6):26-30, 1992.

[275] Proceedings of the IEEE Workshop on Genetic Algorithms, Neural Networks and Simulated Annealing applied to problems in signal and image processing, University of Glasgow (UK), 1990. IEEE.

[276] M. L. Gargano and L. von Gargano. Neural Networks and Genetic Algorithms - Business Applications and Case Studies. International Thomson Publishing, London, 1993.

[277] D. D. Gemmill. Solution to the assortment problem via the genetic algorithm. Mathematical and Computer Modelling, 16(1):89-94, Jan. 1992.

[278] A. Geyer-Schulz and T. Kolarik. Distributed computing with APL. APL Quote Quad, 23(1):60-69, July 1992 (Proceedings of the International Conference on APL 6-10. July 1992 St. Petersburg (Russia)).

[279] A. M. Gillies. Machine learning procedures for generating image domain feature detectors. Ph.D. thesis, University of Michigan, 1985.

[280] A. M. Gillies. Machine learning procedures for generating image domain feature detectors, 1985 (U. S. patent no. 4,821,333. Issued Apr. 11 1989).

[281] D. E. Glover. Experimentation with an adaptive search strategy for solving a key-board design/configuring problem. Ph.D. thesis, University of Iowa, 1986 (University Microfilms No. DA86-22767).

[282] D. E. Glover and H. J. Greenberg. New approaches for heuristic search: A bilateral linkage with artificial intelligence. European Journal of Operations Research, 39(2):119-130, Mar. 1989.

[283] D. E. Goldberg. Computer-aided gas pipeline operation using genetic algorithms and rule learning. Ph.D. thesis, University of Michigan, 1983 (University Microfilms No. 8402282).

[284] D. E. Goldberg. Computer-aided gas pipeline operation using genetic algorithms and rule learning. Part I: Genetic algorithms in pipeline optimization. Engineering with Computers, 3:35-45, 1987.

[285] D. E. Goldberg. Computer-aided gas pipeline operation using genetic algorithms and rule learning. Part II: Rule learning control of a pipeline under normal and abnormal conditions. Engineering with Computers, 3:47-58, 1987.

[286] D. E. Goldberg. Genetic algorithms and Walsh functions: Part I, a gentle introduction. TCGA Report 88006, University of Alabama, 1988 (also as [288]).

[287] D. E. Goldberg. Genetic algorithms and Walsh functions: Part II, deception and its analysis. TCGA Report 89001, University of Alabama, 1988 (also as [289]).

[288] D. E. Goldberg. Genetic algorithms and Walsh functions: Part I, a gentle introduction. Complex Systems, 3:129-152, 1989.

[289] D. E. Goldberg. Genetic algorithms and Walsh functions: Part II, deception and its analysis. Complex Systems, 3:153-171, 1989.

[290] D. E. Goldberg. Genetic Algorithms in Search, Optimization, and Machine Learning. Addison-Wesley, Reading, MA, 1989.

[291] D. E. Goldberg. A note on Boltzmann tournament selection for genetic algorithms and population-oriented simulated annealing. Complex Systems, 4:445-460, Aug. 1990.

[292] D. E. Goldberg. A note on Boltzmann tournament selection for genetic algorithms and population-oriented simulated annealing. TCGA Report 90003, University of Alabama, 1990 (also as [291]).

[293] D. E. Goldberg. Probability matching, the magnitude of reinforcement, and classifier system bidding. Machine Learning, 5:407-425, 1990 (also TCGA Report No. 88002).

[294] D. E. Goldberg. Real-coded genetic algorithms, virtual alphabets, and blocking. IlliGAL Report 90001, University of Illinois at Urbana-Champaign, 1990 (also as [296]).

[295] D. E. Goldberg. Construction of high-order deceptive functions using low-order Walsh coefficients. Annals of Mathematics and Artificial Intelligence, 5(1):35-48, Apr. 1992.

[296] D. E. Goldberg. Real-coded genetic algorithms, virtual alphabets and blocking. Complex Systems, 5(2):139-167, 1992.

[297] D. E. Goldberg. Making genetic algorithms fly: A lesson from the Wright brothers. Advanced Technology for Developers, 2:1-8, Feb. 1993.

[298] D. E. Goldberg. A Wright-brothers theory of genetic-algorithm flight. Journal of the Institute of Systems, Control, and Information Engineers (Japan), 37(8):450-458, 1993.

[299] D. E. Goldberg. Genetic and evolutionary algorithms come of age. Communications of the ACM, 37(3):113-119, Mar. 1994.

[300] D. E. Goldberg and C. L. Bridges. An analysis of a reordering operator on a GA-hard problem. Biological Cybernetics, 62:397-405, 1990 (also TCGA Report No. 88005).

[301] D. E. Goldberg, K. Deb, and J. H. Clark. Genetic algorithms, noise, and the sizing of populations. Complex Systems, 6(4):333-362, 1992 (also TCGA Report No. 91010).

[302] D. E. Goldberg, K. Deb, and B. Korb. Messy genetic algorithms revisited: Studies in mixed size and scale. Complex Systems, 4(4):415-444, Aug. 1990.

[303] D. E. Goldberg, K. Deb, and D. Thierens. Towards a better understanding of mixing in genetic algorithms. IlliGAL Report 92009, University of Illinois at Urbana-Champaign, 1992 (also as [304]).

[304] D. E. Goldberg, K. Deb, and D. Thierens. Toward a better understanding of mixing in genetic algorithms. Journal of the Society of Instrument and Control Engineers, 32(1):10-16, 1993.

[305] D. E. Goldberg and J. H. Holland. Genetic algorithms and machine learning. Machine Learning, 3:95-99, 1988.

[306] D. E. Goldberg, B. Korb, and K. Deb. Messy genetic algorithms: Motivation, analysis, and first results. Complex Systems, 3:493-530, 1989 (also TCGA Report 89003).

[307] D. E. Goldberg and C. H. Kuo. Genetic algorithms in pipeline optimization. Journal of Computing in Civil Engineering, 1(2):128-141, Apr. 1987.

[308] D. E. Goldberg, K. Milman, and C. Tidd. Genetic algorithms: A bibliography. IlliGAL Report 92008, University of Illinois at Urbana-Champaign, 1992.

[309] D. E. Goldberg and W. M. Rudnick. Genetic algorithms and the variance of fitness. Complex Systems, 5(3):265-278, June 1991.

[310] D. E. Goldberg and W. M. Rudnick. Genetic algorithms and the variance of fitness. IlliGAL Report 91001, University of Illinois at Urbana-Champaign, 1991.

[311] E. D. Goodman. Adaptive behavior of simulated bacterial cells subjected to nutritional shifts. Ph.D. thesis, University of Michigan, Ann Arbor, 1972.

[312] M. Gordon. Probabilistic and genetic algorithms for document retrieval. Communications of the ACM, 31(10):1208-1218, Oct. 1988.

[313] M. Gordon. User-based document clustering by redescribing subject descriptions with a genetic algorithm. Journal of the American Society for Information Science, 42(5):311-322, 1991.

[314] M. Gorges-Schleuter. Genetic Algorithms and Population Structures | A Massively Parallel Algorithm. Ph.D. thesis, University of Dortmund, 1990.

[315] T. Goto, H. Ase, M. Yamagishi, Y. Hirota, and S. Fujii. Application of GA, neural network and AI to planning problems. NKK Technical Report (Japan), (144):78-85, 1993 (in Japanese).

[316] A. Gottvald. Optimal magnet design for NMR. IEEE Transactions on Magnetics, 26(2):399-401, 1990.

[317] A. Gottvald, K. Preis, C. A. Magele, O. Biro, and A. Savini. Global optimization methods for computational electromagnetics. IEEE Transactions on Magnetics, 28(2):1537-1540, Mar. 1992.

[318] J. Graf and H. G. Wagemann. Evolutionsstrategie in der halbleitertechnikurf die charakterisierung von MOS-bauelementen (application of evolution strategy in semiconductor modeling for the characterization of MOS-devices). Archivurf Elektrotechnik, 76(2):155-160, 1993.

[319] S. M. L. Grand. The application of the genetic algorithm to protein tertiary structure prediction. Ph.D. thesis, The Pennsylvania State University, 1993.

[320] S. M. L. Grand and K. M. Merz, Jr. The application of genetic algorithms to the minimization of potential energy functions. Journal of Global Optimization, 3:49-66, 1993.

[321] D. G. Green and T. Bossomaier, editors. Complex Systems: from Biology to Computation. IOS Press, Amsterdam, 1993.

[322] D. P. Greene and S. F. Smith. Competition-based induction of decision models from examples. Machine Learning, 13(2-3):229-257, Nov.-Dec. 1993.

[323] J. J. Grefenstette, editor. Proceedings of the First International Conference on Genetic Algorithms and Their Applications, Pittsburgh, PA, 24-26 July 1985. Lawrence Erlbaum Associates: Hillsdale, New Jersey.

[324] J. J. Grefenstette. Optimization of control parameters for genetic algorithms. IEEE Transactions on Systems, Man, and Cybernetics, SMC-16(1):122-128, Jan./Feb. 1986.

[325] J. J. Grefenstette, editor. Genetic Algorithms and their Applications: Proceedings of the Second International Conference on Genetic Algorithms and Their Applications, MIT, Cambridge, MA, 28-31 July 1987. Lawrence Erlbaum Associates: Hillsdale, New Jersey.

[326] J. J. Grefenstette. Credit assignment in rule discovery systems based on genetic algorithms. Machine Learning, 3(2/3):225-246, Oct. 1988.

[327] J. J. Grefenstette. Genetic algorithms. IEEE Expert, 8(5):5-8, 1993.

[328] J. J. Grefenstette. Special issue on genetic algorithms. Machine Learning, 13:157-319, Nov./Dec. 1993.

[329] J. J. Grefenstette, C. L. Ramsey, and A. C. Schultz. Learning sequential decision rules using simulation models and competition. Machine Learning, 5(4):355-381, 1990.

[330] D. E. Grierson and W. H. Pak. Optimal sizing, geometrical and topological design using a genetic algorithm. Struct. Optim. (Germany), 6(3):151-159, 1993.

[331] E. K. U. Gross and R. M. Dreizler. Thomas-Fermi approach to diatomic systems, I Solution of the Thomas-Fermi and Thomas-Fermi-Weizsacker equations. Physical Review A, 20(5):1798-1815, 1979.

[332] M. Gross. Untersuchungenuber dieogmlichkeit der automatischen entwicklung von algebraischen Formeln aus Daten mit hilfe der Evolutionsstrategie. Ph.D. thesis, Technische Universitat der Berlin, 1979.

[333] P. B. Grosso. Computer simulation of genetic adaptation: Parallel subcomponent interaction in a multilocus model. Ph.D. thesis, University of Michigan, 1985 (University Microfilms No. 8520908).

[334] F. C. Gruau and D. Whitley. Adding learning to the cellular development process: a comparative study. Evolutionary Computation, 1(3):213-233, 1993.

[335] Y. V. Guliaev, V. F. Krapivin, and I. L. Bukatova. On the way towards evolutionary informatics. Soviet Journal of the Academy of Sciences of the USSR, 11:53-61, 1987.

[336] Z. Guo. Nuclear power plant fault diagnostics and thermal performance studies using neural networks and genetic algorithms. Ph.D. thesis, University of Tennessee, 1992.

[337] J. N. D. Gupta and C. N. Potts. Editorial. European Journal of Operations Research, 70(3):269-271, Nov. 1993.

[338] M. C. Gupta, Y. P. Gupta, and A. Kumar. Minimizing flow time variance in a single machine system using genetic algorithm. European Journal of Operations Research, 70(3):289-303, Nov. 1993.

[339] J. Haata ja. Geneettisten algoritmien simulointi Matlab 4.0:lla [Simulating genetic algorithms with Matlab 4.0]. SuperMenu, (2):21-25, 1993 (in Finnish).

[340] J. Haata ja. Menetelmia ja ohjelmisto ja globaaliin optimointiin [Methods and programs for global optimization]. SuperMenu, (4):9-12, 1993 (in Finnish).

[341] J. Haata ja and M. Ryynanen. Synkrotronisateilylahteen optimointi geneettisella algoritmilla [Optimization of synchrotron radiation source by using a genetic algorithm]. SuperMenu, (4):12-15, 1993 (in Finnish).

[342] K. Haefner, editor. Evolution of Information Processing Systems, An Interdisciplinary Approach to a New Understanding of Nature and Society. Springer-Verlag, Berlin, 1992.

[343] M. Haggerty. Evolution by esthetics. IEEE Computer Graphics and Applications, 11(2):5- 9, Mar. 1991.

[344] P. Ha jela. Genetic search I an approach to the nonconvex optimization problem. AIAA Journal, 28(7):1205-1210, July 1990.

[345] C. Hampel. Ein Vergleich von Optimierungsverfahrenurf die zeitdiskrete Simulation. Ph.D. thesis, Technische Universitat der Berlin, 1981.

[346] P. J. B. Hancock. Coding strategies for genetic algorithms and neural nets. Ph.D. thesis, University of Stirling, Department of Computing Science and Mathematics, 1992.

[347] B. Hartke. Global geometry optimization of clusters using genetic algorithms. The Journal of Physical Chemistry, 97(39):9973-9976, 1993.

[348] D. Hartmann. Optimierung balkenartiger Zylinderschalen aus Stahlbeton mit elastischem und plastischem Werkstoffverhalten. Ph.D. thesis, University of Dortmund, 1974.

[349] D. Hartmann. Optimierung flacher hyperbolischer Paraboloidschalen. Beton- und Stahlbetonbau, 9:216-222, 1977.

[350] D. Hartmann and G. Hartmann. Identification of material parameters for inelastic constitutive models using principles of biological evolution. J. of Eng. Mater. Technol. Trans. ASME, 111(3):299-305, July 1989.

[351] D. F. Hartmann. Identifikationsstrategien zur Rissformbestimmung an Rotoren. Zeitschrift urf angewandte Mathematik und Mechanik, 71(4):T139-T141, 1991.

[352] I. Harvey and P. Newquist. The life and death of new AI techniques. AI Expert, 8(12):39-40, Dec. 1993.

[353] A. T. Hatjimihail. Optimization of alternative quality control procedures using genetic algorithms [abstract]. Clinical Chemistry, 38(6):1019-1020, 1992 (in Proceedings of the 44th National Meeting of the American Association for Clinical Chemistry, Chicago, IL, 19-23 July 1992).

[354] A. T. Hatjimihail. Genetic algorithms-based design and optimization of statistical quality control procedures. Clinical Chemistry, 39(9):1972-1978, 1993 (in Proceedings of the 25th Annual Oak Ridge Conference on Advanced Analytical Concepts for the Clinical Laboratory, Knoxville, TN, 22-24 Apr. 1993).

[355] U. Hegde and B. Ashmore. A feasibility study of genetic placement. Texas Instrument Technology Journal, 9(6):72-82, Nov.-Dec. 1992.

[356] M. Heidari and P. C. Heigold. Determination of hydraulic conductivity tensor using a nonlinear least squares estimator. Water Resources Bulletin, 29(3):415-424, June 1993.

[357] A. Hemker. Ein wissensbasierter genetischer Algorithmus zur Rekonstruktion physicalischer Ereignisse. Ph.D. thesis, Gesamthochschule Wupperthal, 1992.

[358] M. Herdy. The number of offspring as strategy parameter in hierachically organized evolution strategies. SIGBIO Newsletter, 13(2):2-7, 1993.

[359] R. Herrmann. Evolutionsstrategische Regressionanalyse. Nobel Hefte, 49(1/2):44-54, 1983.

[360] J. Hesslich and P. J. Kuntz. A diatomics-in-molecules model for singly-ionized argon clusters. Zeitschrifturf Physik D — Atoms, Molecules and Clusters, 2:251-252, 1986.

[361] D. B. Hibbert. Generation and display of chemical structures by genetic algorithms. Chemometrics and Intelligent Laboratory Systems, 20(1):35-43, Aug. 1993.

[362] D. B. Hibbert. Genetic algorithm for the estimation of kinetic parameters. Chemometrics and Intelligent Laboratory Systems, 19(3):319-329, July 1993.

[363] D. B. Hibbert. Genetic algorithms in chemistry. Chemometrics and Intelligent Laboratory Systems, 19(3):277-293, July 1993.

[364] T. Higuchi. Towards flexible mechanisms for association — evolvable hardware with genetic learning. Denshi Gijutsu Sogo Kenkyusho Iho, 57(12):55-60, 1993.

[365] T. Higuchi and H. Kitano. Genetic algorithms. Joho Shori (Japan), 34(7):871-883, July 1993 (in Japanese).

[366] P. Hilgers. Der Einsatz eines Mikrorechners zur hybriden Optimierung und Schwingungsanalyse. Ph.D. thesis, Ruhruniversitat Bochum, 1978.

[367] A. Hill and C. J. Taylor. Model-based image interpretation using genetic algorithms. Image and Vision Computing, 10(5):295-300, June 1992.

[368] W. D. Hillis. Optimization problems. Nature, 337:27-28, 1987.

[369] G. E. Hinton and S. J. Nowlan. How learning can guide evolution. Complex Systems, 1:495-502, 1987.

[370] A. Hofler. Formoptimierung von Leichtbaufachwerken durch Einsatz einer Evolutionsstrategies. Ph.D. thesis, Technische Universitat der Berlin, 1976.

[371] J. H. Holland. Outline for a logical theory of adaptive systems. Journal of the Association for Computing Machinery, 3:297-314, 1962.

[372] J. H. Holland. Genetic algorithms and the optimal allocations of trials. SIAM Journal of Computing, 2(2):88-105, 1973.

[373] J. H. Holland. Adaptation in Natural and Artificial Systems. The University of Michigan Press, Ann Arbor, 1975.

[374] J. H. Holland. Adaptive algorithms for discovering and using general patterns in growing knowledge-bases. International Journal of Policy Analysis and Information Systems, 4(3):245-268, 1980.

[375] J. H. Holland. Searching nonlinear functions for high values. Applied Mathematics and Computation, 32:255-274, 1989.

[376] J. H. Holland. Adaptation in Natural and Artificial Systems. MIT Press, Cambridge, 1992.

[377] J. H. Holland. Complex adaptive systems. Daedalus, 121(1):17-30, Winter 1992.

[378] J. H. Holland. Genetic algorithms. Scientific American, 267(1):44-50, 1992.

[379] J. H. Holland and A. W. Burks. Adaptive computing system cabable of learning and discovery, 1985 (U. S. patent no. 4,697,242. Issued Sep. 29 1987).

[380] J. H. Holland and A. W. Burks. Method of controlling a classifier system, 1989 (U.S. Patent 4,881,178. Issued Nov. 14 1989).

[381] J. H. Holland, K. J. Holyoak, R. E. Nisbett, and P. R. Thagard. Induction: Processes of Inference, Learning, and Discovery. MIT Press, Cambridge, MA, 1986.

[382] R. B. Hollstien. Artificial genetic adaptation in computer control systems. Ph.D. thesis, University of Michigan, Ann Arbor, 1971 (University Microfilms No. 71-23,773).

[383] C. W. Holsapple, V. S. Jacob, R. Pakath, and J. S. Zaveri. A genetic-based hybrid scheduler for generating static schedules in flexible manufacturing contexts. IEEE Transactions on Systems, Man, and Cybernetics, 23(4):953-972, 1993.

[384] A. Homaifar, S. Guan, and G. E. Liepins. Schema analysis of the traveling salesman problem using genetic algorithms. Complex Systems, 6(6):533-552, Dec. 1992.

[385] J. S. Hong. Genetic approach to bearing estimation with sensor location uncertainties. Electronics Letters, 29(23):2013-2014, Nov. 1993.

[386] R. Hong. Neurocontrols and vision for Mars robots. Advanced Technology for Developers, 1(2):1, June 1992.

[387] E. Horiuchi and K. Tani. Architecture and implementation issues about learning for a group of mobile robots with a distributable genetic algorithm. Kikai Gijutsu Kenkyusho Shoho, 47(6):247-256, Nov. 1993.

[388] A. Horner, J. Beauchamp, and L. Haken. Machine tongues XVI. genetic algorithms and their application to FM matching synthesis. Comput. Music J., 17(4):17-29, Winter 1993.

[389] A. Horner, J. Beauchamp, and L. Haken. Methods for multiple wavetable synthesis of musical instrument tones. Journal of Audio Engineers Society, 41(5):336-356, May 1993.

[390] C. M. Hosage and M. F. Goodchild. Discrete space location-allocation solutions from genetic algorithms. Annals of Operations Research, 6:35-46, 1986.

[391] R. Huang. Systems control with the genetic algorithm and the nearest neighbour classification. CC-AI, 9(2-3):225-236, 1992.

[392] R. Huang and T. C. Fogarty. Learning prototype control rules for combustion control with the genetic algorithm. Journal of Modeling, Measurement and Control, C, 38(4):55-64, 1992.

[393] B. A. Huberman, editor. The Ecology of Computation. North-Holland, New York, 1988.

[394] M. Hughes. Why nature knows best about design. The Guardian Newspaper, 14 Sept. 1989.

[395] M. Hughes. Improving products and processes — nature's way (genetic algorithms). Industrial Management + Data Systems, 90(6):22-25, 1990.

[396] S.-L. Hung. Neural network and genetic learning algorithms for computer-aided design and pattern recognition. Ph.D. thesis, The Ohio State University, 1992.

[397] C. L. Huntley and D. E. Brown. Parallel heuristics for quadratic assignment problems. Computers & Operations Research, 18(3):275-289, 1991.

[398] P. Husbands. An ecosystems model for integrated production planning. International Journal on Computer Integrated Manufacturing, 6(1&2):74-86, 1993.

[399] H.-S. Hwang, S.-K. Oh, and K.-B. Woo. Fusion of genetic algorithms and fuzzy inference system. Trans. Korean Inst. Electr. Eng. (South Korea), 41(9):1095-1103, 1992. (in Korean).

[400] K.-Y. Hwang. Part selection for predefined configurations using genetic search based algorithms. Ph.D. thesis, The University of Utah, 1993.

[401] W.-R. Hwang. Intelligent control based on fuzzy algorithms and genetic algorithms. Ph.D. thesis, New Mexico State University, 1993.

[402] H. Iba and T. Sato. Bugs: a bug-based search strategy using genetic algorithms. Journal of Japanese Society for Artificial Intelligence, 8(6):786-796, Nov. 1993.

[403] K. Iba. Reactive power planning in large power systems using genetic algorithms. Transactions of the Institute of Electrical Engineers of Japan B, 113-B(8):865-872, Aug. 1993 (in Japanese).

[404] IEEE. Proceedings of ICCI94/Fuzzy Systems, Orlando, FL, 26 June-2. July 1994. IEEE.

[405] IEEE. Proceedings of ICCI94/Neural Networks, Orlando, FL, 26 June-2. July 1994. IEEE.

[406] IEEE. Proceedings of the First IEEE Conference on Evolutionary Computation, volume 1, Orlando, FL, 27-29 June 1994. IEEE.

[407] IEEE. Proceedings of the First IEEE Conference on Evolutionary Computation, volume 2, Orlando, FL, 27-29 June 1994. IEEE.

[408] H. Iima and N. Sannomiya. Genetic algorithm approach to a production ordering problem. Transactions of the Society of Instrument and Control Engineers (Japan), 28(11):1337- 1344, Nov. 1992 (in Japanese).

[409] H. Iima and N. Sannomiya. A solution of modified flowshop scheduling problem by using genetic algorithm. Transaction of Systems, Control and Information, 6(10):437-445, Oct. 1993 (in Japanese).

[410] T. Ikegami and K. Kaneko. Genetic fusion. Physical Review Letters, 65(26):3352-3355, 24 Dec. 1990.

[411] L. Ingber and B. Rosen. Genetic algorithms and very fast simulated annealing: A comparison. Mathematical and Computer Modelling, 16(11):87-100, Nov. 1992.

[412] Y. E. Ioannidis, T. Saulys, and A. J. Whitsitt. Conceptual learning in database design. ACM Transactions on Information Systems, 10(3):265-293, 1992.

[413] H. Ishibuchi, K. Nozaki, N. Yamamoto, and H. Tanaka. Selection of fuzzy if-then rules by a genetic method. Transaction of the Institute of Electronics, Information and Communication Engineers A (Japan), J76-A(10):1465-1473, Oct. 1993. (in Japanese).

[414] H. S. Ismail and K. K. B. Hon. New approaches for the nesting of two-dimensional shapes for press tool design. International Journal of Production Research, 30(4):825-837, Apr. 1992.

[415] B. Jacob, E. K. U. Gross, and R. M. Dreizler. Solutions of the Thomas-Fermi equations for triatomic systems. Journal of Physics B - Atom. Molec. Phys., 11(22):3795-3802, 1978.

[416] C. Z. Janikow. Inductive learning of decision rules from attribute-based examples: A knowledge-intensive genetic algorithm approach. Ph.D. thesis, University of North Carolina at Chapel Hill, 1991.

[417] C. Z. Janikow. A knowledge-intensive genetic algorithm for supervised learning. Machine Learning, 13(2-3):189-228, Nov.-Dec. 1993.

[418] R. A. Jarvis. Adaptive global search by the process of competitive evolution. IEEE Transactions on Systems, Man, and Cybernetics, 5(3):297-311, 1975.

[419] J. N. R. Jeffers. Rule induction methods in forestry research. AI Applications, 5(2):37-44, 1991.

[420] W. M. Jenkins. Structural optimization with the genetic algorithm. The Structural Engineer, 69(24):418-422, Dec. 1991.

[421] W. M. Jenkins. Towards structural optimization via the genetic algorithm. Computers & Structures, 40(5):1321-1327, May 1991.

[422] W. M. Jenkins. Plane frame optimum design environment based on genetic algorithm. Journal of Structural Engineering - ASCE, 118(11):3103-3112, Nov. 1992.

[423] E. D. Jensen. Topological structural design using genetic algorithms. Ph.D. thesis, Purdue University, 1992.

[424] M. Jervis, P. L. Stoffa, and M. K. Sen. 2-D migration velocity estimation using a genetic algorithm. Geophysical Research Letters, 20(14):1495-1498, July 1993.

[425] L.-M. Jin and S.-P. Chan. Analogue placement by formulation of macrocomponents and genetic partitioning. International Journal of Electronics, 73(1):157-173, July 1992.

[426] L.-M. Jin and S.-P. Chan. A genetic approach for network partitioning. International Journal Computers and Mathematics, 42(1-2):47-60, 1992.

[427] S. Jin and R. Madariaga. Background velocity inversion with a genetic algorithm. Geophysical Research Letters, 20(2):93-96, Jan. 1993.

[428] P. Jog, J. Y. Suh, and D. V. Gucht. Parallel genetic algorithms applied to the traveling salesman problem. SIAM Journal on Optimization, 1(4):515-529, 1991.

[429] R. C. Johnson. Defining artificial life leads to tough goals. Electronic Engineering Times, 80(3):37,41, 1990.

[430] R. C. Johnson. Machine-age natural selection: Finding solutions is in the genes. Electronic Engineering Times, 80(2):33-34, 1990.

[431] G. Jones, A. M. Robertson, and P. Willett. The use of genetic algorithms for identifying equifrequent groupings and for searching databases of flexible molecules. Information Research News, 4(2):2-11, 1993.

[432] K. A. D. Jong. Analysis of the Behaviour of a Class of Genetic Adaptive Systems. Ph.D. thesis, University of Michigan, 1975. (University Microfilms No. 76-9381).

[433] K. A. D. Jong. Adaptive system design: A genetic approach. IEEE Transactions on Systems, Man, and Cybernetics, SMC-10(9):566-574, 1980.

[434] K. A. D. Jong. Learning with genetic algorithms: An overview. Machine Learning, 3(2/3):121-138, 1988.

[435] K. A. D. Jong. Genetic algorithms. Machine Learning, 5(4):351-353, Oct. 1990.

[436] K. A. D. Jong. Editorial introduction. Evolutionary Computation, 1(1), 1993.

[437] K. A. D. Jong and W. M. Spears. A formal analysis of the role of multi-point crossover in genetic algorithms. Annals of Mathematics and Artificial Intelligence, 5(1):1-26, Apr. 1992.

[438] K. A. D. Jong, W. M. Spears, and D. F. Gordon. Using genetic algorithms for concept learning. Machine Learning Journal, 13(2-3):161-188, Nov.-Dec. 1993.

[439] R. S. Judson. Teaching polymers to fold. The Journal of Physical Chemistry, 96(25):10102, 1992.

[440] R. S. Judson, M. E. Colvin, J. C. Meza, A. Huffer, and D. Gutierrez. Do intelligent configuration search techniques outperform random search for large molecules? International Journal of Quantum Chemistry, 44(2):277-290, 1992.

[441] R. S. Judson, E. P. Jaeger, and A. M. Treasurywala. A genetic algorithm based method for docking flexible molecules. THEOCHEM, 114:191-206, 10. May 1994.

[442] R. S. Judson, E. P. Jaeger, A. M. Treasurywala, and M. L. Peterson. Conformation searching methods for small molecules II: A genetic algorithm approach. Journal of Computational Chemistry, 14(11):1407-1414, 1993.

[443] R. S. Judson and H. Rabitz. Teaching lasers to control molecules. Physical Review Letters, 68(10):1500-1503, 1992.

[444] N. Kadaba. Xroute: A knowledge-based routing system using neural networks and genetic algorithms. Ph.D. thesis, North Dakota State University of Agriculture and Applied Sciences, Fargo, 1990.

[445] R. R. Kampfner. Computational modeling of evolutionary learning. Ph.D. thesis, University of Michigan, Ann Arbor, 1981 (University Microfilms No. 81-25143).

[446] A. Kanarachos. A contribution to the problem of designing optimum performance bearings. Transactions of the ASME, pages 462-468, 1977.

[447] A. Kanarachos. Zur Anwendung von Parameteroptimierungsverfahren in der rechnergestutzten Konstruktion. Konstruktion, 31(5):177-182, 1979.

[448] J. J. Kanet and V. Sridharan. Progenitor: A genetic algorithm for production scheduling. Wirtschaftsinformatik, 33(4):332-336, Aug. 1991.

[449] J. J. Kanet and V. Sridharan. Progenitor: a genetic algorithm for production scheduling (reply). Wirtschaftsinformatik, 34(2):256, Apr. 1992.

[450] A. Kapsalis, V. J. Rayward-Smith, and G. D. Smith. Solving the graphical Steiner tree problem using genetic algorithms. Journal of the Operational Research Society, 44(4):397- 406, Apr. 1993.

[451] C. L. Karr. Analysis and optimization of an air-injected hydrocyclone. Ph.D. thesis, University of Alabama, 1989 (also TCGA Report No. 90001).

[452] C. L. Karr. Applying genetics to fuzzy logic. AI Expert, 6(3):38-43, Mar. 1991.

[453] C. L. Karr. Genetic algorithms for fuzzy controllers. AI Expert, 6(2):26-33, Feb. 1991.

[454] C. L. Karr. Adaptive process control with fuzzy logic and genetic algorithms. Sci. Comput. Autom. (U.S.A.), 9(10):23-24,26,28-30, 1993.

[455] C. L. Karr and E. J. Gentry. Fuzzy control of pH using genetic algorithms. IEEE Transactions on Fuzzy Systems, 1(1):46-52, 1993.

[456] C. R. Karr, S. K. Sharma, W. J. Hatcher, and T. R. Harper. Fuzzy control of an exothermic chemical reaction using genetic algorithms. Engineering Applications of Artificial Intelligence, 6(6):575-582, Dec. 1993.

[457] M. Kasper. Shape optimization by evolution strategy. IEEE Transactions on Magnetics, 28(2):1556-1560, Mar. 1992.

[458] T. Kawakami and Y. Kakazu. Study on an autonomous robot navigation problem using a classifier system. Nippon Kikai Gakkai Ronbunshu C Hen, 59(564):2339-2345, Aug. 1993.

[459] T. Kawakami and M. Minagawa. Automatic tuning of 3-D packing strategy and rule-base contruction using GA. Trans. Inf. Process. Soc. Jpn. (Japan), 33(6):761-768, 1992.

[460] S. A. Kennedy. Five ways to a smarter genetic algorithm. AI Expert, 8(12):35-38, Dec. 1993.

[461] B. L. N. Kennett and M. S. Sambridge. Earthquake location — genetic algorithms for teleseisms. Physics of the Earth and Planetary Interiors, 75(1-3):103-110, 1992.

[462] J. O. Kephart, T. Hogg, and B. A. Huberman. Dynamics of computational ecosystems. Physical Review A, 40:404-421, 1989.

[463] A. R. Khoogar. Kinematic motion planning for redundant robots using genetic algorithms. Ph.D. thesis, University of Alabama, 1989.

[464] L. Kierman and K. Warwick. Adaptive alarm processor for fault diagnosis on power transmission networks. Intelligent Systems Engineering, 2(1):25-37, 1993.

[465] S.-W. Kim, H.-K. Jung, and S.-Y. Hahn. Optimal design of capasitor-driven coil gun. Trans. Korean Inst. Electr. Eng. (South Korea), 41(12):1379-1386, Dec. 1992 (in Korean).

[466] Y. Kim, Y. Jang, and M. Kim. Stepwise-overlapped parallel annealing and its application to floorplan design. Computer Aided Design, 23(2):133-144, Mar. 1991.

[467] Y. C. Kim and Y. S. Hong. A genetic algorithm for task allocation in multiprocessor systems. J. Korea Inf. Sci. Soc. (South Korea), 20(1):43-51, 1993.

[468] B. Kirste. Least-squares fitting of EPR spectra by Monte Carlo methods. Journal of Magnetic Resonance, 73:213-224, 1987.

[469] B. Kirste. Methods for automated analysis and simulation of electron paramagnetic resonance spectra. Analytica Chimica Acta, 265(2):191-200, Aug. 1992. (6th CIC Workshop on Software Development in Chemistry, Bergakad Freiberg (Germany), 20-22 Nov. 1991).

[470] M. Kishimoto, K. Sakasai, and K. Ara. Estimation of current distribution from magnetic fields by combination method of genetic algorithm and neural-network. Transactions of the Institute of Electrical Engineers of Japan C, 113-C(9):719-727, Sept. 1993. (in Japanese).

[471] H. Kitano. Designing neural networks using genetic algorithms with graph generation system. Complex Systems, 4(4):461-476, 1990.

[472] H. Kitano. Genetic algorithms. Journal of Japanese Society for Artificial Intelligence, 7, Jan. 1992.

[473] H. Kitano. Continuous generation genetic algorithms. Journal of the Society of Instrument and Control Engineers, 32(1):31-38, 1993.

[474] H. Kitano. Genetic algorithm. Sangyo Tosho K.K., Tokyo, 1993.

[475] C. C. Klimasauskas. An Excel macro for genetic optimization of a portfolio. Advanced Technology for Developers, 1(8):11-17, Dec. 1992.

[476] C. C. Klimasauskas. Genetic function optimization for time series prediction. Advanced Technology for Developers, 1(3), July 1992.

[477] C. C. Klimasauskas. Gray codes. Advanced Technology for Developers, 1:18-19, Nov. 1992.

[478] C. C. Klimasauskas. Hybrid neuro-genetic approach to trading algorithms. Advanced Technology for Developers, 1(7):1-8, Nov. 1992.

[479] C. C. Klimasauskas. Genetic algorithm optimizes 100-city route in 21 minutes on a PC! Advanced Technology for Developers, 2:9-17, Feb. 1993.

[480] A. Knijnenburg, E. Matthaus, and V. Wenzel. Concept and usage of the interactive simulation system for ecosystems. Ecological Modelling, 26:51-76, 1984.

[481] S. Koakutsu, Y. Sugai, and H. Hirata. Block placement by improved simulated annealing based on genetic algorithm. Transactions of the Institute of Electronics, Information and Communication Engineers (Japan), J73A(1):87-94, Jan. 1990 (in Japanese).

[482] S. Koakutsu, Y. Sugai, and H. Hirata. Floorplanning by improved simulated annealing based on genetic algorithms. Transactions of the Institute of Electrical Engineers of Japan C, 112-C(7):411-416, July 1992 (in Japanese).

[483] D. Kobelt and G. Schneider. Optimierung im Dialog unter verwendung von Evolutionsstrategie und Einflussgrossenrechnung. Chemie-Technik, 6:369-372, 1977.

[484] G. J. Koehler. Linear discriminant functions determined by genetic search. ORSA Journal on Computing, 3(4):345-357, 1992.

[485] H. M. Kohler. Adaptive genetic algorithm for the binary perceptron problem. Journal of Physics A — Mathematical and General, 23(23):L1265-L1271, 1990.

[486] A. Konagaya. New topics in genetic algorithm research. New Generation Computing, 10(4):423-427, 1992.

[487] A. Konagaya. A stochastic approach to genetic information processing. Journal of Japanese Society for Artificial Intelligence, 8(4):427-438, July 1993 (in Japanese).

[488] H. Kopfer. Genetic algorithms concepts and their application to freight minimization in commercial long distance freight transportation. OR Spektrum, 14(3):137-147, 1992 (in German).

[489] H. Kopfer. Progenitor — a genetic algorithm for production scheduling. Wirtschaftsinformatik, 34(2):255-256, Apr. 1992.

[490] M. Kouchi, H. Inayoshi, and T. Hoshino. Optimization of neural-net structure by genetic algorithm with diploidy and geographical isolation model. Journal of Japanese Society for Artificial Intelligence, 7(3):509-517, 1992 (in Japanese).

[491] J. R. Koza. Non-linear genetic algorithms for solving problems, 1990. (U. S. patent no. 4,935,877. Filed May 20 1988 and issued June 19, 1990).

[492] J. R. Koza. Non-linear genetic algorithms for solving problems by finding a fit composition of functions, 1990 (U.S. patent application filed Mar. 28 1990).

[493] J. R. Koza. A non-linear genetic algorithms for solving problems, 1991 (Australian patent 611,350. Issued Sept. 21, 1991).

[494] J. R. Koza. Genetic Programming: On Programming Computers by Means of Natural Selection and Genetics. The MIT Press, Cambridge, MA, 1992.

[495] J. R. Koza. A non-linear genetic algorithms for solving problems, 1992 (Canadian patent 1,311,561. Issued Dec. 15, 1992).

[496] J. R. Koza, editor. Artificial Life at Stanford. Stanford University Bookstore, Stanford, CA, 1993.

[497] J. R. Koza, editor. Genetic Algorithms at Stanford. Stanford University Bookstore, Stanford, CA, 1993.

[498] J. R. Koza. Genetic programming as a means for programming computers by natural selection. Stat. Comput. (UK), 4(2):87-112, June 1994.

[499] J. R. Koza and J. P. Rice. Non-linear genetic process for use with co-evolving populations, 1990 (U.S. patent application filed Sept. 18, 1990).

[500] J. R. Koza and J. P. Rice. Non-linear genetic process for use with plural co-evolving populations, 1990. (U. S. patent 5,148,513. Filed Sept. 18, 1990. Issued Sept. 15, 1992).

[501] J. R. Koza and J. P. Rice. A non-linear genetic process for data encoding and for solving problems using automatically defined functions, 1992 (U.S. patent application filed May 11, 1992).

[502] J. R. Koza and J. P. Rice. Non-linear genetic process for data encoding and for solving problems using automatically defined functions, 1992 (U. S. patent Application. Filed May 11, 1992).

[503] J. R. Koza and J. P. Rice. A non-linear genetic process for problem solving using spontaneously emergent self-replicating and self-improving entities, 1992 (U.S. patent application filed June 16, 1992).

[504] J. R. Koza and J. P. Rice. Non-linear genetic process for problem solving using spontaneously emergent self-replicating and self-improving entities, 1992 (U. S. patent Application. Filed Jun. 16, 1992).

[505] J. R. Koza, J. P. Rice, and J. Roughgarden. Evolution of food foraging strategies for the Caribbean anolis lizard using genetic programming. Adaptive Behavior, 1(2):47-74, 1992.

[506] V. Kreinovich, C. Quintana, and O. Fuentes. Genetic algorithms: what fitness scaling is optimal? Cybernetics and Systems, 24(1):9-26, Jan.-Feb. 1993.

[507] K. Krishnakumar and D. E. Goldberg. Control system optimization using genetic algorithms. Journal of Guidance, Control, and Dynamics, 15(3):735-739, May-June 1991 (Proceedings of the 1991 AIAA Guidance, Navigation and Control Conference).

[508] K. Krishnakumar and D. E. Goldberg. Control system optimization using genetic algorithms. Journal of Guidance Control and Dynamics, 15(3):735-740, May-June 1992.

[509] K. Kristinsson and G. A. Dumont. System identification and control using genetic algorithms. IEEE Transactions on Systems, Man, and Cybernetics, 22(5):1033-1046, 1992.

[510] B. Korger. Elegant tiefstabeln. MC, 5:72-88, 1991.

[511] W. Kuhn and A. Visser. Identification der Systemparameter 6-achsiger Gelenkarmroboter mit hilfe der Evolutionsstrategie. Robotersysteme, 8(3):123-133, 1992.

[512] J. Kulkarni and H. R. Parsaei. Information resource matrix for production and intelligent manufacturing using genetic algorithm techniques. Computers & Industrial Engineering, 23(1-4):483-485, 1992 (14th Annual Conference on Computers and Industrial Engineering).

[513] L. Kuncheva. Genetic algorithm for feature selection for parallel classifiers. Information Processing Letters, 46(4):163-168, June 1993.

[514] P. J. Kuntz and J. Valldorf. A dim model for homogeneous noble gas ionic clusters. Zeitschrifturf Physik D — Atoms, Molecules and Clusters, 8:195-208, 1988.

[515] F. Kursawe. Evolution strategies: simple "models" of natural processes? Rev. Int. Syst. (Fra), 7(5):627-642, 1993.

[516] W. Kwasnicki and H. Kwasnicka. Market, innovation, competition an evolutionary model of industrial dynamics. Journal of Economic Behaviour and Organization, 19(3):343-368, 1992.

[517] J. E. Labossiere and N. Turrkan. On the optimization of the tensor polynomial failure theory with a genetic algorithm. Transactions of the Canadian Society for Mechanical Engineering, 16(3-4):251-265, 1992.

[518] A. Lane. Programming with genes. AI Expert, 8(12):16-19, Dec. 1993.

[519] C. G. Langton, editor. Artificial Life, The Proceedings of an Interdisciplinary Workshop on the Synthesis and Simulation of Living Systems. Addison-Wesley, Reading, MA, 1989.

[520] C. G. Langton, C. Taylor, J. D. Farmer, and S. Rasmussen, editors. Artificial Life II, Proceedings of the Workshop on Artificial Life Held February, 1990 in Santa Fe, New Mexico, Proceedings Volume X, Santa Fe Institute Studies in the Sciences of Complexity. Addison-Wesley, Reading, MA, 1992.

[521] C. G. Langton, C. Taylor, J. D. Farmer, and S. Rasmussen, editors. Artificial Life III, Santa Fe, NM, 15-19 June 1993. Addison-Wesley, Redwood City, CA.

[522] J. E. Lansberry, L. Wozniak, and D. E. Goldberg. Optimal hydrogenerator governor tuning with a genetic algorithm. IEEE Transactions on Energy Conversion, 7(4):623-630, Dec. 1992 (1992 Winter Meeting of the IEEE/Power Engineering Soc., New York, 26-30 Jan.).

[523] M. Lawo. Automatische Bemessungurf Stochastische Dynamische Belastung. Ph.D. thesis, Universitat-Gesamthochschule Essen, Fachbereich Bauwesen, 1981.

[524] G. Lawton. Genetic algorithms for schedule optimization. AI Expert, 7(5):23-27, May 1992.

[525] R. Leardi, R. Boggia, and M. Terrile. Genetic algorithms as a strategy for feature selection. Journal of Chemometrics, 6(5):267-281, Sept.-Oct. 1992.

[526] B. Lee. Three new algorithms for exact D-optimal design problems. Ph.D. thesis, The Ohio State University, 1993.

[527] J. Lee. Tolerance optimization using genetic algorithm and approximated simulation. Ph.D. thesis, University of Michigan, 1992.

[528] J. Lee and G. E. Johnson. Optimal tolerance allotment using a genetic algorithm and truncated Monte-Carlo simulation. Computer Aided Design, 25(9):601-611, Sept. 1993.

[529] P. L. Lee, editor. Nonlinear Process Control: Applications of Genetic Model Control. Advances in Industrial Control. Springer-Verlag, Berlin, 1993.

[530] M. Lei. Automated acquisition of knowledge for an intelligent system. Zhongguo Jixie Gongcheng, 4(1):4-6, Feb. 1993.

[531] L. Lemarchand, A. Plantec, B. Pottier, and S. Zanati. An object-oriented environment for specification and concurrent execution of genetic algorithms. SIGPLAN OOPS Messenger, 4(2):163-165, Apr. 1993 (addentum to the proceedings of OOPSLA'92).

[532] R. Lerch. Simulation von Ultraschall-wandlern. ACOUSTICA, 57:205-217, 1985.

[533] M. C. Leu and H. Wong. Planning of component placement/insertion sequence and feeder setup in PCB assembly using genetic algorithm. Transactions of ASME, Journal of Electronics Packaging, 115(4):424-432, Dec. 1993.

[534] G. Levitin and J. Rubinovitz. Genetic algorithm for linear and cyclic assignment problem. Computers & Operations Research, 20(6):575-585, Aug. 1993.

[535] S. Levy. Artificial Life: The Quest for new Creation. Pantheon, New York, 1992.

[536] D. R. Lewin. Feedforward control design for distillation systems aided by disturbance cost contour maps. Comput. Chem. Eng., 18(SUPPL):S421-S426, 1994 (Proceedings of the 25th European Symposium of the Working Party on Computer Aided Process Engineering3, Graz (Austria), Jul. 5-7, 1993).

[537] T.-H. Li, C. B. Lucasius, and G. Kateman. Optimization of calibration data with a dynamic genetic algorithm. Analytica Chimica Acta, 268(1):123-134, Oct. 1992.

[538] Y. Li. Heuristic and exact algorithms for the quadratic assignment problem. Ph.D. thesis, The Pennsylvania State University, 1992.

[539] J. Liebowitz. Roll your own hybrids. BYTE, 18(7):113-115, July 1993.

[540] G. E. Liepins. Comparison of neural classifier system approaches to the multiplexer problem. Neural Networks, 1(1):196, 1988 (Proceedings of International Neural Network Society 1988 First Annual Meeting, Boston, MA, 6-10 Sept.).

[541] G. E. Liepins and M. R. Hilliard. Genetic algorithms: Foundations and applications. Annals of Operations Research, 21(1-4):31-58, Nov. 1989.

[542] G. E. Liepins and M. R. Hilliard. Credit assignment and discovery in classifier systems. International Journal of Intelligent Systems, 6:55-69, 1991.

[543] G. E. Liepins and M. D. Vose. Representational issues in genetic algorithms. Journal of Experimental and Theoretical Artificial Intelligence, 2:101-115, 1990.

[544] G. E. Liepins and M. D. Vose. Representational issues in genetic optimization. Journal of Experimental and Theoretical Artificial Intelligence, 2(2):4-30, 1990.

[545] G. E. Liepins and M. D. Vose. Polynomials, basis sets, and deceptiveness in genetic algorithms. Complex Systems, 5(1):45-64, 1991.

[546] G. E. Liepins and M. D. Vose. Characterizing crossover in genetic algorithms. Annals of Mathematics and Artificial Intelligence, 5(1):27-34, 1992.

[547] C.-Y. Lin. Genetic search methods for multicriterion optimal design of viscoelastically damped structures. Ph.D. thesis, University of Florida, 1991.

[548] F.-T. Lin, C.-Y. Kao, and C.-C. Hsu. Applying the genetic approach to simulated annealing in solving some NP-hard problems. IEEE Transactions on Systems, Man, and Cybernetics, 23(6):1752-1767, Dec. 1993.

[549] J.-L. Lin. An analysis of genetic algorithm behavior for combinatorial optimization problems. Ph.D. thesis, The University of Oklahoma, 1993.

[550] X. Liu, A. Sakamoto, and T. Shimamoto. Restrictive channel routing with evolution programs. IEICE Transactions on Fundamentals of Electronics Communications and Computer Sciences, E76-A(10):1738-1745, Oct. 1993.

[551] R. Lohmann. Bionische Verfahren zur Entwicklung visueller Systeme. Ph.D. thesis, Technische Universitat der Berlin, 1991.

[552] S. Louis, G. McGraw, and R. O. Wyckoff. Case-based reasoning assisted explanation of genetic algorithm research. Journal of Experimental and Theoretical Artificial Intelligence, 5(1):21-37, Jan.-Mar. 1993.

[553] S. J. Louis. Genetic algorithms as a computational tool for design. Ph.D. thesis, Indiana University, 1993.

[554] C. B. Lucasius. GATES towards evolutionary large-scale optimization: A software-oriented approach to genetic algorithms. II. toolbox description. Comput. Chem., 18(2):137-156, June 1994.

[555] C. B. Lucasius, L. M. C. Buydens, and G. Kateman. Genetic algorithms for optimization problems in chemometrics. Trends in Analytical Chemistry, 1990.

[556] C. B. Lucasius, A. P. Deweijer, L. M. C. Buydens, and G. Kateman. Cfit — a genetic algorithm for the survival of the fitting. Chemometrics and Intelligent Laboratory Systems, 19(3):337-341, July 1993.

[557] C. B. Lucasius and G. Kateman. Genetic algorithms for large-scale optimization problems in chemometrics — an application. Trac-Trends in Analytical Chemistry, 10(8):254-261, Sept. 1991.

[558] C. B. Lucasius and G. Kateman. Understanding and using genetic algorithms. 1. concepts, properties and context. Chemometrics and Intelligent Laboratory Systems, 19(1):1-33, May 1993.

[559] C. B. Lucasius and G. Kateman. GATES towards evolutionary large-scale optimization: A software-oriented approach to genetic algorithms. I. general perspectives. Comput. Chem., 18(2):127-136, June 1994.

[560] T. A. Ly and J. T. Mowchenko. Applying simulated evolution to high level synthesis. IEEE Transactions on Computer-Aided Design of Integrated Circuits and Systems, 12(3):389- 409, Mar. 1993.

[561] D. Maclay and R. Dorey. Application of genetic search techniques to drivetrain modeling. In Proceedings of the 1992 IEEE International Symposium on Intelligent Control, pages 542-547, Glasgow (Scotland), 11-13 Aug. 1992. IEEE.

[562] D. Maclay and R. Dorey. Applying genetic search techniques to drivetrain modeling. IEEE Control Systems Magazine, 13(3):50-55, 1993.

[563] D. Maclay and R. Dorey. Drivetrain modelling with genetic search techniques. Automotive Engineer, 18(2):47-48, Apr./May 1993.

[564] C. A. Magele, K. Preis, W. Renhart, R. Dyczij-Edlinger, and K. R. Ritcher. Higher order evolution strategies for the global optimization of electromagnetic devices. IEEE Transactions on Magnetics, 29(2):1775-1778, Mar. 1993.

[565] S. W. Mahfoud. An analysis of Boltzmann tournament selection. IlliGAL Report 91007, University of Illinois at Urbana-Champaign, 1991 (also as [566]; anonymous ftp at site gal4.ge.uiuc.edu file /pub/papers/IlliGALs/91007.ps.Z).

[566] S. W. Mahfoud. Finite Markov chain models of an alternative selection strategy for the genetic algorithm. Complex Systems, 7(2):155-170, Apr. 1993.

[567] U. Mahlab, J. Shamir, and H. J. Caulfield. Genetic algorithms for optical pattern recognition. Optics Letters, 16(9):648-650, May 1991.

[568] A. Z. Maksymowicz, J. E. Galletly, M. S. Magdon, and I. L. Maksymowicz. Genetic algorithm approach for Ising-model. Journal of Magnetism and Magnetic Materials, 133(13):40-41, 1993 (11th International Conference on Soft Magnetic Materials, Venice, Italy, Sept. 19 - Oct. 1 1993).

[569] V. R. Mandava, J. M. Fitzpatrick, and I. David R. Pickens. Adaptive search space scaling in digital image registration. IEEE Transactions on Medical Imaging, 8(3):251-262, Sept. 1989.

[570] M. Mangel. Evolutionary optimization and neural network models of behaviour. Journal of Mathematical Biology, 28(3):237-256, 1990.

[571] R. Marimon, E. McGrattan, and T. Sargent. Money as a medium of exchange in an economy with artificially intelligent agents. Journal of Economic Dynamics and Control, 14, 1990.

[572] F. J. Marin, F. Garcia, and F. Sandoval. Genetic algorithms: a strategy for search and optimization. Informatica y Automatica (Spain), 25(3-4):5-15, Nov. 1992. (in Spanish).

[573] R. E. Marks. Breeding hybrid strategies: Optimal behavior for oligopolists. Journal of Evolutionary Economics, 2:17-38, 1992.

[574] R. M. L. Marques, P. J. Schoenmakers, C. B. Lucasius, and G. Kateman. Modelling chromatographic behaviour as a function of pH and solvent composition in RPLC. Chromatographia, 36:83-95, 1993 (in the Proceedings of the 19th International Symposium on Chromatography, Aix-en-Provence (France), 13-18 Sept. 1992).

[575] N. Martin. Convergence properties of a class of probabilistic schemes called reproductive plans. Ph.D. thesis, University of Michigan, Ann Arbor, 1973.

[576] T. Maruyama. Parallel graph partitioning algorithm using a genetic algorithm. JSPP, pages 71-78, 1992 (in Japanese).

[577] A. J. Mason. Genetic Algorithms and Job Scheduling. Ph.D. thesis, University of Cambridge, Department of Engineering, 1992.

[578] K. Mathias. Delta coding strategies for genetic algorithms. Ph.D. thesis, Colorado State University, Fort Collins, 1991.

[579] K. Matsuura, H. Shiba, Y. Nunokawa, and H. Shimizu. Calculation of optimal strategies for fermentation processes by genetic algorithm. Sebutsu-Kogaku Kaishi — Journal of the Society for Fermentation and Bioengineering, 71(3):171-178, 1993.

[580] R. A. J. Matthews. The use of genetic algorithms in cryptanalysis. Cryptologia, 17(2):187- 201, Apr. 1993.

[581] S. Matwin, T. Szapiro, and K. Haigh. Genetic algorithms approach to a negotiation support system. IEEE Transactions on Systems, Man, and Cybernetics, 21(1):102-114, Jan.-Feb. 1991.

[582] A. C. W. May and M. Johnson. Protein structure comparisons using a combination of a genetic algorithm, dynamic programming and least-squares minimization. Protein Engineering, 7(4):475-485, Apr. 1994.

[583] J. S. McCaskill. A stochastic theory of macromolecular evolution. Biological Cybernetics, 50:63-73, 1984.

[584] D. B. McGarrah and R. S. Judson. An analysis of the genetic algorithm method of molecular conformation determination. Journal of Computational Chemistry, 14(11):1385- 1395, 1993.

[585] R. S. McGowan. Recovering articulatory movement from formant frequency trajectories using task dynamics and a genetic algorithm: preliminary model tests. Speech Communications, 14(1):19-48, Feb. 1994.

[586] K. Messa and M. Lybanon. Improved interpretation of satellite altimeter data using genetic algorithms. Telematics and Informatics, 9(3-4):349-356, 1992.

[587] J.-A. Meyer and S. W. Wilson, editors. Proceedings of the First International Conference on Simulation of Adaptive Behavior: From animals to animats, Paris, 24-28 Sept. 1991. A Bradford Book, MIT Press, Cambridge, MA.

[588] W. Michaeli. Materials processing | a key factor. Angewandte Chemie, Advanced Materials, 28(5):660-665, 1989.

[589] Z. Michalewicz. Genetic algorithm for statistical database security. IEEE Bulletin on Database Engineering, 13(3):19-26, Sept. 1990.

[590] Z. Michalewicz. Genetic Algorithms + Data Structures = Evolution Programs. Artificial Intelligence. Springer-Verlag, New York, 1992.

[591] Z. Michalewicz, C. Z. Janikow, and J. R. Krawczyk. A modified genetic algorithm for optimal control problems. Computers & Mathematics with Applications, 23(12):83-94, 1992.

[592] Z. Michalewicz, G. A. Vignaux, and M. F. Hobbs. A nonstandard genetic algorithm for the nonlinear transportation problem. ORSA Journal on Computing, 3(4):307-316, 1991.

[593] E. Michielsen et al. Design of lightweight, broad-band microwave absorbers using genetic algorithms. IEEE Transaction on Microwave Theory and Techniques, 41:1024-1031, 1993.

[594] E. Michielssen, S. Ranjithan, and R. Mittra. Optimal multilayer filter design using real coded genetic algorithms. IEE Proceedings — J Optoelectronics, 139(6):413-420, Dec. 1992.

[595] J. A. Miller, W. D. Potter, R. V. Gandham, and C. N. Lapena. An evaluation of local improvement operators for genetic algorithms. IEEE Transactions on Systems, Man, and Cybernetics, 23(5):1340-1351, Sept./Oct. 1993.

[596] J. F. Miller, H. Luchian, P. V. G. Bradbeer, and P. J. Barclay. Using a genetic algorithm for optimizing fixed polarity Reed-Muller expansions of Boolean functions. International Journal of Electronics, 76(4):601-609, Apr. 1994.

[597] M. Mitchell. Complexity: Imitating life. New Scientist, 137(1860):12-13, 13 Feb. 1993.

[598] A. K. Mitra and H. Brauer. Optimization of a two phase co-current flow nozzle for mass transfer. Verfahrenstechnik, 7(4):92-97, 1973.

[599] Y. Miyamoto, T. Miyatake, S. Kurosaka, and Y. Mori. A parameter tuning for dynamic simulation of power plants using genetic algorithms. Transactions of the Institute of Electrical Engineers of Japan C, 113-D(12):1410-1415, Dec. 1993 (in Japanese).

[600] E. Mjolsness, D. H. Sharp, and B. K. Alpert. Scaling, machine learning, and genetic neural nets. Advances in Applied Mathematics, 10(2):137-163, Dec. 1989.

[601] S. Mohan and P. Mazumder. Wolverines: standard cell placement on a network of workstations. IEEE Transactions on Computer-Aided Design of Integrated Circuits and Systems, 12(9):1312-1326, Sept. 1993.

[602] F. Montoya and J.-M. Dubois. Darwinian adaptive simulated annealing. Europhysics Letters, 22(2):79-84, 10 Apr. 1993.

[603] K. Mori, M. Tsukiyama, and T. Fukuda. Immune algorithm with searching diversity and its application to resource allocation problem. Transactions of the Institute of Electrical Engineers of Japan C, 113-C(10):872-878, Oct. 1993.

[604] K. Morikawa, T. Nakayama, T. Furuhashi, and Y. Uchikawa. LSI assembly line scheduling using a genetic algorithm. Transactions of the Institute of Electrical Engineers of Japan C, 113-D(12):1416-1422, Dec. 1993 (in Japanese).

[605] R. Morin. A look at genetic algorithms. SUNEXPERT Magazine, pages 43-46, 1990.

[606] M. Morrow. Genetic algorithms. Dr. Dobb's Journal, 16(4):26,28,30,32,86,88-89, Apr. 1991.

[607] H. Muhlenbein. Darwin's continent cycle theory and its simulation by the prisoner's dilemma. Complex Systems, 5(5):459-478, 1992.

[608] H.Muhlenbein, M. Gorges-Schleuter, and O. Kramer. New solutions of the mapping problem of parallel systems — the evolution approach. Parallel Computing, 4:269-279, 1987.

[609] H.Muhlenbein, M. Gorges-Schleuter, and O. Kramer. Evolution algorithms in combinatorial optimization. Parallel Computing, 7:65-85, Apr. 1988.

[610] H.Muhlenbein and D. Schlierkamp-Voosen. Predictive models for the breeder genetic algorithm. Evolutionary Computation, 1(1):25-49, 1993.

[611] H.Muhlenbein, M. Schomisch, and J. Born. The parallel genetic algorithm as function optimizer. Parallel Computing, 17:619-632, Sept. 1991.

[612] H.Muller and H. Hofmann. Kinetische untersuchung zur heterogen-katalytischen dehydrochloririerung von 1,1-difluor-1-chlorethan. Chemiker-Zeitung, 114(3):93-100, 1990.

[613] H. Muller and G. Pollhammer. Evolutionsstrategische Lastflussoptimierung. E und M, pages 613-614, 1984.

[614] K. D.Muller. Optimieren mit der Evolutionsstrategie in der Industrie anhand von Beispielen. Ph.D. thesis, Technische Universitat der Berlin, Fachbereich Verfahrenstechnik, 1986.

[615] T. Muntean and E.-G. Talbi. Methodes de placement statique des processus sur architectures paralleles. Technique et Science Informatique TSI, 10(5):355 373, Nov. 1991.

[616] L. J. Murphy, A. R. Simpson, and G. C. Dandy. Design of a pipe network using genetic algorithms. Water, pages 40-42, Aug. 1993.

[617] M. Muselli and S. Ridella. Global optimization of functions with the interval genetic algorithm. Complex Systems, 6(3):193-212, June 1992.

[618] C. Muth. Einfuhrung in die Evolutionsstrategie. Regelungstechnik, 30:297-303, 1982.

[619] J. H. Nachbar. Evolution in the finitely repeated prisoner's dilemma. Journal of Economic Behaviour and Organization, 19(3):307-326, 1992.

[620] T. Nagao, T. Agui, and H. Nagahashi. Extraction of straight lines using a genetic algorithm. Transaction of the Institute of Electronics, Information and Communication Engineers D-II (Japan), J75D-II(4):832-834, 1992 (in Japanese).

[621] T. Nagao, T. Agui, and H. Nagahashi. Structural evolution of neural networks by a genetic method. Transaction of the Institute of Electronics, Information and Communication Engineers D-II (Japan), J76D-II(3):557-565, 1993 (in Japanese).

[622] T. Nagao, T. Agui, and H. Nagahashi. Structural evolution of neural networks having arbitrary connection by a genetic method. IEICE Transactions on Information and Systems, E76-D(6):689-697, June 1993.

[623] S. Nagendra, R. T. Haftka, and Z.Gurdal. Stacking sequence optimization of simply supported laminates with stability and strain constraints. AIAA Journal, 30(8):2132- 2137, Aug. 1992.

[624] Y. Nakanishi and S. Nakagiri. Representation of topology by boundary cycle and its application to structural optimization (a formulation to combine algebraic topology with genetic algorithm). Nippon Kikai Gakkai Ronbunshu A Hen, 59(567):2783-2788, Nov. 1993.

[625] K. Nara, A. Shiose, M. Kitagawa, and T. Ishihara. Implementation of genetic algorithm for distribution systems loss minimum re-configuration. IEEE Transactions on Power Systems, 7(3):1044-1051, Aug. 1992.

[626] S. Nara and W. Banzhaf. Pattern search using a genetic algorithm. Japanese Journal on Condensed Matter Research, 56:235-238, 1991.

[627] M. N. Narayanan and S. B. Lucas. A genetic algorithm to improve a neural network to predict a patient's response to Warfarin. Methods of Information in Medicine, 32(1):55-58, Feb. 1993.

[628] J. T. Ngo and J. Marks. Physically realistic motion synthesis in animation. Evolutionary Computation, 1(3), 1993.

[629] Y. Nishikawa and H. Tamaki. A genetic algorithm as applied to the jobshop scheduling. Transactions of the Society of Instrument and Control Engineers (Japan), 27(5):593-599, May 1991 (in Japanese).

[630] V. Nissen. Evolutionare Algorithmen, Darstellung, Beispiele, betriebswirtschaftliche Anwendungmoglichkeiten. DUV Deutscher Universitats Verlag, Wiesbaden, 1994.

[631] A. E. Nix and M. D. Vose. Modeling genetic algorithms with Markov chains. Annals of Mathematics and Artificial Intelligence, 5(1):79-88, Apr. 1992.

[632] W. Nooss. Konnen Rechenautomaten durch Optimierungsprogramme Neues entdecken? Burotechnik + Automation, 11:214-221, 1970.

[633] W. Nooss. Automatische Synthese von Viergelenkgetrieben durch Digitalrechner. Feinwerktechnik, 75(4):165-168, 1971.

[634] W. Nooss. Ein Universell anwendbares Rechner-Unterprogrammurf Entwurf und Optimierung. Angewandte Informatik, 13:123-129, 1971.

[635] H. G. Nurnberg and G. Vossius. Evolutionsstrategie-ein Regelkonzeptufr die funktionelle Elektrostimulation gelahmter Gliedmassen. Biomedizinische Technik, 31:52-53, Sept. 1986.

636] J. T. Nutter and Y. Ding. Bridging the gap: combining high and low level representations for knowledge retention with genetic algorithms. International Journal of Expert Systems Research and Applications, 4(3):249-280, 1991.

[637] W. Oberdieck, B. Richter, and P. Zimmermann. Evolutionsstrategie I Ein Hilfsmittel bei derosLung fahrzeugtechnischer Aufgaben. Automobiltechnische Zeitschrift, 84(7/8):331- 337, 1982.

[638] J. Oda, N. Matsumoto, and A. lin Wang. Selection method of control members for adaptive truss structures using genetic algorithms (GA). Nippon Kikai Gakkai Ronbunshu C Hen, 60(570):513-518, Feb. 1994 (in Japanese).

[639] J. Oda, N. Matsumoto, and A. Wang. Design method of homologous structures using genetic algorithms (ga). Nippon Kikai Gakkai Ronbunshu A Hen, 59(568):3056-3061, Dec. 1993 (in Japanese).

[640] M. J. O'Dare and T. Arslan. Generating test patterns for VLSI circuits using a genetic algorithm. Electronics Letters, 30(10):778-779, 12 May 1994.

[641] J. Oliver. Finding decision rules with genetic algorithms. AI Expert, 9(3):33-39, Mar. 1994.

[642] A. W. O'Neill. Genetic based training of two-layer, optoelectronic neural network. Electronics Letters, 28(1):47-48, Jan. 1992.

[643] J. Onoda and Y. Hanawa. Actuator placement optimization by genetic and improved simulated annealing algorithms. AIAA Journal, 31(6):1167-1169, June 1993.

[644] N. H. Packard. A genetic learning algorithm for the analysis of complex data. Complex Systems, 4(5):543-572, Oct. 1990.

[645] S. E. Page and D. W. Richardson. Walsh functions, schema variance, and deception. Complex Systems, 6(2):125-135, Apr. 1992.

[646] K. F. Pal. Genetic algorithms for the traveling salesman problem based on a heuristic crossover. Biological Cybernetics, 69(5-6):539-549, 1993.

[647] S. K. Pal, D. Bhandari, and M. K. Kundu. Genetic algorithms for optimal image enchancement. Pattern Recognition Letters, 15(3):261-271, Mar. 1994.

[648] F. Papentin. A Darwinian evolutionary system — ii. experiments on protein evolution and evolutionary aspects of the genetic code. Journal of Theoretical Biology, 39:417-430, 1973.

[649] F. Papentin. A Darwinian evolutionary system — iii. experiments on the evolution of feeding patterns. Journal of Theoretical Biology, 39:431-445, 1973.

[650] S. H. Park, Y. H. Kim, K. B. Sim, and H. T. Jeon. Auto-generation of fuzzy rule base using genetic algorithms. Journal of Korean Institute of Telematics and Electronics, 29B(2):60- 68, Feb. 1992 (in Korean).

[651] I. Parmee and P. Booker. Applying the genetic algorithm to design problems: Progress at the Plymouth Engineering Design Center. Engineering Designer, 19(3):17-18, May/June 1993.

[652] S. Parry. Fittest filters in real world. New Electronics (UK), 26(3):15-16, Mar. 1993.

[653] A. W. R. Payne and R. C. Glen. Molecular recognition using a binary genetic search algorithm. Journal of Molecular Graphics, 11(2):74-91, June 1993.

[654] Z. A. Perry. Experimental study of speciation in ecological niche theory using genetic algorithms. Ph.D. thesis, University of Michigan, Ann Arbor, 1984 (University Microfilms No. 8502912).

[655] T. K. Peters, H.-E. Koralewski, and E. W. Zerbst. Search for optimal frequencies and amplitudes of therapeutic electrical carotid sinus nerve stimulation by application of the evolution strategy. Artificial Organs, 13(2):133-143, 1980.

[656] T. K. Peters, H.-E. Koralewski, and E. W. Zerbst. The evolution strategy — a search strategy used in individual optimization of electrical parameters for therapeutic carotid sinus nerve stimulation. IEEE Transactions on Biomedical Engineering, 36(7):668-675, July 1991.

[657] U. Petersohn, K. Voss, and K. H. Weber. Genetische Adaptation — ein stochastisches Suchverfahrenufr diskrete Optimierungsprobleme. Matematische Operationsforschung und Statistik, 5(7,8):555-571, 1974.

[658] C. Peterson. Parallel distributed approaches to combinatorial optimization: benchmark studies on traveling salesman problem. Neural Computation, 2:261-269, 1990.

[659] I. Peterson. Natural selection for computers. Science News, 136:346-348, 1989.

[660] G. Pettersson. Evolutionary optimization of the catalytic efficiency of enzymes. European Journal of Biochemistry, 206(1):289-295, May 1992.

[661] C. C. B. Pettey. An analysis of a parallel genetic algorithm. Ph.D. thesis, Vanderbilt University, Nashville, 1990 (University Microfilms No. 90-26497).

[662] E. J. Pettit and M. J. Pettit. Analysis of the performance of a genetic algorithm-based system for message classification in noisy environments. International Journal of ManMachine Studies, 27(2):205-220, Aug. 1987.

[663] D. T. Pham and D. Karaboga. Optimum design of fuzzy logic controllers using genetic algorithms. Journal of Systems Engineering, 1(2):114-118, 1991.

[664] D. T. Pham and H. H. Onder. A knowledge-based system for optimizing workplace layouts using a genetic algorithm. Ergonomics, 35(12):1479-1497, 1992.

[665] D. T. Pham and Y. Yang. A genetic algorithm based preliminary design system. Proceedings of the Institution of Mechanical Engineers, Part D, (Journal of Automobile Engineering), 207(D2):127-133, 1993.

[666] E. E. Pichler, J. D. Keeler, and J. Ross. Comparison of self-organization and optimization in evolution and neural networks models. Complex Systems, 4:75-106, 1990.

[667] W. E. Pinebrook. Drag minimization on a body of revolution. Ph.D. thesis, University of Houston, Texas, 1982 (University Microfilms No. 82-19517).

[668] W. E. Pinebrook. The evolution strategy applied to drag minimization on a body of revolution. Mathematical Modelling, 4:439-450, 1983.

[669] W. E. Pinebrook and C. H. Dalton. Drag minimization on a body of revolution through evolution. Computer Methods in Applied Mechanics and Engineering, 39(2):179-197, 1983.

[670] T. W.-S. Plum. Simulation of a cell-assembly model. Ph.D. thesis, University of Michigan, Ann Arbor, 1972.

[671] H. J. Poethke and H. Kaiser. A simulation approach to evolutionary game theory: The evolution of time-sharing behavior in a dragonfly mating system. Behavioural Ecology and Sociobiology, 18:155-163, 1985.

[672] P. W. Poon. Genetic algorithms and fuel cycle optimization. Nuclear Engineer, 31(6):173- 177, Nov.-Dec. 1990.

[673] J. Popplau. Die Anwendung einer (fl =ae; )-Evolutionsstrategie zur direkten Minimierung eines nicht-linearen Funktionals unter Vervendung von FE-Ansatzfunktionen am Beispiel des Brachistochronenproblems. Zeitschrift urf Angewandte Mathematik und Mechanik, 61:T305-T307, 1981.

[674] B. Porter and A. H. Jones. Genetic tuning of PID controllers. Electronics Letters, 28(9):843-844, 23. Apr. 1992.

[675] B. Porter and S. S. Mohamed. Genetic design of minimum-time controllers. Electronics Letters, 29(21):1897-1898, Oct. 1993.

[676] W. D. Potter, J. A. Miller, B. E. Tonn, R. V. Gandham, and C. N. Lapena. Improving the reliability of heuristic multiple fault diagnosis via the EC-based genetic algorithm. International Journal of Artificial Intelligence, 2(1):5-23, July 1992.

[677] D. J. Powell. Inter-GEN: A hybrid approach to engineering design optimization. Ph.D. thesis, Rensselaer Polytechnic Institute, Troy, New York, 1990.

[678] K. Preis, O. Biro, M. Friedrich, A. Gottvald, and C. A. Magele. Comparison of different optimization strategies in the design of electromagnetic devices. IEEE Transactions on Magnetics, 27(5):4145-4147, 1991.

[679] K. Preis, C. A. Magele, and O. Biro. FEM and evolution strategies in the optimal design of electromagnetic devices. IEEE Transactions on Magnetics, 26(2):2181-2183, 1990.

[680] K. Preis and A. Ziegler. Optimal design of electromagnetic devices with evolution strategies. Compel — The International Journal for Computations and Mathematics in Electrical and Electronic Engineering, 9(Supplement A):119-122, 1990.

[681] X. Qi. Analysis and Application of Darwinian optimization Algorithms in the Multidimensional Spaces. Ph.D. thesis, The University of Connecticut, 1993.

[682] N. Queipo, R. Devarakonda, and J. A. C. Humphrey. Genetic algorithms for thermosciences research: Application to the optimized cooling of electronic components. Int. J. Heat Mass Transfer, 37(6):893-908, Apr. 1994.

[683] J. R. Quinlan. An empirical comparison of genetic and decision-tree classifiers. Machine Learning, 5:135-141, 1990.

[684] R. Rada. Evolution and gradualness. BioSystems, 14:211-218, 1981.

[685] R. Rada. Evolutionary structure and search. Ph.D. thesis, 1981. University Microfilm No. 81-14463.

[686] A. Radcliffe. A problem solving technique based on genetics. Creative Computing, 3(2):78-81, Apr. 1981.

[687] N. J. Radcliffe. Equivalence class analysis of genetic algorithms. Technical Report TR-9003, Edinburgh Parallel Computing Centre, 1990 (published also as [689]; anonymous ftp at site ftp.epcc.ed.ac.uk file /pub/tr/90/tr9003.ps.Z).

[688] N. J. Radcliffe. Genetic neural networks on MIMD computers. Ph.D. thesis, University of Edinburgh, Theoretical Physics, 1990.

[689] N. J. Radcliffe. Equivalence class analysis of genetic algorithms. Complex Systems, 5(2):183-205, 1991.

[690] N. J. Radcliffe. Genetic set recombination and its application to neural network topology optimization. Technical Report TR-91-21, Edinburgh Parallel Computing Centre, 1991 (published also as [691]; anonymous ftp at site ftp.epcc.ed.ac.uk file/pub/tr/91/tr9121.ps.Z).

[691] N. J. Radcliffe. Genetic set recombination and its application to neural network topology optimization. Neural Computing and Applications, 1(1):67-90, 1993.

[692] N. J. Radcliffe and G. Wilson. Natural solutions give their best. New Scientist, 126:47-50, 14 Apr. 1990.

[693] S. Rahman. Artificial intelligence in electric power systems: a survey of the Japanese industry. IEEE Transactions on Power Systems, 8(3):1211-1218, Aug. 1993.

[694] S. Rajeev and C. S. Krishnamoorthy. Discrete optimization of structures using genetic algorithms. Journal of Structural Engineering — ASCE, 118(5):1233-1250, May 1992.

[695] S. Rajeev and C. S. Krishnamoorthy. Discrete optimization of structures using genetic algorithms (closure). Journal of Structural Engineering — ASCE, 119(8):2495-2496, Aug. 1993.

[696] R. P. Rankin. Considerations for rapidly converging genetic algorithms designed for application to problems with expensive evaluation functions. Ph.D. thesis, University of Missouri-Rolla, 1993.

[697] S. S. Rao, T.-S. Pan, and V. B. Venkayya. Optimal placement of actuators in actively controlled structures using genetic algorithms. AIAA Journal, 29(6):942-943, June 1991.

[698] B. Ravichandran. Two-dimensional and three-dimensional model-based matching using a minimum representation criterion and a hybrid genetic algorithm. Ph.D. thesis, Rensselaer Polytechnic Institute, Troy, NY, Department of Electrical, Computer and Systems Engineering, 1993.

[699] G. J. E. Rawlins, editor. Foundations of Genetic Algorithms, Indiana University, 15-18 July 1990 1991. Morgan Kaufmann: San Mateo, CA.

[700] I. Rechenberg. Evolutionsstrategie: Optimierung technischer Systeme nach Prinzipien der biologischen Evolution. Ph.D. thesis, Technische Universitat der Berlin, 1971.

[701] I. Rechenberg. Bionik, evolution und Optimierung. Naturwissenschaftliche Rundschau, 11(26):465-472, 1973.

[702] I. Rechenberg. Evolutionsstrategie: Optimierung technisher Systeme nach Prinzipien der biologischen Evolution. Frommann-Holzboog Verlag, Stuttgart, 1973 (2nd edition 1993).

[703] I. Rechenberg. Problemlosungen mit Evolutionsstrategien. Proceedings in Operations Research, 9:499, 1980.

[704] R. D. Recknagel and W. A. Knorre. Anwendung biologischer Evolutionsprinzipien zur Optimierung von Fermentationsprozessen. Zeitschrifturf allgemeine Mikrobiologie, 24(7):479-483, 1984.

[705] J. Reed, R. Toombs, and N. A. Barricelli. Simulation of biological evolution and machine learning. Journal of Theoretical Biology, 17:319-342, 1967.

[706] B. Reetz. Greedy solutions to the traveling sales person problem. Advanced Technology for Developers, 2:8-14, May 1993.

[707] C. R. Reeves, editor. Modern Heuristic Techniques for Combinatorial Problems. Blackwell Scientific Publications, Oxford, 1993.

[708] C. Reiter. Toy universes. Science '86, 7(5):55-59, 1986.

[709] R. G. Reynolds and J. I. Maletic. The use of version space controlled genetic algorithm to solve the Boole problem. International Journal of Artificial Intelligence Tools, Architectures, Languages and Algorithms (Singapore), 2(2):219-234, June 1993.

[710] G. G. Richards and H. Yang. Distribution system harmonic worst case design using a genetic algorithm. IEEE Transactions of Power Delivery, 8(3):1484-1491, July 1993.

[711] G. G. Richards, H. Yang, P. K. Kalra, S. C. Srivastava, S. K. Mishra, R. Adapa, and P. Ribeiro. Distribution-system harmonic worst-case design using a genetic algorithm. IEEE Transactions on Power Delivery, 8(3):1484-1491, 1993 (in Proceedings of 1992 Summer Meeting of IEEE/Power-Engineering-Society, Seattle, WA, 12-16 July).

[712] R. L. Riche and R. T. Haftka. Optimization of laminate stacking sequence for buckling load maximization by genetic algorithm. AIAA Journal, 31(5):951-956, May 1993.

[713] H. J. Riedel. Einsatz rechnergestutzter optimierung mittels der Evolutionsstrategie zur oslung galvanotechnischer Probleme. Ph.D. thesis, Technische Universitat der Berlin, Fachbereich Verfahrenstechnik, 1984.

[714] L. Riekert. Moglichkeiten und Grenzen deduktiven Vorgehens bei der Entwicklung technischer Katalysatoren. Chem.-Ing.Tech., 53(12):950-954, 1981.

[715] R. L. Riolo. Empirical studies of default hierarchies and sequences of rules in learning classifier systems. Ph.D. thesis, University of Michigan, Department of Computer Science and Engineering, 1988 (University Microfilms No. 89-07143).

[716] R. L. Riolo. Survival of the fittest bits. Scientific American, 267(1):89-91, July 1992.

[717] B. J. Ritzel, J. W. Eheart, and S. Ranjithan. Using genetic algorithms to solve a multiple objective groundwater pollution containment problem. Water Resources Research, 30(5):1589-1603, May 1994.

[718] M. Rizki and M. Conrad. Evolve III: A discrete events model of an evolutionary ecosystem. BioSystems, 18:121-133, 1985.

[719] M. Rizki and M. Conrad. Computing the theory of evolution. Physica D, 22:83-99, 1986.

[720] J. Roberts. Structure-based drug design ten years on. Nature-Structural Biology, 1(6), 1994.

[721] S. M. Roberts and B. Flores. An engineering approach to the travelling salesman problem. Man. Sci., 13:269-288, 1966.

[722] G. G. Robertson. Population size in classifier systems. Machine Learning, 5:142-152, 1990.

[723] G. G. Robertson and R. L. Riolo. A tale of two classifier systems. Machine Learning, 3(2/3):139-160, Oct. 1988.

[724] D. Rock and J. Hirsh. Will GAs breed with aerospace? AI Expert, 8(12):28-34, Dec. 1993.

[725] R. Rodloff and H. Neuhauser. Application of an evolution strategy to calculate statistic and dynamic dislocation group configurations. Physica Status Solidi (a), 37:K93-K96, 1976.

[726] L. L. Rogers. Optimal groundwater remediation using artificial neural networks and the genetic algorithm. Ph.D. thesis, Stanford University, 1992.

[727] J. P. Ros. Learning Boolean functions with genetic algorithms: A PAC analysis. Ph.D. thesis, University of Pittsburgh, 1992.

[728] R. S. Rosenberg. Simulation of genetic populations with biochemical properties. Ph.D. thesis, University of Michigan, Ann Arbor, 1967 (University Microfilm No. 67-17,836).

[729] R. S. Rosenberg. Simulation of genetic populations with biochemical properties: I. the model. Mathematical Biosciences, 7:223-257, 1970.

[730] R. S. Rosenberg. Simulation of genetic populations with biochemical properties: II. selection of crossover probabilities. Mathematical Biosciences, 8:1-37, 1970.

[731] W. M. Rudnick. Genetic algorithms and fitness variance with an application to the automated design of artificial neural networks. Ph.D. thesis, Oregon Graduate Institute of Science and Technology, 1992.

[732] R. Ruthen. Trends in nonlinear dynamics: Adapting to complexity. Scientific American, 268(1):110-117, Jan. 1993.

[733] J. Ryan. Review of: D. E. Goldberg, 1989 genetic algorithms in search, optimization and machine learning. ORSA Journal on Computing, 3(2):176, 1991.

[734] Y. G. Saab and V. B. Rao. Combinatorial optimization by stochastic evolution. IEEE Transactions on Computer-Aided Design of Integrated Circuits and Systems, 10(4):525- 535, 1991.

[735] L. Saarenmaa. Induktiivinen oppiminen metsanviljelyn tietokannan tulkinnassa. Ph.D. thesis, University of Helsinki, Department of Forest Ecology, 1992.

[736] J. Sakamoto and J. Oda. Technique for determination of optimal truss layout using genetic algorithm. Nippon Kikai Gakkai Ronbunshu A Hen, 59(562):1568-1573, June 1993.

[737] S. Sakane, T. Kuruma, T. Omata, and T. Sato. Planning focus of attention with consideration of time varying aspect-search of the best plan by using a genetic algorithm. Transactions of the Society of Instrument and Control Engineers (Japan), 28(9):1111-1117, Sept. 1992 (in Japanese).

[738] M. S. Sambridge and G. Drijkoningen. Genetic algorithms in seismic waveform inversion. Geophysical Journal International, 109(2):323-342, May 1992.

[739] A. V. Sannier, II. A computational theory of learning in distributed systems. Ph.D. thesis, Michigan State University, 1988.

[740] N. Saravanan and D. B. Fogel. A bibliography of evolutionary computation & applications. Technical Report FAU-ME-93-100, Florida Atlantic University, Department of Mechanical Engineering, 1993 (available via anonymous ftp at magenta.me.fau.edu /pub/ep-list/bib/EC-ref.ps.Z).

[741] J. D. Schaffer. Some experiments in machine learning using vector evaluated genetic algorithms. Ph.D. thesis, Vanderbilt University, Nashville, TN, 1984 (University Microfilms No. 85-22492).

[742] J. D. Schaffer, editor. Proceedings of the Third International Conference on Genetic Algorithms, George Mason University, 4-7 June 1989. Morgan Kaufmann Publishers, Inc.

[743] J. D. Schaffer and A. Morishima. Adaptive knowledge representation: A content sensitive recombination mechanism for genetic algorithms. International Journal of Intelligent Systems, 3:229-246, 1988.

[744] A. Scheel. Ein beitrag zur Theorie der Evolutionsstrategie. Ph.D. thesis, Technische Universitat der Berlin, 1985.

[745] L. Schmid. Discrete optimization of structures using genetic algorithms (discussion). Journal of Structural Engineering-ASCE, 119(8):2494-2496, Aug. 1993.

[746] H. Schmiedl. Anwendung der Evolutionsoptimierung bei Microwellenschaltungen. Frequenz, 35(11):306-310, 1981.

[747] K. Schneider. Evolving the best solution. Industrial Solutions, 222(19):27-28, 1989.

[748] A. Schober, M. Thuerk, and M. Eigen. Optimization by hierarchical mutant production. Biological Cybernetics, 69(5-6):493-501, 1993.

[749] P. Scholz. Die darwinische Evolution als Strategie-modellurf die numerische Optimierung von Parametern nichtlinearer Regressionsfunktionen. EDV in Medizin und Biologie, 13(2):36-43, 1982.

[750] E. Schoneburg and F. Heinzmann. Perplex: Produktionsplanung nach dem Vorbild der Evolution. Wirtschaftsinformatik, 34(2):224-232, Apr. 1992.

[751] N. N. Schraudolph and R. K. Belew. Dynamic parameter encoding for genetic algorithms. Machine Learning, 9(1):9-21, June 1992.

[752] L. Schreiber. Parametrization of mass models with the evolution strategy. Zeitschrifturf Angewandte Mathematik und Mechanik, 73(4-5):T343-T345, 1993 (in German).

[753] P. A. Schrodt. Short-term prediction of international behavior using a Holland classifier. Mathematical and Computer Modelling, 12(4/5):589-600, 1989.

[754] R. Schultheis, R. Rautenbach, and G. Bindl. Entwicklung von Ventrikelmodellen nach dem Prinzip der biologischen Evolution. Biomedizinische Technik, 21E:197-198, 1976.

[755] A. C. Schultz, J. J. Grefenstette, and K. A. D. Jong. Test and evaluation by genetic algorithms. IEEE Expert, 8(5):9-14, 1993.

[756] H.-P. Schwefel. Evolutionsstrategie und numerische Optimierung. Ph.D. thesis, Technische Universitat der Berlin, 1975.

[757] H.-P. Schwefel. Numerische Optimierung von Computer-Modellen mittels der Evolutionsstrategie. Birkhauser Verlag, Basel and Stuttgart, 1977 (in German; in English as [758]).

[758] H.-P. Schwefel. Numerical Optimization of Computer Models. John Wiley, Chichester,1981 (also as [757]).

[759] H.-P. Schwefel. Evolution strategies: A family of non-linear optimization techniques based on imitating some principles of organic evolution. Annals of Operations Research, 1:165- 167, 1984.

[760] H.-P. Schwefel. Systems analysis, systems design, and evolutionary strategies. Systems Analysis - Modeling - Simulation, 7(11/12):853-864, 1990.

[761] H.-P. Schwefel, editor. Proceedings of PPSN3, Israel, 9-14 Oct. 1994. Springer-Verlag.

[762] H.-P. Schwefel and T. Back. Kunstliche Evolution — eine intelligente Problemlosungsstrategie? KI -Kunstliche Intelligenz, 6(2):20-27, June 1992.

[763] A. M. Segre. Applications of machine learning. IEEE Expert, 7(3):30-34, 1992.

[764] M. T. Semertridis, S. Hazout, and J.-P. Mornon. A computer based simulation with artificial adaptive agents for predicting secondary structure from the protein hydrophobicity [abstract]. Protein Science, 2(Suppl. 1):66, July 1993 (Proceedings of the Seventh Symposium of the Protein Society, San Diego, CA, July 24-28).

[765] M. T. Semertzidis. Developement de etmhodes besees sur les mathematiques, l'informatique et l'intelligence artificielle pour l'alignement de sequences et la eprdiction de structures proteines [Development of mathematical, computing and artificial intelligence methods for the protein secondary structure prediction]. Ph.D. thesis, University of Paris 7, 1994 (in French).

[766] M. K. Sen and P. L. Stoffa. Rapid sampling of model space using genetic algorithms: Examples from seismic waveform inversion. Geophysical Journal International, 108(1):281+, Jan. 1992.

[767] D. Seniw. A genetic algorithm for the traveling salesman problem. Ph.D. thesis, University of North Carolina at Charlotte, 1991.

[768] D. Shafer. Global optimization in optical design. Computers in Physics, 8(2):188-195, Mar./Apr. 1994.

[769] K. Shahookar and P. Mazumder. A genetic approach to standard cell placement using meta-genetic parameter optimization. IEEE Transactions on Computer-Aided Design, 9(5):500-511, May 1990.

[770] K. Shahookar and P. Mazumder. VLSI cell placement techniques. ACM Computer Surveys, 23(2):143-220, June 1991.

[771] T. Shibata and T. Fukuda. Path planning using genetic algorithms (2nd report, selfish planning and coordinative planning for multiple robot systems). Nippon Kikai Gakkai Ronbunshu C Hen, 59(560):1134-1141, Apr. 1993 (in Japanese).

[772] T. Shibata and T. Fukuda. Coordination in evolutionary multi-agent-robotic system using fuzzy and genetic algorithm. Control Engineering Practice, 2(1):103-111, Jan. 1994 (Proceedings of 1993 IEEE Workshop on Neuro-Fuzzy Control: Instrumentation and Control Applications, Muroran (Japan)).

[773] L. Shu. The impact of data structures on the performance of genetic-algorithm-based learning. Ph.D. thesis, University of Alberta, Canada, 1992.

[774] W. Siedlecki and J. Sklansky. A note on genetic algorithms for large scale feature selection. Pattern Recognition Letters, 10(5):335-347, Nov. 1989.

[775] R. Sikora. Learning control strategies for chemical processes, a distributed approach. IEEE Expert, 7(3):35-43, 1992.

[776] A. R. Simpson and S. D. Priest. The application of genetic algorithms to optimization problems in geotechnics. Computers and Geotechnics, 15(1):1-19, 1993.

[777] K. Sims. Artificial evolution for computer graphics. Computer Graphics, 25(4):319-328, July 1991.

[778] K. Sims. Interactive evolution of equations for procedural models. The Visual Computer, 9:466-476, 1993.

[779] S. R. F. Sims and B. V. Dasarathy. Automatic target recognition using a passive multisensor suite. Optical Engineering, 31(12):2584-2593, Dec. 1992.

[780] M. Sinclair. Comparison of the performance of modern heuristics for combinatorial optimization on real data. Computers & Operations Research, 20(7):687-695, Sept. 1993.

[781] R. E. Smith. Default hierarchy formation and memory exploitation in learning classifier systems. Ph.D. thesis, University of Alabama, 1991 (also TCGA Report No. 91003).

[782] R. E. Smith and D. E. Goldberg. Diploidy and dominance in artificial genetic search. Complex Systems, 6(3):251-285, June 1992.

[783] R. E. Smith and D. E. Goldberg. Reinforcement learning with classifier systems: Adaptive default hierarchy formation. Applied Artificial Intelligence, 6(1):79-102, 1992 (also TCGA Report No. 90002).

[784] R. W. Smith. Energy minimization in binary alloy models via genetic algorithms. Computer Physics Communications, 71(2):134-146, Aug. 1992.

[785] S. F. Smith. A learning system based on genetic adaptive algorithms. Ph.D. thesis, University of Pittsburgh, 1980 (University Microfilms No. 81-12638).

[786] H. Sonnenschein. A modular optimization calculation method of power station energy balance and plat efficiency. Journal of Engineering for Power, 104:255-259, 1982.

[787] B. Soucek and the IRIS Group, editors. Dynamic, Genetic, and Chaotic Programming. Sixth Generation Computer Technologies. John Wiley & Sons, New York, 1992.

[788] R. Spillman. Cryptanalysis of knapsack ciphers using genetic algorithms. Cryptologia, 17(4):367-377, Oct. 1993.

[789] R. Spillman. Genetic algorithms. Dr. Dobb's Journal, 18(2):26,28,30,90-93, Feb. 1993.

[790] R. Spillman, M. Janssen, B. Nelson, and M. Kepner. Use of a genetic algorithm in the cryptanalysis of simple substitution ciphers. Cryptologia, 17(1):31-44, Jan. 1993.

[791] J. L. Sponsler. Genetic algorithms applied to the scheduling of the Hubble space telescope. Telematics and Informatics, 6(3-4):181-190, 1989.

[792] T. J. Starkweather. Optimization of sequencing problems using genetic algorithms. Ph.D. thesis, Colorado State University, 1993.

[793] R. M. Stein. Real artificial life. BYTE, pages 289-298, Jan. 1991.

[794] J. Stender, editor. Parallel Genetic Algorithms. IOS Press, Amsterdam, 1993.

[795] D. J. Stockton and L. Quinn. Identifying economic order quantities using genetic algorithms. International Journal of Production Management, 13(11):92-103, 1993.

[796] P. L. Stoffa and M. K. Sen. Nonlinear multiparameter optimization using genetic algorithms — inversion of plane wave seismograms. Geophysics, 56(11):1794-1810, Nov. 1991.

[797] D. Suckley. Genetic algorithm in the design of FIR filters. IEE Proceedings, Part G: Electronic Circuits and Systems, 138(2):234-238, Apr. 1991.

[798] H. Sugimoto, B. L. Lu, and H. Yamamoto. Study on an improvement of reliability of GA for the discrete structural optimization. Doboku Gakkai Rombun Hokokushu, (471):67-76, July 1993.

[799] H. Sugimoto, H. Yamamoto, T. Sasaki, and J. Mitsuo. On design optimization of design of retaining wall structures by genetic algorithm. Doboku Gakkai Rombun Hokokushu, (474):105-114, 1993.

[800] B. H. Sumida. Genetics for genetic algorithms. SIGBIO Newsletter, 12(2):44-46, 1992.

[801] B. H. Sumida, A. I. Houston, J. M. McNamara, and W. D. Hamilton. Genetic algorithms and evolution. Journal of Theoretical Biology, 147(1):59-84, Nov. 1990.

[802] S. Sun. Reduced representation model of protein structure prediction: Statistical potential and genetic algorithms. Protein Science, 2(5):762-785, May 1993.

[803] Y. Takahashi. Convergence of the genetic algorithm to the type I two bit problem. Transaction of the Institute of Electronics, Information and Communication Engineers A (Japan), J76-A(3):556-559, 1993.

[804] M. Takeuchi and A. Sakurai. A genetic algorithm with self-formation mechanism of genotype-to-phenotype mapping. Transaction of the Institute of Electronics, Information and Communication Engineers D-I (Japan), J76D-I(6):229-236, June 1993 (in Japanese).

[805] E.-G. Talbi. Etude experimentale d'algorithmes de placement de processus. Lettre du Transputer et des Calculateurs Distribues, 15:7-26, Sept. 1992 (in French).

[806] E.-G. Talbi. Allocation de processus sur les architectures parallelesaemmoire distribuee. Ph.D. thesis, l'Institut National Polytechnique de Grenoble, May 1993. (in French).

[807] E.-G. Talbi and P. Bessiere. A parallel genetic algorithm applied to the mapping problem. SIAM News, 24(4):12-27, July 1991.

[808] S. N. Talukdar, P. S. de Souza, and S. Murthy. Organizations for computer-based agents. Int. J. Eng. Intell. Syst., 1(2):75-87, Sept. 1993.

[809] K. Y. Tam. Genetic algorithms, function optimization, and facility design. European Journal of Operational Research, 63(2):322-346, Dec. 1992.

[810] H. Tamura, A. Hirahara, I. Hatono, and M. Umano. An approximate solution method for combinatorial optimization — a hybrid approach of genetic algorithm and Lagrange relaxation method. Transactions of the Society of Instrument and Control Engineers (Japan), 30(3):329-336, Mar. 1994 (in Japanese).

[811] M. Tanaka, T. Hattori, and T. Tanino. Jump detection and identification of linear systems by the genetic algorithm. Transactions of the Society of Instrument and Control Engineers (Japan), 28(11):1383-1385, Nov. 1992 (in Japanese).

[812] R. Tanese. Distributed genetic algorithms for function optimizations. Ph.D. thesis, University of Michigan, Department of Electrical Engineering and Computer Science, 1989 (University Microfilms No. 90-01722).

[813] N. Taniguchi, X. Liu, A. Sakamoto, and T. Shimamoto. An approach to channel routing using genetic algorithm. Bulletin of Faculty of Engineering, Tokushima University (Japan), (38):99-112, 1993.

[814] N. Taniguchi, X. Liu, A. Sakamoto, and T. Shimamoto. An attempt to solve channel routing using genetic algorithm. Transaction of the Institute of Electronics, Information and Communication Engineers A (Japan), J76-A(9):1376-1379, Sept. 1993 (in Japanese).

[815] S. R. Thangiah. Gideon: A genetic algorithm system for vehicle routing with time windows. Ph.D. thesis, North Dakota State University of Agriculture and Applied Sciences, Fargo, 1991.

[816] E. Thro. Artificial Life Explorer's Kit. Sams Publishing, 11711 N. College Ave., Carmel, IN 46032, 1993.

[817] P. Tian and Z. Yang. An improved simulated annealing algorithm with genetic characteristics and the traveling salesman problem. J. Inf. Optim. Sci. (India), 14(3):241-255, Sept. 1993.

[818] P. M. Todd. The evolution of learning: Simulating the interaction of adaptive processes. Ph.D. thesis, Stanford University, Psychology Department, 1992.

[819] P. M. Todd. Book review: Stephanie Forrest, ed., emergent computation: Self-Organizing, collective, and cooperative phenomena in natural and artificial computing networks. Artificial Intelligence, 60(1):171-183, 1993.

[820] P. M. Todd. Parental guidance suggested: How parental imprinting evolves through sexual selection as an adaptive learning mechanism. Adaptive Behavior, 2(1):5-47, 1993.

[821] S. Todd and W. Latham. Evolutionary Art and Computers. Academic Press, London, 1992.

[822] S. Tokinaga and A. B. Whinston. Applying adaptive credit assignment algorithm for the learning classifier system based upon the genetic algorithm. IEICE Transactions on Fundamentals of Electronics Communications and Computer Sciences, E75-A(5):568-577, May 1992.

[823] S. S. Tong and B. A. Gregory. Turbine preliminary design using artificial intelligence and numerical optimization. Transactions of the ASME, 90-GT-148, 1990.

[824] B. H. V. Topping and A. I. Khan, editors. Neural Networks and Combinatorial Optimization in civil and Structural Engineering, Edinburgh (UK), 17-19 Aug. 1993. Civil Comp. Press, Edingburgh.

[825] D. S. Touretzky, editor. Advances in Neural Information Processing Systems 2, Proceedings of the Neural Information Processing Systems (NIPS), Denver, CO, 1990. Morgan Kaufmann Publishers.

[826] J. Z. Tu. Genetic algorithms in machine learning and optimization. Ph.D. thesis, University of Cincinnati, 1992.

[827] P. Tuffrey, C. Etchebest, S. Hazout, and R. Lavery. A new approach to the rapid determination of protein side chain comformations. Journal of Biomolecular Structure & Dynamics, 8(6):1267-1289, 1991.

[828] P. Tuffrey, C. Etchebest, S. Hazout, and R. Lavery. A critical comparison of search algorithms applied to the protein side-chain comformations. Journal of Computational Chemistry, 14:790-798, 1993.

[829] S. Uckun, S. Bagchi, K. Kawamura, and Y. Miyabe. Managing genetic search in job shop scheduling. IEEE Expert, 8(5):15-24, Oct. 1993.

[830] S. Ulam and R. Schrandt. Some elementary attempts at numerical modelling of problems concerning rates of evolutionary processes. Physica D, 22:4-12, 1986.

[831] R. Unger and J. Moult. Genetic algorithms for protein folding simulations. Journal of Molecular Biology, 231(1):75-81, May 1993.

[832] P. Urwin and P. Alison. Genetic selection of information. Systems Science, 17(1):105-109, 1991.

[833] R. J. M. Vaessens, E. H. L. Aarts, and J. H. van Lint. Genetic algorithms in coding theory — a table for A3 (n; d). Discrete Applied Mathematics, 45(1):71-87, Aug. 1993.

[834] M. Valenzuela-Rendon. Two analysis tools to describe the operation of classifier systems. Ph.D. thesis, University of Alabama, Tuscaloosa, 1989 (also TCGA report No. 89005).

[835] P. van Bommel. A randomised schema mutator for evolutionary database optimization. Aust. Comput. J. (Australia), 25(2):61-69, 1993.

[836] J. Vancza and A Markus. Genetic algorithms in process planning. Computers in Industry, 17(2-3):181-184, Nov. 1991.

[837] F. J. Varela and P. Bourgine, editors. Toward a Practice of Autonomous System: Proceedings of the First European Conference on Artificial Life, Paris, 11.-13. Dec. 1991. MIT Press, Cambridge, MA.

[838] A. Varsek, T. Urbancic, and B. Filipic. Genetic algorithms in control design and tuning. IEEE Transactions on Systems, Man, and Cybernetics, 23(5):1330-1339, Sept./Oct. 1993.

[839] V. Venkatasubramanian, K. Chian, and J. M. Caruthers. Computer-aided molecular design using genetic algorithms. Computers in Chemical Engineering, 18(9):833-844, 1994.

[840] V. Venugopal and T. T. Narendran. A genetic algorithm approach to the machine component grouping problem with multiple objectives. Computers & Industrial Engineering, 22(4):469-480, Oct. 1992.

[841] P. F. M. J. Verschure. Formal minds and biological brains: AI and Edelman's extended theory of neuronal group selection. IEEE Expert, 8(5):66-75, Oct. 1993.

[842] G. A. Vignaux and Z. Michalewicz. A genetic algorithm for the linear transportation problem. IEEE Transactions on Systems, Man, and Cybernetics, 21(2):445-452, 1991.

[843] H.-M. Voigt. Evolution und Optimierung: Ein populationsgenetischer Zugang zu kombinatorischen Optimierungsproblemen. Dr. sc. techn., Academy of Sciences, Berlin, 1987.

[844] H.-M. Voigt. Evolution and Optimization: An Introduction to Solving Complex Problems by Replicator Networks. Akademie-Verlag, Berlin, 1989.

[845] H.-M. Voigt. Optimization by selection pressure controlled replicator networks. Syst. Anal. Model. Simul., 6(4):267-278, 1989.

[846] H.-M. Voigt, H. Muhlenbein, and H.-P. Schwefel, editors. Evolution and Optimization '89, Selected Papers on Evolution Theory, Combinatorial Optimization, and Related Topics, Wartburg Castle, Eisenach (Germany), 2-4 Apr. 1989. Akademie-Verlag, Berlin.

[847] K. von Falkenhausen. Optimierung regionaler Entsorgungssysteme mit der Evolutionsstrategie. Proceedings in Operations Research, 9:46-51, 1980.

[848] G. von @ Ein parallel genetischer Algorithmusurf das Graph-Partitionierungsproblem. Ph.D. thesis, University of Bonn, 1990.

[849] J. von Neumann. Theory of self-reproducing automata. University of Illinois Press, Urbana, 1966 (edited and completed by A. W. Burks).

[850] M. D. Vose. Generalizing the notion of schema in genetic algorithms. Artificial Intelligence, 50(3):385-396, 1991.

[851] M. D. Vose and G. E. Liepins. Punctuated equilibria in genetic search. Complex Systems, 5(1):31-44, Feb. 1991.

[852] K.-N. Wada, H. Doi, C.-I. Tanaka, and Y. Wada. A neo-Darwinian algorithm: Asymmetrical mutations due to semiconservative DNA-type replication promote evolution. Proceedings of the National Academy of Sciences of the United States of America, 90(24):11934-
11938, Dec. 1993.

[853] R. L. Wainwright. A family of genetic algorithm packages on a workstation for solving combinatorial optimization problems. SIGICE Bulletin, 19(3):30-36, Feb. 1994.

[854] C. T. Walbridge. Genetic algorithms: What computers can learn from Darwin. Technol. Rev., 92(1):46-48, Jan. 1989.

[855] V. W. Waldmann and T. Gerhaard. Kurvenanpassung und Lastflussoptimierung mittels Evolutionsstrategie. E und M, page 518, 1985.

[856] M. Walk and J. Niklaus. Some remarks on computer-aided design of optical lens systems. Journal of Optimization Theory and Applications, 59(2):173-181, 1988.

[857] C. Walnum. Adventures in Artificial Life. Que Corporation, 11711 N. College Ave., Carmel, IN 46032, 1993.

[858] D. C. Walters, G. B. Sheble, and M. E. El-Hawary. Genetic algorithm solution of economic dispatch with valve point loading. IEEE Transactions on Power Systems, 8(3):1325-1332, 1993 (Proceedings of the 1992 Summer Meeting of the Power-Engineering-Society of IEEE, Seattle, WA, 12-16 July. 1992).

[859] Q. Wang. Optimization by simulating molecular evolution. Biological Cybernetics, 57:95-101, 1987.

[860] Q. J. Wang. The genetic algorithm and its application to calibrating conceptual rainfall runoff models. Water Resources Research, 27(9):2467-2471, Sept. 1991.

[861] T. L. Ward, P. A. S. Ralston, and K. E. Stoll. Intelligent control of machines and processes. Computers & Industrial Engineering, pages 205-209, 12-14 Mar. 1990 (Proceedings of the 12th Annual Conference on Computers and Industrial Engineering).

[862] T. Warwick. Genetic algorithms. Computing (UK), pages 18-19, 8 Aug. 1991.

[863] H. Watabe and N. Okino. An evolutional shape design by genetic algorithm. J. Jpn. Soc. Precision Eng., 59(9):1471-1476, Sept. 1993 (in Japanese).

[864] K. Watanabe, Y. Ikeda, S. Matsuo, and T. Tsuji. Improvement of genetic algorithm and its applications. Memoirs of the Faculty of Engineering, Fukui University, 40(1):133-149, 1992 (in Japanese).

[865] P. Wayner. Genetic algorithms. BYTE, 16(1):361-368, Jan. 1991.

[866] H. Wechsler. A perspective on evolution and the Lamarckian hypothesis using artificial worlds and genetic algorithms. Rev. Int. Syst. (France), 7(5):573-592, 1993.

[867] R. Wehrens, C. B. Lucasius, L. M. C. Buydens, and G. Kateman. HIPS, a hybrid self-adapting expert-system for nuclear-magnetic-resonance spectrum interpretation using genetic algorithms. Analytica Chimica Acta, 277(2):313-324, May 1993.

[868] R. Wehrens, C. B. Lucasius, L. M. C. Buydens, and G. Kateman. Sequential assignment of 2D-NMR spectra of proteins using genetic algorithms. Journal of Chemical Information and Computer Sciences, 33(2):245-251, Mar.-Apr. 1993.

[869] H. H. Weiland. Optimierung von Saugkopfeinlaufen zur gewinnung mariner lockermaterialien mit hilfe der evolutionsstrategischen Experimentiertechnik. Ph.D. thesis, Technische Universitat der Berlin, 1986.

[870] R. Weinberg. Computer simulation of a living cell. Ph.D. thesis, University of Michigan, Ann Arbor, 1970.

[871] E. D. Weinberger. A stochastic generalization of Eigen's model of natural selection. Ph.D. thesis, New York University, 1987 (University Microfilms No. 87-22798).

[872] E. D. Weinberger. A more rigorous derivation of some properties of uncorrelated fitness landscapes. Journal of Theoretical Biology, 134:125-129, 1988.

[873] E. D. Weinberger. Correlated and uncorrelated fitness landscapes and how to tell the difference. Biological Cybernetics, 63:325-336, 1990.

[874] M. P. Wellman. A market-oriented programming environment and its application to distributed multicommodity problems. Journal of Artificial Intelligence Research, 1:1-23, 1993.

[875] T. H. Westerdale. An application of Fischer's theorem on natural selection to some reenforcement algorithms for choice strategies. Journal of Cybernetics, 4:31-42, 1974.

[876] T. H. Westerdale. A reward scheme for production systems with overlapping conflict sets. IEEE Transactions on Systems, Man, and Cybernetics, SMC-16(3):369-383, 1986.

[877] K. W. Whitaker, R. K. Prasanth, and R. E. Markin. Specifying exhaust nozzle contours with a neural network. AIAA Journal, 31(2):273-277, Feb. 1993.

[878] D. Whitley. Applying genetic algorithms to neural network problems. Neural Networks, 1(1):230, 1988 (Proceedings of International Neural Network Society 1988 First Annual Meeting, Boston, MA, 6-10 Sept.).

[879] D. Whitley. Deception, dominance and implicit parallelism. Technical Report No. CS91-120, Colorado State University, Department of Computer Science, Fort Collins, 1991 (also as [880]).

[880] D. Whitley. Deception, dominance and implicit parallelism in genetic search. Annals of Mathematics and Artificial Intelligence, 5(1):49-78, 1992.

[881] D. Whitley, editor. Foundations of Genetic Algorithms l 2 (FOGA-92), Vail, CO, 24.29. July 1992 1993. Morgan Kaufmann: San Mateo, CA.

[882] D. Whitley. A genetic algorithm tutorial. Stat. Comput. (UK), 4(2):65-85, June 1994.

[883] D. Whitley, R. Das, and C. Crabb. Tracking primary hyperplane competitors during genetic search. Annals of Mathematics and Artificial Intelligence, 6(4):367-388, 1992.

[884] D. Whitley and T. J. Starkweather. Genitor ii: A distributed genetic algorithm. Journal of Experimental and Theoretical Artificial Intelligence, 2(3):189-214, July-Sept. 1990.

[885] D. Whitley, T. J. Starkweather, and C. Bogart. Genetic algorithms and neural networks: Optimizing connections and connectivity. Parallel Computing, 14(3):347-361, Aug. 1990.

[886] W. Wienholt. Durch zufall zum erfolg: Genetische Algorithmen. Microcomputer Zeitschrift, 3:152-154,156-158,160-163, Mar. 1990 (in German).

[887] D. Wienke, C. B. Lucasius, M. Ehrlich, and G. Kateman. Multicriteria target vector optimization of analytical procedures using a genetic algorithm. 2. polyoptimization of the photometric calibration graph of dry glucose sensors for quantitative clinical analysis. Analytica Chimica Acta, 271(2):253-268, Jan. 1993.

[888] D. Wienke, C. B. Lucasius, and G. Kateman. Multicriteria target vector optimization of analytical procedures using a genetic algorithm. 1. theory, numerical simulations and applications to atomic emission spectroscopy. Analytica Chimica Acta, 265(2):211-225, Aug. 1992 (6th CIC Workshop on Software Development in Chemistry, Bergakad Freiberg (Germany), 20-22 Nov. 1991).

[889] R. Wiggins. Docking a truck: A genetic fuzzy approach. AI Expert, 7(5):28-35, May 1992.

[890] K. Wilmanski and A. N. van Breemen. Competitive adsorption of trichloroethylene and humic substances from groundwater on activated carbon. Water Research, 24(6):773-779, 1990.

[891] V. Wilms. Auslegung von Bolzenverbindungen mit minimalem Bolzengewicht. Konstruktion, 34(2):63-70, 1982.

[892] S. Wilson. How to grow a starship pilot [genetic algorithms for space probes]. AI Expert, 8(12):20-26, Dec. 1993.

[893] S. W. Wilson. Classifier systems and the Animat problem. Machine Learning, 2(3):199- 228, 1987.

[894] S. W. Wilson. Bid competition and specificity reconsidered. Complex Systems, 2(6):705- 723, 1988.

[895] W. G. Wilson and K. Vasudevan. Application of the genetic algorithm to residual statics estimation. Geophysical Research Letters, 18(12):2181-2184, Dec. 1991.

[896] E. Winkler. Optimum design of gamma-irradiation plants by means of mathematical methods. Radiat. Phys. Chem., 26(5):599-601, 1985.

[897] E. Winkler. A mathematical approach to the optimum design of gamma-irradiation facilities. Isotopenpraxis, 22(1):7-11, 1986.

[898] A. Wittmus, R. Straubel, and R. Rosenmuller. Interactive multi-criteria decision procedure for macroeconomic planning. Systems Analysis — Modeling — Simulation, 1(5):411-424, 1984.

[899] S. J. Wodak and M. J. Rooman. Generating and testing protein folds. Current Opinion in Structural Biology, 3(3):247-259, June 1993.

[900] R. L. Wood. A comparison between the genetic algorithm and the function specification methods for an inverse thermal field problem. Eng. Comput. (UK), 10(5):447-457, 1993.

[901] X.-L. Wu. Darwin's ideas applied to magnetic response. The marriage broker. Journal of Magnetic Response, 85:414-420, 1989.

[902] Y. L. L. Xiao and D. E. Williams. Genetic algorithm: a new approach to the prediction of the structure of molecular clusters. Chemical Physics Letters, 215(1-3):17-24, Nov. 1993.

[903] Y. L. L. Xiao and D. E. Williams. Game: Genetic algorithm for minimization of energy, an interactive FORTRAN program for three-dimensional intermolecular interactions. Computers & Chemistry, 18:199-201, 1994.

[904] Y. L. L. Xiao and D. E. Williams. Genetic algorithms for docking of actinomycin D and deoxyguanosine molecules with comparison to the crystal structure of actinomycin D-deoxyguanosine complex. Journal of Physical Chemistry, 98:7191-7200, 1994.

[905] Y. Xiong. Optimization of transportation network design problems using a cumulative genetic algorithm and neural networks. Ph.D. thesis, University of Washington, WA, 1992.

[906] T. Yamagishi and T. Tomikawa. Polygonal approximation of closed curve by GA. Transaction of the Institute of Electronics, Information and Communication Engineers D-II (Japan), J76D-11(4):917-919, 1993 (in Japanese).

[907] M. Yamamura and S. Kobayashi. Combinatorial optimization with genetic algorithms. J. Jpn. Soc. Simul. Technol. (Japan), 12(1):4-10, 1993 (in Japanese).

[908] M. Yamamura, T. Ono, and S. Kobayashi. Character-preserving genetic algorithms for traveling salesman problem. Journal of Japanese Society for Artificial Intelligence, 7(6):1049-1059, Nov. 1992 (in Japanese).

[909] C.-H. Yang. Genetic search and time constrained routing. Ph.D. thesis, North Dakota State University of Agriculture and Applied Sciences, 1992.

[910] G. Yang. Genetic algorithm for the optimal design of diffractive optical elements and the comparison with simulated annealing. Guangxue Xuebao, 13(7):577-584, July 1993 (in Chinese).

[911] J.-J. Yang and S. S. Rich. Linkers: A simulation programming system for generating populations with genetic structure. Computers in Biology and Medicine, 20(2):135-144, 1990.

[912] L. Yao. Parameter estimation for nonlinear systems. Ph.D. thesis, The University of Wisconsin-Madison, 1992.

[913] L. Yao, W. A. Sethares, and D. C. Kammer. Sensor placement for on-orbit modal identification of large space structure via a genetic algorithm. AIAA Journal, 31(10):1922-1928, Oct. 1993.

[914] X. Yao. A review of evolutionary artificial neural networks. International Journal of Intelligent Systems, 8(4):539-567, Apr. 1992.

[915] X. Yao. An empirical-study of genetic operators in genetic algorithms. Microprocessing and Microprogramming, 38(1-5):707-714, 1993.

[916] X. Yao. Evolutionary artificial neural networks. International Journal of Neural Systems (Singapore), 4(3):203-222, Sept. 1993.

[917] C. Yilin, L. Feipeng, and H. Zheng. Displacement estimation by 2-D genetic optimizer algorithm for image sequence coding. Acta Electronica Sinica, 20(1):61-66, Jan. 1992. (in Chinese).

[918] X. Yin and N. Germay. Investigations on solving the load flow problem by genetic algorithms. Electric Power Systems Research, 22(3):151-163, Dec. 1991.

[919] A. Zeyher. Optical packages look for global minima. Computers in Physics, 8(2):137-140, Mar./Apr. 1994.

[920] J. Zhang and P. D. Roberts. Use of genetic algorithms in training diagnostic rules for process fault diagnosis. Knowledge-Based Systems (UK), 5(4):277-288, Dec. 1992.

[921] Y. Zhou. Genetic algorithm with qualitative knowledge enchancement for layout design under continuous space formulation. Ph.D. thesis, University of Illinois at Chicago, 1993.

[922] A. Ziegler and W. Rucker. Die Optimierung der Strahlungscharakteristik linearer Antennengruppen mit hilfe der Evolutionsstrategie. Archiv urf Elektronik und Ubertragungstechnik, 40(1):15-18, 1986.

[923] D. C. Zimmerman. A Darwinian approach to the actuator number and placement problem with non-negligible actuator mass. Mech. Syst. Signal Process. (UK), 7(4):363-374, July 1993.

[924] J. M. Zurada, I. Robert J. Marks, and C. J. Robinson, editors. Computational Intelligence Imitating Life. IEEE Press, New York, 1994.

**Bibliography entry formats**

This documentation was prepared with aL TE X and reproduced from camera-ready copy supplied by the editor. The ones who are familiar with BibTeX may have noticed that the references are printed using abbrv bibliography style and have no difficulties in interpreting the entries.

For those not so familiar with BibTeX are given the following formats of the most common entry types. The optional fields are enclosed by "[]" in the format description. Unknown fields are shown by "?". y after the entry means that neither the article nor the abstract of the article was available for reviewing and so the reference entry and/or its indexing may be more or less incomplete.

**Book**

Author(s), Title, Publisher, Publisher's address, year.

Example

John H. Holland. Adaptation in Natural and Artificial Systems. The University of Michigan Press, Ann Arbor, 1975.

**Journal article**

Author(s), Title, Journal, volume(number): first page - last page, [month,] year.

Example

David E. Goldberg. Computer-aided gas pipeline operation using genetic algorithms and rule learning. Part I: Genetic algorithms in pipeline optimization. Engineering with Computers, 3(?):35-45, 1987.

Note: the number of the journal unknown, the article has not been seen.

**Proceedings article**

Author(s), Title, editor(s) of the proceedings, Title of Proceedings, [volume,] pages, location of the conference, date of the conference, publisher of the proceedings, publisher's address.

Example

John R. Koza. Hierarchical genetic algorithms operating on populations of computer programs. In N. S. Sridharan, editor, Eleventh International Joint Conference on Artificial Intelligence (IJCAI-89), pages 768-774, Detroit, MI, 20.-25. August 1989. Morgan Kaufmann, Palo Alto, CA. .

**Technical report**

Author(s), Title, type and number, Institute, year.

Example

Thomas Back, Frank Hoffmeister, and Hans-Paul Schwefel. Applications of evolutionary algorithms. Technical Report SYS-2/92, University of Dortmund, Department of Computer Science, 1992.

# INDEX

## D



## E

## F



## G



## H

## Q



## R



## S